（京）新登字035號

圖書在版編目（CIP）數據

　現實中的夢想：建築師潘祖堯的心路歷程：1968～1998：中、英文對照／潘祖堯著．北京：中國建築工業出版社，1999

　ISBN　7-112-03897-9

　Ⅰ.現...Ⅱ.潘...Ⅲ.①建築設計-圖集②建築藝術-繪畫-圖集③建築藝術-攝影-圖集Ⅳ.TU2-64

現實中的夢想
The Dream in Reality
潘祖堯　著

出版發行：	中國建築工業出版社（北京西郊百萬莊）
經　　銷：	新華書店
制　　版：	深圳中華商務聯合印刷有限公司制版
印　　刷：	深圳中華商務聯合印刷有限公司印刷
開　　本：	787 × 1092毫米 1/12
印　　張：	35.5
印　　數：	1-1000冊
定　　價：	300.00元

1999年5月第一版　1999年5月第一次印刷

ISBN　7-112-03897-9
　TU · 3030（9254）

版權所有　翻印必究
如有印裝質量問題，可寄本社退換
（郵政編碼 100037）

現實中的夢想

建築師潘祖堯的心路歷程
1968~1998

The Dream in Reality

Ronald Poon 1968~1998

潘祖堯著
by Ronald Poon

中國建築工業出版社
China Architectural & Building Press

目錄

前言 4
回顧 7
建築設計 27
    學生時代作品(1968) 29
        空中庭園 31
    在潘衍壽土木工程師事務所的作品(1968~1972) 33
        邵氏製片廠警衛屋 34
        邵氏大屋 36
        翡翠明珠戲院 40
        蜆殼油站 41
        蕭明天主教中學 42
        培敦中學 44
        筆架山道四間排屋 48
        海外信託銀行大廈 50
    潘祖堯則師事務所(1973~1986) 53
    羅素/潘建築師集團(1976~1986) 53
    潘祖堯顧問有限公司(1986~1998) 53
        石澳村屋 54
        又一村公寓樓 55
        蒲台島度假村 56
        南豐路公寓大樓 58
        山光道公寓大樓 62
        銀禧體育中心 64
        加達樓 66
        弱能兒童院 68
        豐樂閣 72
        紙幣廠 78
        城市理工大學 80
        牡丹山莊 84
        鵬鷹山莊 86
        聰葉山房 90
        華獻度假公寓大樓 92
        北芭提雅度假公寓大樓 94
        清邁山莊度假村 96
        蘇榮特皇路住宅小區 98
        拉曼9街商住大樓 100
        曼谷市42街公寓大樓 102
        曼谷市郊河畔小築 104
        蕭山市金家浜花園 106
        "西西工程"總規劃 110
        2號地商辦大樓 114
        4及5號地商辦大樓 116
        武康路公寓大樓 122
        龍潭湖度假村 124
        都江堰別墅山莊 126
        深圳少年宮 128

室內設計 135
    "起居窩" 136
    香港節詢問處 137
    時裝店 137
    公寓 138
    怡和保險有限公司九龍辦事處 138
    中華文化促進中心 139
    北京人民大會堂香港廳 140
視覺設計 143
攝影 163
繪畫 191
古物收藏 207
學會工作 223
    香港建築師學會 224
    中國建築學會 228
    亞洲建築師協會 237
    英聯邦建築師學會 240
    國際建築師協會 242
    香港科技協進會 243
    建築師協會亞洲會 245
    中建學社 246
    中國建築論壇 247
    現代中國建築創作研究小組 250
社會工作 253
    中國人民政治協商會議全國委員會 255
    香港房屋協會 258
    香港市政局 261
    敏求精舍 264
    香港政府屬下委員會 266
    建築設計評委工作 267
建築教育 271
已發表文章及學術報告選輯 275
有關文章及主編的圖書目錄 343
個人檔案及作品年表 347
附錄 359

Contents

Forward 4
Reminescence 7
Architectural design 27
    Student work (1968) 29
        "Sky Courtyards" 29
    Projects at Peter Y.S. Pun & Associates(1968-1972) 33
        "Guard House", Shaw's Studio 34
        Shaw's residence 36
        Jade and Pearl cinemas 40
        Shell's petrol stations 41
        Siu Ming catholic secondary school 42
        Pooi Tun secondary school 44
        4 terrace houses, Beacon Hill Road 48
        OTB Bank building 50
    Ronald Poon Associates (1973~1986) 53
    Russell/Poon Group Partnership (1976~1986) 53
    Ronald Poon Consultants Ltd. (1986~1998) 53
        Village shop, Shek O 54
        Apartment complex, Yau Yat Chuen 55
        Po Toi Island resort 56
        Apartment complex, Nan Fung Road 58
        Apartment complex, San Kwong Road 62
        Silver Jubilee sports centre 64
        Cactus Mansion 66
        Home for the severely handicapped children 68
        Albron Court 72
        Printing factory for Thomas De La Rue Ltd. 78
        City Polytechnic University 80
        Xian Mountain Inn 84
        Lanzhou Mountain Inn 86
        House, Ilan 90
        Holiday condominium, Hua Hin 92
        Holiday condominium, North Pattaya 94
        Misty Hill holiday homes, Chiang Mai 96
        Housing estate, Survintawongs Road 98
        Commercial/residential complex, Rama 9 Road 100
        Residential condominium, Soi 42 102
        Riverside housing, near Bangkok 104
        Riverside Garden, Xiao Shan 106
        Master plan for "Xi-Xi Project" 110
        Office/commercial complex, no. 2 site 114
        Office/commercial complex, no. 4 & 5 sites 116
        Apartment complex, Wu Kang Road 122
        Holiday villas complex, Lun Tam Lake 124
        Holiday villas, Dujiangyan 126
        Youth centre, Shenzhen 128
Interior design (1969~1998) 135
    "Living Pod" 136
    Festival of Hong Kong kiosk 137
    Malcolm Starr International showroom 137
    Apartment 138
    Jardine Insurance Co. Ltd., Kowloon 138
    Chinese Cultural Centre 139
    "Hong Kong Room", Hall of the People 140
Graphic design 143
Photography 163
Painting 191
Antique collecting 207
Professional institution work 223
    Hong Kong Institute of Architects 224
    Architectural Society of China 228
    Architects Regional Council Asia 237
    Commonwealth Association of Architects 240
    International Union of Architects 242
    Hong Kong Association for the Advancement of Science & Technology 243
    A A Asia 245
    Zhong Jian Society 246
    China Architectural Forum 247
    The Reasearch Group on Modern Chinese Architecture 250
Community work 253
    The National Committee of the Chinese People's Political Consultative Conference 255
    Hong Kong Housing Society 258
    Urban Council, Hong Kong 261
    Min Chiu Society 264
    Hong Kong Government committees 266
    Work of design competition juror 267
Architectural education 271
Published articles & academic papers 275
Related articles & edited publications 343
Biographical chronology 347
Appendix 359

前言
Forward

這本書能面世,連我自己都難以相信。我早於1993年初已有一個夢想,想把我以往的工作來個回顧及編寫一部書。至少可把我堆積如山的舊圖及檔案,付諸一炬,讓出一些空間來。而且當時覺得在設計思路上,有點前路茫茫,有在十字路口的感覺,希望溫故知新。我有了這個構想後與我台北的好友朱祖明建築師交談,他不但極力支持,而且提意書名為"心路歷程"。我雖然主編過數本與建築設計有關的書籍和雜誌特刊以及學術報告,但從未嘗試過寫書,而且當時也不能肯定有出版商會支持,更不肯定有人會對我的工作發生興趣,所以遲遲未能起筆寫作。直至1996年中,獲黑龍江科學技術出版社寄來"當代中國名家建築創作與表現叢書"中由彭一剛教授所寫的《創意與表現》一書及信一封,信中邀請我編寫一本有關我的工作的書,列入叢書系列。但當時適逢工作比較繁忙,所以就提議數月後再進一步商討,直至去年7月出版社也音信全無。此間,我的兩位好友陳世民及程泰寧,相繼把他們寫作的書送給我,使我對出書的意念倍增,所以,去年7月決定動筆,而且為了確保每頁的佈局都能符合我意,我只好親自剪貼及挑選所有圖案及照片。這20多年來我每日沉迷於工作的時間平均有12小時,所以動手寫書後不到3個月已經有一本300多頁的初稿。正好碰上我與好友楊永生合辦了4年的建築論壇每年一次的研討會在重慶召開,所以我日夜趕工,把一本較完整的初稿於10月15日帶到重慶與永生兄商量出版事宜,幸好得到他及中國建築工業出版社的大力支持,這本書才能得以面世。但我始終覺得祖明兄提意的書名有點太直白,恰逢年初五我久別的書法老師好友林悅恒與他的太太到寒舍敘舊,我把將近完成的第二稿給他看,並請他提出一個書名,幸好他提出《現實中的夢想》這個書名正合我意,所以當晚飯後便要求他即席揮毫,替我題寫書名。真有些"踏破鐵鞋無覓處,得來全不費功夫"之感。悅恒兄滿腹文彩,而且寫得一手好字,他的書名就好像沙漠中的一滴水,把整個出書的夢想化為多采的現實。此外我更要多謝黑龍江科學技術出版社出書邀請的鼓勵,使我最終能下決心去完成這個夢想。

在編寫的開端,我花了不少時間去落實書的內容。因我興趣頗廣,除建築設計外,對藝術、攝影、繪畫、古物收藏、學會工作、社會工作、建築教育、學術研究等都有濃厚的興趣。我認為這些在建築設計外的"嗜好"與建築設計是有密切的關係,而且是從事建築設計不可缺少的"維他命"。所以這本書所載的內容記錄了我在打好根基階段30年的經歷,在籌備這本書的過程中,回顧了我以往的工作,使我有機會檢討以往的得失。我覺得每個人都應該在適當的時候作一回顧,從而對以後的工作帶出一個方向,才不會盲目地向前跑。溫故知新,才能度過一個豐盛的人生。

我要多謝我公司的員工黃偉綽、梁國華、黃樹堅、劉淑媛、黎朗、黃智峰及黎華在這艱苦時期,還能全力投入,協助我完成這"溫故知新"的夢想,更要多謝永生兄替我修改中文稿。最後我要多謝與我同甘共苦已20多年的太太——麗華及我的孩子們——正煒、美慈、正輝、美婷及美桐,沒有他們的諒解及支持,我的"夢"難以成真。

潘祖堯
1999年2月
于香港小聰颿樓

I still find it difficult to belief that this book is now in print. Early in 1993 I had a dream and wish to review my past work and produce a book on them. At least I can then throw away all those drawings and files and clear away some much needed storage space in the office. At that time I was a little "lost" in my design approach and feel like "standing in the cross road". I hope by reviewing my past, it can throw some light for the future. I then discussed my idea with Mr. Chu Joe Ming, my architect friend in Taipei. He not only gave me great encouragement but also suggested the name of the book to be "Recollections". Although I have edited several books & magazines concerning architectural design and written several papers, I have not written a book. At the same time I was not sure I can get support from a publisher nor will there be anyone interested in my work. Therefore the idea remained a dream until mid 1996 when the Heilongjiang Science and Technology Press forwarded to me a book called "Creativity and Realization" by Prof. Peng Yigang being one of the books in a series called "Creativity and Realization of the famous contemporary Chinese Architects" together with a letter of invitation to me to write a book on my work which will form part of the series. Unfortunately I was a little busy at the time and I suggested to them to discuss this further in a few months' time. However by July last year I have not heard further from them. During the intervening period I received two books from my friends Mr. Chen Shi Min and Cheng Taining on their own work. These books gave me great encouragement. I decided to make a start on the book in July last year. In order to ensure that each page is properly laid out, I have decided to do the pasting up myself. In the last 20 years, I normally spent an average of 12 hours per day in my work. Therefore after starting on the book for three months, I already have a first draft consisting of some three hundred odd pages. The annual architectural forum founded jointly by Mr. Yang Yong Sang and I in 1995 was scheduled to take place in Chongqing on the 15th October. I decided to take the first draft to Chongqing and discuss with Yong Sang about publication. Luckily both Yong Sang and eventually the Architectural& Building Press agreed to publish the book. At the end I felt that Joe Ming's suggested book title was a little bit "matter of fact". It so happened that my good friend and calligraphy teacher Lin Yue Xing and his wife were visiting us on the eve of the 5th day of the lunar new year. I showed him the second draft of the book and asked him to suggest a title. Luckily he suggested "the Dream in Reality" which was exactly what I wanted. After dinner I asked him to write the title for me with ink and brush. I felt that "Despite a long hard search with worn soles, the title was found by the drop of a hat". Yue Xing is well versed in literary matters and also writes beautiful calligraphy. His suggested title was like a drop of water in the dessert. It brought colourful reality to my "dream". I am also particularly indebted to the Heliongjiang Science and Technology Press for their earlier invitation to me for writing the book which prompted me to make my dream come true.

In the beginning I spent a lot of time in deciding on the contents of the book as my interest is rather broad. Other than architecture, I am also interested in art, photography, painting, antique collecting, professional institutional work, community work, academic research and architectural education etc. I consider these interests closely related to architecture and in fact, are essential "vitamins" of architectural design. Therefore what is recorded here is thirty years of foundation work. In the course of writing this book, I had the opportunity to review my past work. I feel that everyone should do the same at appropriate time so that a direction can be found for the future and one would not blindly march forward as "reviewing the past brings knowledge of the future". Only in this way one can enjoy a rewarding life.

I would like to thank my staff Edward Wong, Peter Leung, Peter Wong, Candy Lau, Lai Long, Wong Chi Fung and Zoyee Lai for their devoted hard work despite the difficult time we are having. Their efforts assisted me in realizing my dream. I would also like to thank Yong Sang for correcting my chinese text. Finally I would like to thank my wife Ellie who has "for better or worse" spent the last twenty odd years with me and my children, Justin, Marissa, Geoffrey, Vanessa & Serena for their understanding and support, without which my dream will not have come true.

Ronald C. Y. Poon
February 1999
at Shao Ting Fan Lau,
Hong Kong.

回 顧

Reminiscence

1949年與大哥祖期及妹憫賢攝於香港淺水灣
Taken in Repulse Bay, Hong Kong with brother Albert and sister Beatrice in 1949

1957年聖誕節與父母及兄妹攝於香港堅道99號大廳
Taken in 99, Caine Road with parent, borther and sister at Christmas, 1957

1942年我在香港出生，當時正是日軍佔港時期。先祖原籍福建省泉州府同安縣積善里明盛鄉，於清乾隆41年(1776年)落籍廣東番禺縣龍溪鄉，即現今廣州市河南海珠區。16歲時先父便安排我到英國就學，所以我的中文根基不好，幸好在小學期間先父特請了一位精通古文的老師到家裏授課，所以至今我還能背出文天祥的"正氣歌"。到1968年畢業回港，開始了打工的生涯，月薪港幣1,800元。幸好有親戚幫助，以港幣500元的租金與妻兒入住一間大房，以港幣1,300元過活，先父沒有給任何津貼。從我開始執業以來，我有些同行誤以為我家底非常富裕，所以不要工作也能過活，實在有苦自己知。對我來說，這是我一生的包袱，不知使我失去多少想僱用我的業主。一個人工作，基本上當然要攢一點錢，但最重要是要對社會有所貢獻及不要辜負先人栽培養育之恩。

1958年夏天坐船經意大利到英國，首先在倫敦進修英文，9月開學時到英國西南部深馬石州入讀一寄宿中學，度過了兩個嚴寒的冬天。我於1960年秋天考進英國西南部的蒲路藝術學院，當時我的志願是想成為一個藝術家。經過一個學期之後，我便申請轉往鄰近城市一所著名的保密夫藝術學院繼續攻讀。在這個時候，我發覺我的興趣集中於人體素描和陶瓷方面，但感到在藝術學院的生活是有點兒過於輕鬆，感到浪費時光，覺得就算成為一個成功的藝術家，對社會的貢獻也是有限度的。我希望能夠進修一門對社會更有直接貢獻的專業。我的乾爹當時是一位在新加坡執業的建築師，或者因此令我投身建築設計方面。

經過一年在倫敦補習入學必需的科目後，便於1962年夏天考進了由英國著名的建築協會主辦的建築學院。從1961年起，我開始將所有從報章、雜誌中剪存下來有興趣的資料貼在剪貼簿上，這興趣一直維持至今，已有剪貼簿10多本。這些剪貼簿就是我的圖畫性日記，記載當時對我有興趣的一事一物，從中可以看出早期的剪貼簿是充滿了初學的心思。可惜中期所剪貼的資料，表現出一些中年人的弱點，就是充滿了與知名人士所合拍的照片及在報章刊登有關我的報導，幸而大部分近期的資料都是關于學術性的事物和有初學精神的資料。

我是在1958年離開香港往英國的，在1962年的夏天我初次返港，炎熱的夏天加上我已被英國的建築協會主辦的建築學院接納的消息，我就買了一大堆油漆一口氣畫了數張大型的油畫及人像。我可算是做了一個短期的藝術家。別人會說我是受積臣·保羅及中國傳統山水畫影響(見193頁)。

在建築學院就讀四年後，經導師推薦到史密夫/沙民/羅建築師事務所實習。在這事務所我從事了一些民居的設計工作。兩個月後我就被倫敦市哈寧基區的政府建築部門錄用在一小組內工作。範圍包括低層的民居方案及地區性購物商場的設計。這是我到現今唯一在政府部門裏工作的經驗，在這段時間中，我最難忘的有兩件事，第一就是每天早上11時及下午3時半硬性規定的茶點時間；第二就是每到下午5時一定要馬上離開工作崗位，不然的話就會被清潔工人驅逐出房。

1967年秋重返建築學院，進修由著名研究熱帶建築學的高力畢加教授主辦的熱帶建築學系，至1968年夏天畢業。我的畢業論文"空中庭園"是關於香港高密度的房屋建設(見31頁)。工地是位於現在香港島西南部的華富村。這方案是將住宅單位用高層的處理方式，集中在橫跨工地的長形建築物上，使居民有更多休憩的地方，住宅單位的設計注意反映公共屋村居民的生活要求，更附帶能夠擴建的可能性，當時老師的評論覺得這高層建築所增加的費用，居民並沒有受到很大的益處。雖然方案建議了在香港這高密度的地方一種新的生活方式，但沒有徹底去解決所有問題，例如，建築費用及對這龐大屋村週圍的環境，及交通的要求等各種問題。現在來看，如果以香港房屋署在這工地所建成的屋村來作比較，我總覺得我的方案最低限度是多了一點休憩的地方。如果密度的要求是每公頃2千人的話，你怎樣去避免運用高層來解決問題呢？

當年夏天隨即回港加入潘衍壽土木工程師事務所工作。我當時是事務所唯一的建築師，所以大部分建築設計都由我主持，對一個剛畢業的學生，實在是一千載難逢的機會。我在潘衍壽土木工程師事務所工作了4年，所負責設計的項目大約有24項。在這個階段，我把在學生時代所敬佩的大師——柯布西埃及路易斯·康的作風都盡量施展到我的設計中，尤其是對路易斯·康的"服務者及被服務者"的思路更多次嘗試採用。使我所設計的建築物，在功能上分佈上更加清楚，但其中也產生了一些冷感。

邵氏電影製片廠藝員宿舍內的警衛樓(見35頁)可算是我個人設計工作中第一座落成的建築物。外表來看那些野獸派的混凝土外牆及其中的圖案當然是受柯布西埃的影響。邵氏大屋(見37頁)的佈局，把建築設備房及佣人的工作間合成一組，集中於樓面的中部而客廳及睡房從這一組服務性的核心向外伸展，使客廳及睡房都能享受到週圍的美景，反映出利用"服務者及被服務者"佈局的一個長處。

我的老板潘衍壽先生是一個有豐富經驗的工程師，他在結構設計上，有獨特的新意。當時在香港，他是在工廠及辦公樓設計上首位利用預製構件、預應力混凝土結構的工程師。我在這方面吸收了不少經驗。在1968年至1969年間我設計了一些小工程，包括電油站、汽車陳列室等(見41頁)。其中在一間預製的電油站，因業主的要求，要在施工過程中繼續營業，所以採用了預製的結構。雖然在施工上要特別編排程序，但最終都能滿足業主的要求。預製預應力混凝土結構最能發揮作用的地方，是在大型多層的工廠大廈及辦公樓。這類的結構樓身兩旁是要有適當的支柱牆或支柱，預製預應力混凝土的橫樑樓板放在其中，而樓梯、電梯及廁所等設施如放在樓身的兩旁，剛好提供了所需要的支柱牆，這便能在樓身中部提供一個無柱的大型空間，配合工廠設備及辦公功能所需要的靈活性。我就是利用了這個構思去設計多處多層工廠大廈(見348頁)，及海外信託銀行總行等項目(見51頁)。

1969年我設計了兩間中學，蕭明天主教中學(見43頁)位於一小山頂上。因為週圍環境比較空曠，所以我把課室大樓及實驗室等高層大樓放在工地的西面，給向南的球場提供了遮太陽的作用。而西面的課室就加設混凝土的遮陽板，建築物的外形充分表現我從熱帶建築學院深造後的成績。另一間中學培敦中學(見45頁)在設計方面比較有挑戰性。因為工地是由兩個不同方向45度斜坡組成，而形狀是長而窄的，在平面設計方面，學校有4個基本單位例如禮堂、課室、實驗室及建築設備軸心。因為想減輕開闢斜坡的工作，所以將禮堂升高，放在課室與實驗室之間。這樣一來，禮堂的下層成為有蓋的運動場。在這兩間學校的設計中，課室的架構表露於外牆的設計中。兩者來說，培敦中學在外形設計上，是比較優秀的。在20年後的今天，我對此建築物的感受是這外牆設計真是充滿複雜與矛盾。

從1970年至今，我每年都設計自己的賀年卡(見157頁)，開始的幾年只用一些有趣的圖案或書法來作題材，但從1984起，我每年都按當時國內或香港的時勢作打油詩一首，加以形容，並從我在這年拍到的照片中挑選適合反映我的意見的照片，希望收到我賀年卡的朋友一笑。多年來有些朋友要求我把以往設計的賀年卡出書，結果我決定把所有的卡都放到這本書裏。

1984年10月我嘗試籌辦一次中、港、台建築師學術研討會，命名"蘭亭會"并設計一個會徽(見154頁)。這個集會聚集了北京、上海、天津、廣州、台北、台中及香港建築界的精英去研討"中國傳統環境觀念與現代體形設計的關係"，我將內地及台灣省學會的徽章用90度的安排及面對面形式的合成一標誌。由於本來是一家人，所以兩會所用的會徽是用中國傳統的斗栱為圖案，只是顏色上略有不同，內地是紅，而台灣省是金色的。為什麼要將徽章放在90度的角度呢?因為我想表達我對這集會的願望，就是希望各參會者能以90度轉變的眼光把這主題引進一個新的方向，而兩徽章之間有一比較粗厚的線條，就是間接表達香港在這事情所發揮的聯結作用，顏色方面當然是用紅金混合色。

室內設計是我的副業。從1969年開始，利用晚上空閒的時間設計了一些項目。每一個項目都用盡心思，務求提供既實際又美觀的設計。例如在香港節路旁詢問處(見137頁)，採用了油上鮮艷顏色的三夾板作為詢問處的門及打開後作為陳列小冊子的標板。在一郊外度假屋內，設計了一個坐地式的起居室(見136頁)，增加了濃厚的度假氣氛。我對傢俬設計也感興趣，在一個市區內的公寓我設計了一套皮製的起居室坐椅及以膠片和染色木做的酒櫃(見138頁)。另外在一保險公司的辦公室內，因為面積小，所以用透明膠片作間牆及比較輕巧的皮椅來爭取更大空間的幻覺(見138頁)。在另一個電影院設計項目，我初次嘗試用染色的技術處理木板間牆，創造出一種有透明感而又鮮艷的色彩，增加電影院大堂內的熱鬧氣氛(見40頁)。在一狹窄的通道裏我利用淺色預製混凝土的雕塑牆版，使通道在視覺上增加廣闊的感覺。我的室內設計是充滿新的嘗試，因為在這個尺度的工作，新的嘗試是比較容易控制而錯處是比較容易修改。

1973年我加入了香港建築師學會為會員，從此至今我不斷地參與各個專業學會的工作，并參加國際建築師聯會(1987)、英聯邦建築師協會(1979)、亞洲建築師學會(1979)、中國建築學會(1980)、香港科技協進會(1985)、英國建築師協會——亞洲會(1990)。我在這些學會的工作，在書中已有詳細的紀實，在此也不多說了。

1973年初我畢業後的第五年，我便創辦了自己的設計公司——潘祖堯則師事務所，至今已有25年了。第一個任務，是在港島石澳灣設計一間用作為商店的鄉村小屋(見54頁)。

這間屋的總面積只有100平方米,雖然很小,但這村屋給了我一個機會去實踐路易斯‧康的"服務者與被服務者"的設計原理。雖然在這很小的面積上,兩個樓梯及廁所設備已經佔去一半的面積,剩下來的空間還可以滿足各種用途的需要,例如房屋中間的光井,不只有通風的作用,並且成為地下廁所日間光線的來源。如果不是這樣,香港建築條例並不許可這個廁所在屋中的位置。屋後地下化糞池之上就是戶外廚房。這小屋的設計可算表達了我對功能分佈及物盡其用的構思,下了一個定論。

我的第二個任務就是為嘉頓麵包廠設計一間加建的廠房。原有的廠房是由香港著名的巴馬丹拿建築及工程有限公司設計,而這建築物的外形甚為美觀。我在設計上想將加建的部分與原來建築物協調,而成為其附屬物。在外牆方面除了必要的通風系統之外,並沒有其它窗戶。這樣形成了一幅很單調的白色角形外牆,和原來建築物的六角外形相配合(見351頁)。

蒲台島是我第一個度假村的項目。佈局依山而建,以角型的平面創造出一個與一般住宅不同的度假居住環境(見57頁)。這個項目的工地位於一個島上,所以全部的結構是以預製的方式來解決有關的問題。

加達樓(見67頁)的設計,是從以前我在1969年在潘衍壽土木工程建築師事務所設計的永勝大廈發展而來。雖然仍然是受建築條例所困擾,但設計可算比永勝大廈略勝一籌。香港市中心區的投機性的住宅樓宇,往往賣出之後,被小業主在外牆加建醜陋的鐵籠,以增加樓面面積,但對外牆設計方面,往往有點兒煞風景。因為這個問題我從較早時一個沒有建成的方案裏,開始研究折衷的辦法。這個方案位於香港西環的高陞街(見351頁),解決的辦法是將外牆用三種顏色在表面上看來是沒有規律地塗上各單元的外牆,形成了如萬花筒的圖案。這樣一來,任何加建都會被遮掩在萬花筒圖案中。1975年我參加了在馬尼拉舉行的國際設計比賽(見352頁)。比賽的題目是為馬尼拉北部的平民區作一整體的規劃及設計一典型的屋村。這個區域的估計未來人口為14萬人,這區是位於馬尼拉北部海邊,名打格籐區。我將整區分為35個分區,每分區有小學、區中心遊樂場地及商店。房屋是分佈在行人小徑的兩旁,其中加了公共聚集的空間,空間有傳統性的水龍頭以保留居民聚集的風味。房屋的結構是由政府建設,而外牆屋頂及內部間格則由居民自己負責。房屋的底層是與地面相隔有1.5米之高,原因是避免水浸,另一用途為飼養家畜之用。參加這個比賽只是爭取經驗,因為回想當時真是有點"不知天高地厚"之感。在資歷上我欠缺在這樣規模的大型區規劃的經驗,與中獎的方案比較真是有一段距離。

在1978年我參加另一比賽,就是香港銀禧體育中心的設計(見65頁)。這中心的設備頗為複雜,而我就利用中心主要的設備在建築體形上表露。中心有四種主要設備:第一是建築設備;第二是大型活動性的體育場地;第三是固定的體育場地,例如游泳池;第四就是運動員宿舍。地形的擺佈是受了戶外的體育場所方向的要求及入口道位置的影響。整個建築群有兩行平行式的樓房,近馬路的一邊是作為屬於"服務者"一類的用途空間,例如更衣室、機房、實驗室、辦事處、教育及訓練中心及商店等。近體育場地的另一邊,就有"被服務者"一類的空間,例如游泳池、體育館等等。兩個平行建築物之間,則為連貫通道,整個屋群是可以向一方面擴展。而運動員宿舍大樓形成整個設計的視覺核心。現在看來,這個設計可能是受了一些現代日本設計的影響。

香港堅道99號可算是我建築設計工作中的一個里程碑(見73頁)。1979年已作一個方案(見349頁),在1981年便交由我的事務所再設計。屋頂上的螺旋形是由香港建築條例限制下產生的,加上有此形狀後可以在屋頂上有多間平台式的單元。此外,在整個大廈的外觀上更有一種特別的風格。整座大廈有35層高。外形的設計不單是反映了結構的形式,同時更表露到室內空間的用途,睡房是獨立的,從而形成建築物的管形外狀。起居安排在各管形之間,形狀是從徹底研究傢俬安排的各種認識後而設計成的。因為要特別表達出螺旋形的架構,所以外牆的顏色是以兩色相間的方式。此外,建築物的中部就以鮮紅色來表達出來,外牆的特點就是這一連串的凸出外牆的窗箱。這設計是取材於西藏的建築物,樓身下部的設計反映了我對西藏布達拉宮的設計的仰慕。有一位建築評論家說,在外形上反映了一些中國傳統建築的壯觀。

弱能兒童院代表我的另一里程碑(見69頁)。這個方案是由我和大衛‧羅素所設計。在這個方案中,業主要求將舊樓的一部分保留,其中包括小教堂,而該部分卻位於工地的中間,所以對設計方面產生很大的障礙。第一是如何將這部分舊樓設計成新樓的一部分,第二是怎樣將兩種不同的用途互相協調。設計弱能兒童院是特別要注意幾項事情:(一)弱能兒童在複雜及擠擁的環境中,往往會覺得不安,所以空間要有簡單形狀的設計,無間牆式的空間是方便兒童的起居及看護人員進行工作。(二)這些兒童要經常被照顧,在日常活動中,例如沐浴等是要人扶助的,有部分因弱能嚴重,要長臥床上,但每天要在水裏接受物理治療。(三)適當的空間,會更加鼓勵殘疾程度低的兒童去多作活動。(四)護理人員要經常做監視的工作,空間要有適當的安排,鼓勵這些護理人員與兒童有更多的接觸,同時在工作時不必常常離開這空間。(五)護理人員的宿舍要特別處理,因為這些人員白天所面對的壓力太大,當他們下班後,應該能夠回到一個優美環境來鬆弛一下。但是這個環境又要接近兒童的睡房,以防萬一。根據以上各點,我們是要設計一個比較硬性化的方案,但同時亦要富有溫暖活動性及家庭式的氣氛。

兒童院近入口的地方有一座行政大樓,整個兒童院的設計是利用庭院式來解決各種環境的要求及各種照顧弱能兒童的設備。斜形的屋頂是為了調節室內溫度而設的,因為斜頂在冬天能夠保溫,夏天產生陽光反射的作用。弱能兒童是特別對冷的方面敏感的,而管理機構因經費有限,不能常用暖氣設備,所以在設計方面我們盡量利用天然能源。這個項目的設計我是比較滿意,因為要保留工地中一幢50年代的舊樓,從舊樓的外形及角度伸展,構成了3個不同環境的庭園,加上屋頂採用了斜型,整個項目成功地反映出地方風格。

1979年我與大衛·羅素建築師合作設計了兩項公寓大樓,一個位於香港北角南豐路,工地北面有一水塘(見59頁)。設計構思盡量利用水塘及周圍的天然條件,把水塘的一部分引入首層平台的一面,使平台花園增加了水面,增加了休憩的氣氛。工地東面有多幢的高層公寓,所以把3幢高層的公寓放在工地的西面貼近山邊,爭取視野的空間。各幢公寓的佈局,是務求提供各單元都有遠景及三面的外牆,方便單元內有自然通風的條件。外形上3幢公寓的頂部反映了周圍的山型,成一斜線,減少3幢高層建築物在視覺上的體型。

紙幣廠的設計是一項難度高的考驗(見79頁),印製紙幣設備的要求非常嚴格,室內的溫度及濕度要控制,印製機器的安裝規定了廠房的佈局。庫房、機房及辦公大樓分佈廠房的兩側,使廠房有一大而空闊的空間。整個佈局包括內外牆身的磚位、門窗等,是根據一個1600x1600的規格,創造出一個有規律外形,反映出印製廠的工業形像。整個項目層高兩層,基本上有4個主要功能的範圍,廠房、庫房、機房及辦公樓。而大多數的外牆都是密封式沒有窗口。恰逢主要入口的範圍內要分為兩區,一為客戶入口,一為員工入口。所以用不同高度的水塔來表達兩個入口的特徵,并給整幢廠房一點標誌的外形。因業主保密的要求,所以不便刊登內部圖紙,而內部佈局也不便多言了。

1983年我與大衛羅·素及英國的拿土敦建築師合作參與香港城市理工大學的設計比賽(見81頁)。在設計工作中,我們聘請了風水專家提供土地佈局的指引,加上實地考察的資料,決定將工地北面的山林保留并引進校園內,而使部分建築物與大自然相連,增加校園的綠化。項目的核心是一中心廣場,通過直線型的行人網絡,把學生帶引到大學的各教學設施,校園內增設了多處戶外休閒的小廣場,提供了清靜學習的氣氛及優美的大自然環境。這個項目與英國著名的拿土敦建築師合作設計,他有設計多間大學的經驗并是倫敦國家歌劇院的設計者,通過與他合作吸收了不少設計的心得。當年我安排他到北京向中國建築學會作了學術報告,而且他的報告後來還在《建築師》雜誌刊登了。

1984年我做了一個室內設計項目,就是中華文化促進中心。在這個室內項目,我大膽地用了中國書法的藝術來作主題,并採用傳統宮庭的色彩作牆身的裝飾,增添中華文化的精神(見139頁)。

1985年開始正式進軍國內,第一個項目在西安市郊臨潼縣,工地位於近華清池的一個山坡上(見85頁)。自從1974年我第一次到國內參觀後至1985年間,國內的單位領導及建築師們都忙於興建滿佈玻璃幕牆的高層飯店。經過多次到國內參觀後,我覺得在西安地區當時最能適應外國遊客及符合地方經濟的飯店建設,就是盡量利用本地建材及富有地方特色的現代化設計風格最為理想,所以在牡丹山莊的設計中,我簡單地引用了當地傳統窰洞的半圓型入口,及單面斜頂的黑灰瓦屋頂,使之在視覺上有地方風味的感覺。而電梯塔的外形是吸收了傳統磚窰,及50年代的工業煙囱的特色演變而成,可惜這個項目最後沒有實現,只是能變成紙上談兵的學術研究。

鵬鷹山莊是繼牡丹山莊後的第二間小型飯店(見87頁),位於蘭州市北側的一個山谷內。當時幸得當地的老專家任震英老先生指點,領略到當地傳統建築及生土建造的心得。我花了點時間去研究當地的建造方法及窰洞的構造,發現多方面傳統的營造方式在平價飯店的建設上,是最適合採用的,所以項目全部非結構的牆,都採用了生土建造方式,而一部分的客房是依山而建,採用了改進的傳統窰洞通風設施。在夏天把涼風通過通風管道帶到每一房間內,省了空調設施,平價客房提供了既有效又經濟的自然通風設施。其他的客房,採用了經過時間考驗的庭園式佈局,務求適應西北的氣候。屋頂有兩個做法,有綠化的做法,增加冬暖夏涼的效果;另一做法可用斜頂裝設太陽能板,提供經濟的熱水供應。整個項目盡量採用適當改進的傳統設施及符合經濟條件的建造方法,把傳統建築現代化。我對這個項目覺得有點滿足感,不單是能夠適當地把傳統空間的構思用於現代化的設計中,同時把傳統的營造方式改良及採用,創造出一個既經濟又富有傳統色彩、更有時代精神的低層飯店,與當時流行而奢侈的高層玻璃幕墻項目,提供了一個參考,希望有關人士能回頭是岸。現代化的設計是要實事求是,配合國情,不可盲目抄襲,要對地方氣候、風土人情有適當的交待。

1988年至1992年因泰國經濟起飛,我便移師泰國發展我的業務,在當地註冊了一間Taipan Development Ltd.的公司,提供開發顧問服務及建築設計。通過在亞洲建築師學會的工作中,認識了不少亞洲各地的建築師,在泰國認識一位泰國建築師學會的前會長,除建築設計工作外,她也有投入開發工作。她的公司在曼谷有多塊工地,結果在1988年她邀請我在一塊市中心的工地上,提供設計服務。當時曼谷市區的交通已經十分擠塞,所以對曼谷市的情況了解後,我便提議建一幢商住式的大樓(見354頁),這樣可向部分在大樓工作的高級人員在同一幢大樓內提供一些住宅單元,減少每天往來上班的交通。如果曼谷市內有多幢這樣的大樓,我肯定交通情況一定有所改善。大樓的前部分是一多層的辦公樓,有獨立的電梯及防火梯,而後部分是一梯4戶的公寓樓。每個公寓是2層式,前面的兩個公寓的起居室通過辦公樓與隔鄰建築物間的空間,可以有遠景。而工地後面有一運河及低層建築物,所以後面的兩個公寓的起居室有更廣闊的視野。公寓有4面外牆,入口通道是露天平台,所以室內有一流的通風條件。晚上市區十分寧靜,所以公寓雖然置身鬧市中,到晚上卻有甚為優美的環境。

1989年分別在泰國海邊承擔了兩個項目:一個位於曼谷西南面一個海邊度假勝地(見93頁);一個位於芭堤雅北面海灘(見95頁)。兩個工地的形狀相同又長又窄,處理方法一樣,把大樓放近馬路,減少汽車深入度假區,把近海灘的一面留給泳池及一些別墅。兩幢皆是提供度假居住用,所以很注重海景及適應熱帶氣候的設施,例如每個單元都有闊大的陽台,及遮陽板和自然通風的條件。同年在泰國北部清邁市郊外有一度假山莊(見97頁),為了要適應各種不同背景的業主,所以山莊有10個不同的別墅設計,有歐陸式的山舍、美式的農莊、中式的庭園及泰式的住屋。所以有時為了適應市場的要求,建築師雖有崇高的志願,也要向現實低頭。為了要實現我們的理想,就要把現代化的建築設計普及化,使業主甚至一般市民了解及欣賞高水平的設計,才能把整個社會的建築水平提高。

1989年應太太的要求,在她的家鄉台灣西北部宜蘭縣買了一塊山坡下的農地,並且設計了一幢有4個睡房的農舍(見91頁)作為度假之用。工地位於高山之腳,有山泥傾瀉的危險,所以整幢農舍是在地面一層興建。宜蘭縣是地震區,農舍的結構採用了鋼筋水泥架及土產磚,外形以斜頂及清簡的外牆和木欄杆為主,構成比較有地方風格的一個組合。

1990年在曼谷有兩個項目。一個位於運河邊的4層平價住宅(見99頁),這個項目主要的要求是提供一些既經濟又實用的家庭式工場,所以每個單元的佈局非常簡單,盡量把固定的設施例如樓梯、電梯、浴室及廚房組合成一角,其餘的樓面由個別業主按他們自己的要求,自行分格,一來減低成本,二來提供比較靈活的工場及住宿的空間。另外一個項目是一高級公寓(見355頁),每層有4個2間睡房的公寓,整個佈局是由1981年在香港設計的豐樂閣(見73頁)伸展而來。在大樓的中部提供了一通風的天井,使各公寓有三面的外牆及天然通風的條件,我認為在熱帶氣候,這樣的佈局比較合理。1991年我在曼谷再有2個項目,一個是9層高的公寓大樓(見103頁),這個項目因為工地面積小,加上市區規劃的條例及高密度的要求,局限了在設計上的靈活性。而當時對泰國傳統建築認識不深,在屋的頂部雖然加添了一些所謂地方色彩鋼管架,但現在看來,覺得膚淺。另一項目的工地地形不太理想,工地近馬路的前部分狹窄,後部分伸展到後街(見101頁),但業主要求設計一幢高級辦公樓及公寓樓,真是雪中加霜。經過多方面的考慮,我提議在工地的前部分做一幢兩層高的玻璃商場及陳列室,把行人吸引到工地的中部,用空中步廊與辦公大樓連接起來,而公寓大樓就放在工地的最近後街的部分,以後街為主要入口。辦公大樓採用了八角形,提供多面的視野,而公寓樓的服務性設施,例如電梯及樓梯,盡量收縮,使每個公寓都有3面半的外牆,提供了良好的通風條件。

"河畔小築"是我在泰國最後一個項目(見105頁)。工地位於曼谷市郊一條小河畔,週圍環境優美。我提供了兩個方案,一個是把河水引入工地,構成水鄉的氣氛;另一個把所有住屋沿河而建,提供比較大的康樂活動場地。所有的房屋都採用了斜頂,一來適應地方氣候,二來配合鄉村風味,在項目核心的會所樓在設計上,特別加上地方色彩,是現代化傳統設計的另一嘗試。

1993適逢杭州好友董孝綸建築師介紹,認識了當地在友好飯店任職的經理吳啟元先生,他提供了一些在杭州及附近的開發及設計項目。其中一個就是金家浜花園(見107頁),位於杭州南面蕭山市,工地三面臨水提供了水鄉的背景。項目中除了有一部分的別墅是沿水邊之外,也有一部分別墅靠運河邊。運河橫穿整個工地,增加水鄉的氣氛,而每幢別墅都有南向的花園,居民可通過綠化的人行道,直達項目的會所。可惜工地被一條公路分割,只好在路的兩旁用4層高的公寓樓分隔,給別墅群比較寧靜的環境。從這項目的佈局,可以看出在別墅群的項目中,如果密度要高的話,佈局便限於排形,比較規律化,幸好在其中加設運河的景觀,軟化了排形的佈局。

1994年6月被北京的敬遠房地產開發有限公司聘請為"西西工程"的規劃總顧問。對北京規劃管理局及北京市建築設計研究院所制定的北京西單北大街西側商業區"西西工程"1~9號地塊土地使用規劃條件進行評估及提供深化規劃設計。我公司組織一個顧問團,成員公司包括:吳享洪建築師有限公司、香港城市規劃有限公司、MVA亞洲運輸策劃顧問公司、黃澤恩顧問工程事務所、利比建築工料測量師及北京市建築設計研究院。規劃地區位於西單北大街西側,佔地20.85公頃。我公司對北京的整體規劃"西西工程"地區的情況及規劃局的規劃條件進行了深入的研究,顧問團對交通、運輸、人流等事項也進行了深入探討。規劃提議將人流、車流分在不同的層面(見111頁),車流主要在路面及地下2層,地下的車流可從西單北大街西側通過地下通道到東側的部分新建設通到東側的地面道綱,形成一個地下過街的車路系統。人流在地面行人路圍繞各地塊及在多處設置廣場及標誌性的雕塑,和有部分現有街道改為行人商業區。除此之外,還提議在二層增設一個跨區的行人綱絡,使行人能夠在二層的步廊通過跨路的行人天橋,走過整條西單北大街兩側,整條西單北大街的地下改為一個地下城的構造,有地下鐵車站、商場、過街車路及行人路,還有一條地下自行車路把自行車從北面街頭到長安街十字路口南面安排在有自然光井的地下車道行駛,避免了自行車在路面產生的交通混亂,可惜這個比較超前的地下自行車系統,被當局的保守意見最後否決。

在建築造型上,我提供了多項基本做法,在最南端的重要地塊,當時已有意賣給中國銀行作為總部大樓,我提意八角形大樓,把行人的視覺帶引進西單北大街。同時提供一個比較有傳統特色的大樓,作為西單商業區的門樓,可惜貝聿銘大師的中銀大廈設計沒有接受我的心意。

1994年10月我等完成了一個規劃評估及深入研究報告,提供了現有規劃的評估、城市設計大綱、建築設計總則、土地判用及分佈構思,行人及零售架構、綠化與社會空間、西西工程實施計劃、城市設計的形態及風格、建議的機動車通道、自行車通道等系統及總體交通組織佈局。這個報告比較深入及完整地把"西西工程"推進了可以實施的階段。

1995年我公司被聘請為2、4及5號地的總設計顧問。3個工地的總建築面積近400,000平方米,所以我公司決定與吳享洪建築師合作設計2號地(見115頁),而4及5號地就單獨由我公司負責。2號地最初方案的東立面有一斜形多層的玻璃步廊塔,主要是表達商場的功能,吸引行人視覺,後來因為2號地是回遷樓而造價低廉,所以業主要求修改外形,以配合造價的低標準。國內的業主對建築師修改圖紙的時間及費用有簡單的看法,而且不太接受我等合理的額外收費要求。為了簡化工作,所以我決定把2號地方案設計以後的工作歸由我公司負責,把玻璃步廊塔的設計簡化。商場內部的佈局十分簡單,電梯、樓梯及廁所等設施放在沿邊的位置,使得中央留出一個比較大的商場空間,辦公層的中央加插了光井,減低辦公室的深度。增設過街樓有兩個好處,一可以與隔鄰的建築物相連;二可以增加建築面積,但對建築物的外貌有一定的影響。外牆的設計,在辦公層的窗是安裝於結構架的內邊,加上遮陽板,使外窗對風雨及烈日有適當的保護。商場的外牆不需要太多的透光要求,整個7層高的外牆只有少量細窗,樓身上下部雖然有不同的處理手法,但在綫條上是有統一的構圖系統。整個外貌清楚地表達了功能的分別,提供了一個簡樸而富有時代精神的設計。

4及5號地的設計,可算我的設計生涯中的另一里程碑(見117頁)。在這個項目的設計工作中,我面對了一個難題。項目的內容主要是大型的商場及辦公樓,這些功能與傳統建築設計是拉不上什麼關係的。但在北京市有富有傳統精神的雄偉建築、大規模的建設,中軸線及居民的生活方式以及北京大部分的建築物,無論是傳統或現代的都是帶有王者味道及雄偉的色彩。一般城市的居民都有在路旁乘涼的習慣,尤其是在夏天,因為路旁總比大部分民居涼快。總括來說,我認為北京的建築傳統精神是它的建築尺度的雄偉,平衡對比的佈局和居民的生活方式。西單北大街是北京最繁華的商業街之一,4及5號工地是區內最主要和最大的地塊,總建築面積有320,000平方米。建築物的高度限制於12層(45米)。4及5號地是把兩幢大樓以過街樓相連為一體,為了適應居民徘徊路旁的習慣,及創造一個社交中心點和大樓的主要入口,我在入口與行人路之間提供了一個相連的空間,這個空間把行人帶引到大樓的入口,每層的商場是從一中軸綫伸展,使每層的商場都有明確的佈局。首層及2層的商場成為室內的十字路口,把樓外的行人通過首層的四面入口及2層的步廊入口引進商場,這是把這兩層的商場與過街的步廊相連變為城市行人路綱的一部分。

大樓的體型是用室外防火梯分割,減低體量及在立面上提供一些垂直綫條,外形是從構思的初稿深化而來。造型仿彿有傳統宮庭建築的雄偉,樓身頂層的遮陽板與樓身下部簡單的外牆處理，不單是表達了不同的功能，更創造出一個豐富和清簡的對比，與傳統宮庭建築的外貌有異曲同工之感。我有意無意之間把傳統建築的精華融化到外形的設計內，並提供了一個相連空間去適應居民的傳統休憩習慣，增加了大樓的吸引力。我發覺在12層高的建築物能夠伸展現代化傳統設計是有限的，但這兩個因素在北京地區是與現代化設計有密切的關係。我在這一大規模項目中，嘗試現代化傳統的設計手法，但我認為這只能作為一個過渡式的產品，成功還有待時日。

"西西工程"是至今在我的生涯中最大規模的一個項目。包括了城市規劃及3個個別地塊的建築設計，總造價為人民幣2,047,000,000元。從1994年6月到1997年我公司集中精力去完成我們的任務，2號地已完成了施工圖設計，有待開工。5號地因業主把整個商場部分賣給別人，我公司做完初步設計階段的工作後，新業主便為了節省我公司及其他香港顧問公司的收費，便將以後階段的工作委託給原本與我公司合作的北京市建築設計院去做。其實新業主此舉並不明智，因為這樣龐大的投資，單是保險的收費都比我等的收費高，而且施工圖及施工期間的工作是每一個項目最容易出錯的階段。因小失大，我等唯有搖頭嘆息。4號地因北京商場過盛，我等做完初步設計階段便應業主要求停工，至今已一年多了。

1997年受全國人大常委霍英東先生邀請，就北京人民大會堂的"香港廳"提交了一個室內設計方案(見141頁)。我把英國傳統的板牆裝飾及中國傳統宮庭的頂部色彩合併於一室，務求創造出一個富有中外傳統色彩的空間。傢俱挑選了明式長櫈及英國傳統的皮椅，及美國著名建築師設計帶有中國味道的小椅，加上以鮮明的櫻木為主的牆版及金底色的牆紙，創造出一個富有朝氣的氣氛，充份表達了特區的精神。

1997年下旬，接到一項在上海武康路的高級公寓項目(見123頁)。樓高7層，每層有3戶有3至4間睡房的公寓。起居室有自然通風條件，而在每層中央的電梯大堂比一般的寬闊，中間有光井，提供了一個豪華的氣氛。公寓樓的外形嘗試採用一些傳統上海建築的色彩。

北京龍潭湖度假村位於龍潭湖公園內(見125頁)。週圍環境清靜，別墅及會所的造型為了要配合週圍園林的環境，採用了斜頂及富有傳統氣氛的庭園佈局，會所位於整個方形地塊的中心點，而其他設施及別墅按日照及視野的要求以45度角伸展，形成了一個以中軸綫為主，而有傳統建築風格的佈局，這是我以現代化傳統建築設計手法的另一新嘗試。

四川都江堰青城別墅山莊面積有850畝，位於成都市郊一高爾夫球場對面(見127頁)。我公司首要的工作是提議一個山莊的規劃，根據地理環境及山莊內的人流及會所的關係，提議把原來業主指定的會所位置西移，使人行的距離在最長的情況下不超過600米。其次把整個山莊以綠化的步行徑貫通，使居民能夠通過步行徑，直達會所及其他娛樂設施，盡量與車流分離。然後對各大小地塊的面積進行分析及研究，根據市場情況及不同的分佈圖，選出最理想的地塊面積。在一標準地塊上研究最理想的別墅位置，並提議山莊內的道路綠化指標及盡量保留原有的大樹群，使山莊演變成一個別墅量高而又有充分的度假氣氛的一個山莊。這個設計使我領略到在別墅山莊規劃中，一方面要有高的別墅量，而另一方面能夠減輕一排排別墅群的規則性，實在十分難辦。

1998年最後一個項目是參與深圳市中心少年宮的公開設計比賽。其實我對參加設計比賽興趣不大，通過我多次擔任這類比賽的評委後，我對設計比賽實在有很多意見。但這次因為有關當局來信邀請，所以卻之不恭，在短短3個星期內，日夜趕工，提供了一個方案。

我對深圳市中心區十分熟悉，因為在這兩年內參加了不下5個區內的大型項目設計比賽的評委工作，包括：購物公園、廣播中心、文化中心及黑川紀章的中軸綫設計和李明義的市民中心設計的顧問工作。而深圳市本身是一個現代化的新城市，所以我認為設計應以完全現代化的風格(見129頁)。恰好項目的內容少年宮本身也隱藏著一些面對將來的意味，所以無論在佈局或個別建築物的處理手法，我都採用了清徹而大膽的作風。科技館以一個大型的飛機庫的外形，提供靈活的內部空間，來適應科技展品的不斷變化的需求。多功能廳及太空館實事求是，把內部空間的造型表露於外形上，形成兩個大型的室外雕塑。少年宮的大堂成為項目的核心地帶，人流的交匯點，半開放式，使之有自然通風，節省空調設施。同時整個項目的造型也對城市景觀提供了交待，科技館的半圓型，創造出在視野上新市區中心東北面的入口標誌。對這個項目設計，我頗為滿意，可惜落選，實在有點失望。

回顧我30年來多元化的探索工作，總有心有餘而力不足的感覺。雖然做過的項目有60多個，但成功實現的只有10多個。通過這30年的磨練，從各項目的設計風格來看，我的設計已有肯定的手法。我喜歡將建築設備、廁所及梯間等與主要的功能空間分離，但我總是覺得這樣的安排是最理想的。同時我喜愛運用六角形作為主要設計的體形，因為六角形能減少不需要的樓面面積，及提供多方面的視野。我在設計中，對地方的風土人情特別注重，從1979年開始投入國際學術交流活動，並且每到海外參加會議，我便趁機往書店搜羅當地的傳統建築及生活資料，尤其是對亞洲的資料更感興趣。最近幾年因公務繁忙，只能淺讀所收集的書籍，所以還未能在設計中產生效果，但我認為，建築設計除了要充分表達風土人情之外，還要採用最適當的科學技術，務求設計既有地方風格，又有實事求是的功效。我覺得今後應多花時間去研究中國傳統建築及世界各地先進的建築科技，務求在創造現代化中國建築設計上多做些工作，希望在設計道路上多建幾個里程碑，提供優美的建築空間，為人民服務。

I was born in Hong Kong in 1942, during the Japanese occupation. My ancestors came from Fujian Province in China and settled in Panyu District in Guangdong Province in 1776 during the forty first year of the reign of the Qing Emporor Qianlong. My home village is now part of Hai Zhu District in Guangzhou City. At 16 I was sent to school in England by my late father. That is one of the reasons why my written Chinese is not as good as it should be. Fortunately my late father arranged to have a home tutor during my primary school days to teach us on classical chinese. Up to now I can still recite the Ode to Righteousness by Wen Tian Zheng. After graduation in 1968, I returned to Hong Kong and started my working life at the monthly salary of HK$1,800. Luckily with the help of a relative I was able to rent a big room for my wife and child for HK$500 and used the remaining HK$1,300 to support the family. My late father did not give me any subsidy. There are some colleagues who mistakenly thought that I came from a very rich family and can survive even without working ever since I started practicing. However only I would know the pain after losing so many potential clients because of that insinuation. One works not only to earn some money but, most importantly, to serve the community so that our parent's and teacher's efforts are not in vain.

In the summer of 1958 I took the boat to England via Italy. In the beginning I took some lessons in English and in September I went to a boarding school in the County of Somerset in the south west of England. There I spent two severe winters. In the autumn of 1960 I was accepted by the Poole College of Art. At that time I wanted to be an artist. After one term, I applied and was accepted by the famous Bournemouth College of Art to continue my studies. In the art college I was interested in nude drawings and ceramics. However after a while I felt that the life at the art college was too relaxing and wasteful and even though I might become a successful artist, the benefit to the society would be limited. I liked to take up a profession which could directly benefit the society. At that time my god father was an architect in Singapore and perhaps because of that I decided to take up architecture.

What followed was one year in London studying subjects that were required for entry into architectural schools. In the summer of 1962, I was accepted by the Architectural Association School of Architecture. As early as 1961, I started collecting cuttings from magazines and newspapers and pasted them in scrapbooks. This habit has been maintained to the present day and I have by now accumulated near to 20 such scrapbooks. The scrapbooks are like a pictorial diary, recording my interests at the time. One can see from the early scrapbooks the materials are full of spirit of youth eager to learn. Unfortunately the material in the later scrapbooks reflect a weakness of middle age man full of photographs taken with well known personalities and newspaper articles about myself. Luckily the more recent scrapbooks now contain matters of academic interests and of youthful spirits. I left for England. in the summer of 1958 and returned to Hong Kong for the first time in the summer of 1962 for a holiday. The hot summer and the good news that I have been accepted by the AA School led me to buy a few pots of household paints and painted several large abstract paintings and portraits. I considered that at least I had been an artist for a short time. People commented that my paintings were influenced by Jackson Pollock and chinese landscape paintings. (see page 193)

After studying four years in the architectural school, my tutor suggested that I took a year out and recommended that I start work at Austin Smith/Salmon/Lord Partnership. In that office I did some medium rise housing design. After two months I left and went to work in the Architect's Department, London Borough of Haringey. I was put into a team of four people responsible for the design of some low rise housing projects and shopping centres. This was my only time spent in a government office to date. During that time there were two unforgetable matters. One was that everyday at 11:30 am. and 3:00 pm. the compulsory tea break took place and the other was that everyone must leave by

5:00 pm. or else you would be chased out by the cleaners. I rejoined the AA School in the autumn of 1967 and entered the post graduate Dept. of Tropical Studies headed by the famous tropical design expert Dr. Otto Koenigsberger until my graduation in the summer of 1968. My thesis "Sky Courtyards" was concerned with high density housing in Hong Kong (see page 31). The site is situated in the south west of Hong Kong named Wah Fu Estate. The concept utilized high rise solution by putting the units in a long block which crossed the width of the site leaving a fair amount of ground recreational areas for the residents. The units were designed based on the living pattern of the inhabitants with a degree of flexibility for change. At the time my tutor's felt that the extra costs of the high rise solution did not benefit the inhabitants greatly and though my thesis proposed a new way of living in the high density situation in Hong Kong, it did not resolve all the associated problems such as high building costs, the effect of such a large block on the surrounding environment and associated traffic problems. Looking at it now, if one compares my solution with that executed by the Hong Kong Housing Authority, at least I provided more ground recreational spaces for the residents. I still think that with a high density of 2000 per hectare, how can one not use high rise as a solution.

The summer of that year, I returned to Hong Kong and immediately was employed by Peter Y.S. Pun & Associates. At that time I was the only qualified architect in the office, therefore most of the project was given to me to design. For a fresh graduate, this was the chance of a life time. I have worked in Peter Y.S. Pun & Associates for four years and was responsible for designing some twenty four projects. During that period I utilized the design principles of Le Corbusier and Lou Kahn, two of my most admired architects during my student days, in my design. Particularly Lou Kahn's "the servant and the served" concept was used many times. This created a clarity in the disposition of different functions within a building, but at the same time produced a 'cool' kind of environment.

The Guard House (see page 35) at the Shaw's studio staff's dormitory can be considered my first built project. The off form concrete and pattern on the external wall treatment obviously reflect an influence from Le Corbusier. The layout at Shaw's residence (see page 37) grouping the services and servants' work room in the centre with the living space and bedrooms spreading from it, allowing views of the surrounding countryside demonstrated one of the advantages of utilizing "the servant and the served" concept in design.

My boss at the time Mr. Peter Y.S. Pun was an experienced engineer. He had an unique & innovative approach to structural design. At that time he was the first engineer in Hong Kong to use precast pre-stressed elements in the design of factories and office projects. I have learnt a great deal from him. In 1968 and 1969 I designed several small projects including petrol filling stations, motorcar showrooms etc., (see page 41). One of them was a precast petrol filling station. The client required that business be continued during the construction of the project. This necessitated using precast design. Although it involved complicated phased construction, we managed to satisfy the client. The best projects to employ precast pre stressed elements are multi storey factory buildings and office blocks. This kind of structure requires structural walls or columns at either end of the building with the precast pre stressed beams and floor planks placed between them. If elements such as stairs, lifts and lavatories are positioned on the two ends of a building layout, they can provide for the necessary core walls support required. This will create a columnless space in the middle of the layout suitable for the flexible layout of factories and offices. I had used this concept to design a number of multi-storey factories (see page 348) and the Oversea Trust Bank headquarters building (see page 51).

In 1969 I designed two secondary schools. Siu Ming catholic secondary school (see page 43) is situated on top of a hill. As the site is rather exposed, I put the classroom and laboratory blocks along the western edge of the site to provide shading of the south facing basketball court. The classroom block facing west is provided with concrete external sun shading panels. This perhaps reflects the result of what I have learnt at the Tropical School. The other school Pooi Tun secondary school (see page 45) is more challenging in design. The site is long and narrow and has two 45°C slopes in two directions. The school has four basic planning units, the assembly hall, classroom block, laboratory block and service core. In order to reduce excavation work to the slope, the assembly hall was raised and positioned between the classroom block and the laboratory block. In this way the floor space under the assembly hall could be used as a covered play space. In the design of both schools, the structural system is expressed clearly in the external elevation. Between the two designs, Pooi Tun secondary school is a better product. Looking back after twenty odd years, I feel that my external elevation design at the time was full of complications and contradictions.

Since 1970 I have the habit of designing my own new year greeting cards (see pages 157). In the early years I used some interesting patterns or calligraphy as a theme, but, from 1984 onwards I made up a poem which reflected the current state of affairs in Hong Kong or China and selected amongst the photographs I had taken during that year to illustrate the poem. I was hoping to generate some smiles when my friends received my cards. All these years some of my friends asked me to produce a book on these cards as a record. Therefore I decided to reproduce them here in this book for them.

In October 1984 I attempted to organize a symposium between architects from China, Taiwan and Hong Kong called the "Lanting Meeting" and had designed a logo (see page 154). This symposium would have gathered leading architects from Beijing, Shanghai, Tianjin, Guangzhou, Taipei, Tai Chun and Hong Kong to discuss "the relationship between the spirit of traditional Chinese environment and modern design". In the design of the logo I used the logos of the two professional bodies in China and Taiwan Provence by putting the two logos together but turning them at 90 degree angle. The two logos both used the traditional chinese architectural "bracket system" as basically all architects were originated from China with the only difference being in their use of colours. The professional body in mainland China used red and that in Taiwan Provence used gold. The reason I put them at 90 degree was to express my aspiration for the symposium which was to hope that those attending would use a 90 degree turn attitude to create a new outlook for the symposium theme. I put a thick line between the two logos to express the Hong Kong position in bringing the two organizations together. The colour I used a blending colour between red and gold.

Interior design is my secondary profession. From 1969 onwards I utilized the spare time in the evening to design a number of interior projects. In every project I have tried my best to provide practical as well as pleasing designs. For example in the design of the information kiosk for the Festival of Hong Kong (see page 137), I have used brightly coloured panels as doors and display panel for the kiosk. In the design of an interior for a country house, I provided a floor seating area for the living room(see page 136) to create a holiday like and relaxing interior. I am also interested in furniture design. In the design of an apartment interior, I provided the design of a set of leather sofas and a drinks cupboard with colour dye panels and perspex stands (see page 138). In another design for an insurance company office, I used perspex panels for the partition and light weight leather chairs to create spaciousness in otherwise a very small office interior (see page 138). In the design of a cinema interior, I attempted for the first time to use dye colour on wall panels to create a translucent and cheerful wall finish for the cinema lobby (see page 40). In a narrow passageway I used light coloured precast concrete sculptural panels to create a much wider space impression. I tend to use a lot of new ideas in interior design work as in this scale of operation, mistakes can be rectified easily.

In 1973 I joined the Hong Kong Institute of Architects as a member. Since then I have continuously participated in the work of other professional bodies as well. Union of International Architects (1987), Commonwealth Association of Architects (1979), Architects Regional Council Asia (1979), Architectural Society of China (1980), Hong Kong Association for the Advancement of Science and Technology (1985), and Architectural Association Asia (1990), The work with these professional bodies are recorded elsewhere in this book.

In 1973 I founded my own design firm Ronald Poon Associates four years after graduation. The first project was to design a shop house in Shek-O (see page 54). The total floor area of the house was about 100m2. Although it was a small project, it gave me an opportunity to practice Lou Kahn's "the servant and served" principle of design. The two staircases and the lavatory occupied half of the total floor area and what was left could still satisfy the various functions such as a light well in the middle. This light well not only helped to ventilate but also provided daylight to the ground floor lavatory which normally could not be located in the middle of the house according to the building regulations. The external kitchen is placed above the septic tank. This little shop house seems to have expressed well my attitude towards a clear distribution of different functions and the maximizing of use of space in design.

My second project was to design an extension to the Garden Bakery. The existing bakery was designed by Hong Kong's famous Palmer and Turner office and the external shape of the building is quite beautiful. In my design I wanted the extension to blend in with the existing building and act as part of it. The external wall has no window opening and is solid except for some minor ventilation areas. This produces a plain white wall which blends well with the existing building. (see page 351)

Po Toi Island was my first holiday village project. (see page 57) The layout of the buildings follows the contour of the site. I used triangular shapes to differentiate this development from a normal residential shape and to create a holiday atmosphere. The site is situated on an island therefore a precast system of construction was used.

The design of Cactus Mansion (see page 67) was developed from the design of Winsing Building executed in the office of Peter Y.S. Pun & Associates in 1969. Although it is very much affected by the building regulation, it is somewhat better designed. In the speculative type of building development in urban Hong Kong ,after the individual owners bought their units, they would normally erect additional ugly looking cage extensions on their external walls to have some additional floor space. Unfortunately these extensions ruin the external elevation. In one project I produced an elevation design to counteract this problem (see page 351). This project is located in Ko Shing Street, Western District. The different units on the elevation were painted randomly in differenct colours producing a random colour pattern on the elevation. In this way any unsightly outside addition would be lost in the colour pattern.

In 1975 I entered an international design competition in Manila (see page 352). The subject of the competition was the master plan for development of a settlement district north of Manila which included the proposal for a proto type sector design. The estimate future population was 140,000. The site is at the Torndo Foreshore District of Dagat Dagatan. In my proposal I divided the district into 350 sectors. Each sector had their own primary school, community centre and shops. The houses were spread along two sides of a pedestrian path with public open space in between. The open space had a public water tap to retain a long time habit of the inhabitants. The government was responsible for the structural framework of the house and the inhabitants were responsible to construct the external cladding, roof and internal partitions. The house was raised 1.5m above the ground partly to avoid flooding and partly to provide a place for the poultry. This was my first competition entry and mainly my participation was to gain experience. Looking back I was a bit "big headed" as at that time I had no experience in this scale of work and compare to the winning entry, my proposal still had a long way to go.

In 1978 I entered another design competition and this time it was to design the Silver Jubilee Sports Centre in Hong Kong (see page 65). The centre had a complicated schedule of accommodation. I purposely expressed the major function of the centre in the external bulk design. There are four major elements in the centre, namely, the building services facilities, the large multi-purpose halls, the fixed use halls and the trainees' dormitory. The general planning was dictated by the orientation required for the outdoor fields and entry routes. There are two rows of building in the complex. The one near the road contains the "servant" type of facilities such as changing rooms, machine room, laboratory, offices, training centre and shop etc.. The one row near the outdoor fields contains the "served" type of facilities such as swimming pools, the training halls etc. In between the two rows of building lie the inter connecting routes. The whole complex was design to extent in one direction. The high rise dormitory block forms the centre of the complex. Looking back the whole design was very much influenced by some Japanese design at the time.

99 Caine Road, Hong Kong can be considered a milestone in my design career. This project had another earlier proposal back in 1979 (see page 73). The project was assigned to my office in 1981. (see page 349) The spiral shape on the top part of the building is a response to street lighting requirements in the building regulation. The whole bulk of the building has an unique character. It is 35 storey high and the external design not only reflects the structural framework, but also expresses the function of the interior spaces. The bedroom block is independent forming the external tubular shape with the living room placed in between two bedroom cores. The shape of the living room is based on the best furniture layout associated with the lifestyle of the potential inhabitants. In order to accentuate the spiral effect of the building, the external walls of the tubular form were tiled in alternate colours with the central service core painted bright red. The projecting window box design was derived from traditional Tibetan design and the lower part of the podium reflected my admiration of the design of the Potala Palace in Tibet. An architectural critic once remarked that the external bulk of the building reflected the majestic image of chinese traditional architecture.

Home for the severely handicapped children represented another milestone in my design career (see page 69). The project was designed jointly by David Russell and myself. The client required that a part of the existing building be retained in the middle of the site. This part contains a chapel which created restrictions in the layout of the project. The problem being how to integrate the old part with the new buildings and how to blend two different functions into a cohesive whole. In the design of handicapped children's home one had to pay particular attention to many aspects. The first aspect is that handicapped children normally feel uncomfortable in complicated and congested environment. Therefore the space has to be simple and open to facilitate a healthy living condition for the children and a good working environment for the staff. Secondly the children require constant care in their daily activities including bathing. Some of the seriously handicapped children have to lie in bed all day and require therapy treatment. Thirdly suitable environment will encourage the less severely handicapped children to participate in more activities. Fourthly the staff has to carry out regular supervisory work and suitable space will encourage the staff to work closely with the children and not to leave the area of work and lastly the living quarters of the staff will have to be specially designed as the staff are under pressure in their daily work and after work they need to return to a relaxing and nice environment. However their accommodation has to be near to the children's bedroom in case of emergency. Based on the above aspects one has to design a rather disciplined layout and yet the environment has to be full of warmth and family like atmosphere.

The administrative office is situated near the main entrance to the complex. The whole complex used basically the courtyard layout to resolve all the space requirement and facility requirement for looking after the children. The sloping roof was used to provide ideal weather protection. In winter it can keep the warmth and in summer it will reflect the sun light. The handicapped children are particularly sensitive to the cold and as the client had limited budget and could not afford heating all the time, we had to resort to using natural resources. I was quite satisfy with the design of this project. The requirement to preserve the old house belonging to the 50's period in the middle of the site inspired us to use the angle and shape of the old building to produce three courtyards of different environment and together with the sloping roof created a local identity for the whole complex.

1979 David Russell and I jointly designed two apartment blocks. One of them is situated in Nan Fung Road, North Point. The site has a reservoir in the north edge (see page 59) and the design utilized the reservoir and the surrounding natural setting to the maximum by extending the reservoir into the site on the podium level increasing the water element in the podium garden and created a more restful environment. There are a number of high rise blocks in the east of the site therefore the three blocks of the apartments were sited along the west edge near the hill side to provide a widen visual effect for the inhabitants. The layout of the apartments aimed to provide three external sides to facilitate through ventilation. The three blocks were designed with different heights sloping away from the hills to reduce the bulk effect of the multi storey blocks. The design of the printing factory for the Thomas De La Rue Ltd. was a test of high calibre (see page 79). The requirements for printing bank notes are highly demanding. The interior humidity and temperature has to be strictly controlled and the machine layout dictates the general layout of the factory. The vault, the machine room and the administrative offices were positioned on two sides allowing a large space for the main factory. The whole layout including the block line of the internal and external walls, the windows etc., was based on a planning grid of 1600 x 1600 creating a regimental expression to reflect the nature of the printing factory. The whole complex is two storey high and basically there are four major functions, the printing factory, the vault, the machine rooms and the offices and most of the accommodation are windowless. It so happened that the main entrance had to have two separate areas, one for the visitors and the other for the staff. Therefore we used different height of the water towers to express the two entrance function and to create a focal point for the complex. The client had requested confidentially in our design , therefore we are not at liberty to disclose any plans of the complex including any internal layout arrangement.

In 1983 David Russell and I together with Denis Lasdun from England participated in the design competition for the City Polytechnic University (see page 81). We employed the services of a "Fung Shiu" expert in the site planning and together with site investigation information, we decided to retain most of the natural slope on the north side of the site and included it as part of the campus area making part of the buildings surrounded by natural greenery to create a campus in the landscape environment. The centre of the campus is the central square and through straight line pedestrian routes lead the students to all the educational facilities. There are many rest gardens within the campus to facilitate quiet studies amongst nature. We collaborated with the famous Mr. Denis Lasdun in this competition and he had vast experience in university design beside being responsible for the design of the National Theater in London. I have personally learnt a great deal working with him. In the same year, I arranged for him to deliver a talk in Beijing for the Architectural Society of China and the talk was reprinted in the magazine "Architect" afterwards.

In 1984 I did an interior project for the Chinese Cultural Centre. In this project I used chinese calligraphy as a theme for all the wall decorations together with the colours of the traditional imperial palace to create an atmosphere full of the essence of the chinese culture (see page 139).

In 1985 I undertook my first design project in China. It is located outside the city of Xian on a slope above the Hua Qing Spring (see page 85). My first visit to China was in 1974 and until 1985 the city leaders and architects were busy putting up multi storey curtain wall hotels. After many study tours in China, I felt that at that time the most suitable form of hotel for foreign visitors in the area around Xian was one that used local material and technology with design reflecting local characteristics. Therefore in the design of the Xian Mountain Inn I used the simple half moon shape associated with most traditional cave dwellings and one direction sloped roofs with dark grey tiles to express the essence of local identity. The lift tower design was developed from the design of the traditional brick oven and the industrial chimney of the fifties. Unfortunately this project was not realized and the work remains a paper study. Lanzhou Mountain Inn (see page 87) was the other small hotel after Xian Mountain Inn. It is situated in a valley north of the city of Lanzhou. At that time I was lucky to have the help of an expert Mr. Ren Zheng Ying to familiarize myself with the local traditional architecture and ram earth construction. I spent some time in studying the local construction method and formation of the cave dwelling and discovered that traditional construction in small and low cost hotels was most suitable. Therefore all non structural walls in the project were to be constructed by the ram earth method. Part of the accommodation was positioned against the slope using improved traditional cave ventilation system. In the summer this system brings natural air into every room and eliminated the need to provide air-conditioning. The other bedrooms are planned in a time tested courtyard concept suitable for the climate of the north west of China. There were two ways to design the roof. One used turf to provide coolness in summer and warmth in winter. The other used solar panels to supply hot water. The whole complex utilized improved traditional techniques and economical construction methods to provide modern vernacular design. I was quite satisfied with the design of this project. It not only managed to use the traditional concept of space in a modern design but also facilitated the improved traditional construction method to create an economical solution with a design full of the essence of traditional architecture and the spirit of modern design. This provided a contrasting example compare with the extravagant high rise curtain wall solution which was popular at the time. Lets hope that those concerned will reconsider their design approach as modern design has to be practical and suitable to the economical and social conditions of the country. One cannot blindly imitate but has to take into account the climate and the social aspect of the locality.

The economy of Thailand began to boom between 1988 and 1992. I took the opportunity and started to developed my work prospects in Thailand by forming a company there call Taipan Development Ltd. offering development consultancy and architectural design work. Through the work with ARCASIA, I came to know architects in most countries in Asia. In Thailand I knew the past President of the Association of Siamese Architects. She was an architect as well as a developer and her company in Bangkok owned many sites. In 1988 she invited me to provide my services on a project in Bangkok. At that time the car traffic in Bangkok was already conjested. After familiarizing myself with Bangkok, I proposed a commercial/residential solution (see page 354). In this way some of the senior office staff can live in some of the apartments to reduce in a small way the traffic load. If more of this type of solutions were built in Bangkok, the traffic condition would certainly improve. The front part of the complex is a multi-storey office block with independent lifts and fire stairs, the back part is an apartment block with four apartments on each floor. The apartments are all split level design and the two apartments facing the front part of the complex can have distant view through the space between the office block and the adjacent building. The two apartments at the back enjoy the view of a canal and low rise buildings. Each apartment has four external walls and the entrance side is through an open access giving excellent natural ventilation through the apartment. At night the district is normally very quiet and provides a peaceful environment to live in.

In 1989 I designed two projects by the sea. One of them is situated at the south west of Bangkok in a seaside resort (see page 93). One is located at the north beach of Pattaya. Both sites are long and narrow and the layout are basically the same with the main block sited near the road minimizing the penetration of car into the site and leaving the beach side for swimming pool and some houses. Both projects are for holiday accommodation, therefore, emphasis was placed on sea view and provisions for the tropical climate such as each apartment had large verandah, sun shading and natural ventilation facilities. In the same year I designed a holiday resort in Chiang Mai, north of Thailand (see page 97). In order to design according to market demand, I designed ten different house types to suit potential purchaser from different cultures. The ten types included european mountain chalets, american farm house, chinese courtyard house and thai house etc.. I realized then that although we have an ideal as an architect, we sometimes have to bow to popular demand. Unless the clients and the general public begin to appreciate high quality design, we cannot raise the general standard of design in society. In 1989 at the request of my wife, we bought a piece of farm land near a mountain slope in Ilan, north west of Taiwan and designed a modern farm house with four bedrooms (see page 91) as a holiday house. The site is situated at the foot of a high mountain and in order to avoid landslide, I raised the building one storey above ground. Ilan has earth quakes, therefore, the house has reinforced concrete frame and in fill brick walls. The external expression mainly composed of sloping roof, simple wall and wooden railings creating a house with some local characteristics.

In 1990 I had two projects in Bangkok. One is sited along the canal with four storey of low cost housing (see page 99). This project aimed to provide low cost family workshop type of accommodation. The layout for each unit is very simple, putting the fixed elements such as stairs, lift, bathroom and kitchen on one side leaving a large space for the occupiers to partition themselves according to their own requirement. The other project is a luxury apartment block (see page 355). Each floor has four units of two bedroom each and the general layout was developed from the Albron Court designed in 1981 (see page 73). The middle part of the block has open light wells to enable the apartments to have three external walls facilitating natural cross ventilation. I maintain that in the tropics this kind of layout is most suitable. In 1991 I had a further two projects in Bangkok. One is a nine storey apartment complex (see page 103). The project had a very small site and the restrictive planning and building regulation hindered flexibility in the design. At that time I was not so conversant with traditional thai architecture and the tubular design on the roof was intended to provide some form of local identity but looking back it fails miserably. The other project had a difficult site with a very narrow area near the road and the rest stretched far towards the rear lane (see page 101) and to make matter worse the client insisted on designing a high class office complex with luxury apartment block. After considering the various aspects I proposed that a two storey glass showroom and shopping mall be put at the narrow area near the road to attract the passer by into the complex linked to the main office tower by an elevated walkway. The apartment block was placed near the rear lane using that as the main entrance area. The office block took on a hexagonal shape to facilitate multi view from the office and the service elements in the apartment block were kept to a minimum area so that the apartments could each have three external walls to facilitate cross ventilation. Riverside housing was my last project in Thailand (see page 105). The site is situated in the outskirt of Bangkok near a river with beautiful surroundings. I proposed two solutions. One was to extend the river into the site to create a village in the river environment and the other with the houses all sited on the river edge providing more space for recreational facilities. All the houses have sloping roofs mainly to suit the climatical condition and provide some form of local identity. The design for the club house which is the focal point of the village, reflects a concern for the local vernacular architecture.

In 1993 it so happened that my good friend Mr. Dong Xiao Lun in Hangzhou introduced me to his friend Mr. Wu Qi Yuan who was the General Manager of the Friendship Hotel in Hangzhou. Mr. Wu suggested to me several potential projects. One of them was the Riverside Garden (see page 107) situated in Xiao Shan, south of Hangzhou. The site is surrounded on three sides by the river, and this provided an ideal setting for a village by the river project. Some of the houses are sited on the water edge and some adjacent to the system of canals. The canals were introduced throughout the site to create a village in the river environment. Every house has south facing gardens and the inhabitants can reach the club house via landscaped paths. Unfortunately the site is cut by a road in the middle and four storey apartment blocks were positioned along the road to provide a quiet environment for the rest of the site. One can see from the layout of this project that high density layout will inevitably produce regimental rows of houses but luckily the canal system somewhat managed to soften the situation. In June, 1994 I was appointed lead consultant by the Beijing Jing Yuan Property Development Co. Ltd. for the "Xi-Xi Project". The work involved a review of the parameters set by the Beijing City Planning Bureau and the Beijing Institute of Architectural Design and Research on the western commercial district in Xidan North Street sites 1 to 9 and to develop a master plan for the district. I organized a consultant team for the work including Anthony Ng Architects Ltd., Urban Design Group (H.K.) Ltd., MVA Asia Ltd., Greg Wong & Associates Ltd., Levett & Bailey and Beijing Institute of Architectural Design and Research. The plan area is situated west of Xidan North Street and has an area of 20.85 hectares. We carried out a detail study of the general planning of Beijing, the "Xi-Xi Project" area and the planning regulations. The consultant team also carried further studies of the traffic, transportation and pedestrian flow of the district. Our proposal suggests segregation of pedestrian and vehicular traffic (see page 111). The main vehicular traffic is restricted to the ground level and basement two. The vehicular traffic at basement two can cross under Xidan North Street from both directions forming an underground road network. Pedestrian routes are provided at ground pavement around all sites with many open squares with sculptural nodes and part of the existing streets is designated pedestrian shopping streets. In addition there is a district scale pedestrian link on the first floor level with cross street pedestrian bridge across Xidan North Street and pedestrian can walk the whole length of Xidan North Street on the first floor pedestrian system. The whole ground under Xidan North Street is proposed as an underground city with mass transit station, shopping malls, cross street vehicular roads and pedestrian routes with an underground bicycle path running the length of Xidan North Street with light wells providing natural lighting. This resolved the constant problem of bicycle traffic on the ground. Unfortunately this proposal was a little ahead of the time and the relevant authority turned it down.

We had proposed basic approach to architectural forms for the whole district. The southern most site was about to be sold to the Bank of China for the construction of their headquarters building. I suggested a hexagonal shape building to visually lead the pedestrian into Xidan North Street and at the same time to provide a building with traditional forms as the gate building for the commercial district. Unfortunately Mr. I.M. Pei, the famous architect for the bank headquarters design did not share my aspiration.

We completed a planning report and in-depth study in October, 1994, which contains an assessment of the existing planning parameters, planning and urban design principles, architectural principles, land use and layout concept, pedestrian and retail framework, open space and landscape proposal, development programme and phasing, urban design form and character, vehicle circulation, overall transport requirements etc. This report provided comparatively in depth and complete studies bringing the "Xi-Xi Project" to a step closer to being realized.

In 1995 I was appointed lead and architectural consultant for the design of sites 2, 4 and 5 of the "Xi-Xi Project". The total floor area for the three sites are about 400,000m2. Therefore I decided to ask Anthony Ng's office to deal with the preliminary architectural design of site no. 2 (see page 115). I could then devote my energy to sites 4 and 5. In the preliminary design for site 2, there was a multi storey sloping glass tower enclosing the shoppers' ramp connecting the shopping floors. Mainly to provide a visual attraction for the passer by. Unfortunately no. 2 site was for resettlement accommodation therefore there was a tight budget control and the client requested revision to the glass tower to comply with the budget. The clients in mainland China normally do not fully appreciate the cost of revising drawings and in order to simplify management work I decided to deal with the design of site no. 2 after the preliminary stage and simplified the glass tower design. The layout of the shopping mall in site no. 2 is very simple with the lifts, stairs and lavatories put along the two edges of the building leaving a large space in the middle for the shopping activities. A large light well was added on the office floors to reduce the depth of the office. The advantages of the additional building across the street linking the two sites are firstly to link both buildings together and secondly to increase the total floor area, but there is a definite effect on the external bulk design. In the design of the external wall, the windows were installed on the inside of the structural frame for the office floors with the additional sun shade giving adequate protection against sun and rain. The external wall of the shopping mall did not require too many openings within the seven storey height. Although the external design for the office and shopping mall floors differed, but all the openings followed the same grid control. This gave a clear external expression of the functions on different floors and provided a simple design full of the spirit of the time.

The design for sites 4 & 5 can be consider another milestone in my design career (see pages 117). In the design of this project I was faced with a dilemma. The project consists mainly of shopping malls and multi storey offices. These functions do not exist in the traditional vernacular concepts. What exist however, are the grandeur, the scale, the symmetry and people's habits in the capital city of Beijing. I find even amongst the buildings in Beijing whether modern or traditional, there is an air of supremacy and grandeur. In most cities of China people gathers in the pavement very often. In the summer, particularly, as the pavement is cooler at night than their homes. Therefore I find the grandeur in the proportion of buildings, symmetry and peoples' activities are the most relevant vernacular aspects in the design of buildings in Beijing. Xidan North Street is one of the most busy commercial districts in Beijing. Sites 4 and 5 are the most prominent sites in the district with a total floor area of 320,000m2. The height limit is 12 storeys (45m). Sites 4 and 5 are designed as two buildings linked by bridge buildings making it into one hugh mass. In order to facilitate the peoples' habit of waiting, mingling on the pavement and to create a social node and entry point to the building, I have designed an intermediate space between the entrance doors and the pavement. This atrium serves as an introductory space to the complex, pooling the people off the pavement and into the shopping mall. The shopping mall is symmetrically laid out to facilitate ease of orientation within each floor. The ground and first floor acting as a cross road in the shopping activities as it is accessible from all directions, with pedestrian ground and upper level links to the pedestrian street at the back and to 1st floors of all the surrounding buildings. This is extending the building into the pedestrian matrix of the city.

The bulk of the building is divided by exposed staircases which I am very fond of using on my buildings. They serve to break up the building into more reasonable proportion as well as giving vertical eccentricities to the overall bulk. The external bulk of the building was developed from the initial sketch concept expressing a degree of grandeur reminiscent of the palace structures. This impression is further reinforced by the regimental display of partial sun shading on the office floors on the upper part of the building giving a contrast of complexity and simplicity between the upper and lower part of the building not only to express the different functions but also to reflect a traditional treatment method of the palace structures. I have purposely attempted to capture the essence in the discipline of traditional palace design in the external treatment of my building and have recreated an intermediate space to facilitate a time tested habit of the people to enhance the attractiveness of the buildings. These are the two aspects which I can find valid in the Beijing situation as modern vernacular in a building of 12 storey high has very limited application. This is my first attempt of this scale in modern vernacular. I can only treat it as a product of a transition period as I think I still has a long way to go.

"Xi-Xi Project" is my largest project to-date which includes master planning and architectural design for three sites. The total construction cost amounts to RMB￥2,047,000,000. From June 1994 to 1997 we concentrated our efforts to complete the work. The working drawings for site no. 2 has been completed and the building is awaiting construction. My client has sold the shopping mall in site no. 5, therefore when we completed the development drawings, the new owner in order to make saving of our fees and the other Hong Kong consultants' fees, appointed Beijing Design and Research Institute to continue the work. This I consider unwise of the new owner as in such scale of project, the cost of insurance alone is more than our fees and yet working drawing and supervision of work on site are the stages where mistakes can easily be made. They might have made a little saving to start but the possible lost caused by any mistakes may be more costly at the end. As for site no. 4 because of over supply of shopping malls at the time, work was suspended after we have completed the development drawings and it has been suspended for over a year and a half already.

In 1997 Mr. Fok Ying Tung, member of the executive committee of the People's Congress invited me to provide a design for the interior of the "Hong Kong Room" in the Hall of the People in Beijing (see page 141). I used traditional english wall panels and traditional chinese palace colours on the ceiling to create a blending of colours and textures from east and west within one space. The furniture was a mixture of chinese ming style tables and english leather chesterfield range of chairs and a small chair with chinese characteristics designed by a famous american architect together with brightly coloured burlwood wall panel and gold coloured wall paper created an environment full of bright spirit expressing the aspiration of the Special Administrative Region.

In the later part of 1997 I was appointed to design a luxury apartment block in Wu Kang Road, Shanghai (see page 123). The building is seven storey high with three apartments per floor of three to four bedroom each. The living room has natural ventilation. The lift lobby on each floor has a rather large space with an oval shape light wall in the middle providing an air of luxury. The external treatment is an attempt to reflect the colour of Shanghai's traditional architecture.

The Lun Tam Lake holiday villa complex is situated in the Lun Tam Lake Park in Beijing (see page 125). The surrounding is rather quiet and peaceful and in order to blend in with the environment, the design of the club house and villas used the sloping roofs and the traditional courtyard layout. The club house is located in the centre of the near square site and the other facilities and villas spread in a 45 °C angle from it depending on orientation and visual requirement forming a central axis layout full of traditional spirit. This is another attempt in modern vernacular design. The holiday villas project in Dujiangyan, Sichuan Provence has an area of 850 acres, located opposite a golf course outside of the city of Chengdu (see page 127). The first part of our work was to produce a master plan. I had taken into account the geographical condition and the pedestrian traffic flow to the club house and proposed that the original location for the club house suggested by the client be resited westwards resulting in the longest travel distance reduced to 600m. In addition I proposed landscaped pedestrian paths be provided between the villas and the club house so that the inhabitants in accessing the recreational facilities would be segregated away from the car traffic. I then proceeded to analyze the different sizes of the house plot according to market demand to arrive at the best size of plot. Further studies were carried out to select the best location for the house within a house plot. At the same time landscape guidelines were produced for the road system and suggested the preservation of all mature trees within the site making the project a place full of holiday environment despite the high density housing layout. This project made me realized that in the planning of high density villa complex, it is difficult to avoid the regimental layout.

My last project in 1998 was to participate in the design competition of the Youth Centre at Shenzhen. Frankly speaking I am not very much interested in entering competitions, mainly because through having acted as jury member for many design competitions, I have developed many reservations on design competitions. However this time I received an invitation from the organizers to participate in the competition and had difficulty in refusing. Therefore I submitted the entry after working day and night for the short period of three weeks.

I am very familiar with Shenzhen mainly through participating as jury member in not less than five large scale projects in the area including the Shopping Park, the Broadcast Centre, the Culture Centre, the Central Axis designed by Kisho Kurokawa and the Citizen's Centre by John Lee. Shenzhen is a modern city and therefore one should use modern design as an approach (see page 129). It so happened that the Youth Centre as a theme also reflects a spirit for the future. Therefore I have purposely used simple and bold design approach in both the layout and in individual buildings. The Science Museum is in a shape of a large aircraft hanger providing flexible space to cater for the ever changing display requirement. The shape of the interior space of the multi purpose hall and the Space Museum are expressed in their external bulk creating two large scale sculptures. The main lobby is the focal point of the complex and meeting point of the people. It is semi open permitting natural ventilation to flow through, thus saving on the provision of air-conditioning. The shapes of the buildings also blends in with the general townscape. The half round shape of the Science Museum created an entry form for the central area in the north east corner of the town. I was quite satisfied with the entry and was a little disappointed when it was not selected.

Looking back at my multi facet work in the last thirty years, I cannot help feeling that I have the heart but not the ability to design well. Although I have designed more than sixty odd projects, but only less than one third of them got built. Through my work of the last thirty years, I can see that my design approach has taken shape. I like to separate the major space with the services, lavatories and stairs as I do feel that it is the best way. At the same time I am fond of using hexagonal shapes as a design form as hexagon cuts out a lot of useless space and provide multi visual effect. I am particularly interested in the local vernacular and from 1979 onwards I started to involve myself in international academic exchanges. Whenever I was participating in oversea conferences, I always took the opportunity to visit the book stores to collect books on traditional architecture and material on lifestyles, particularly concerning the Asian region. In the last few years work has prevented me from doing too much reading and that is why one cannot yet see any result in my design. I maintain that architectural design other than being able to express a sense of local identity should also be able to utilize the latest in technology so that design can have both identity as well as being practical. I feel that from now on I shall spend more time in studying traditional chinese architecture and the latest architectural technology around the world in order to reinforce my quest for a modern chinese architectural design and hope to create more milestones in my design work to provide better designed spaces for the people.

建築設計
Architectural design

Taken near Nice, South of France, 1961. 在法國南部尼斯市近郊拍攝

學生時代作品

Student work.

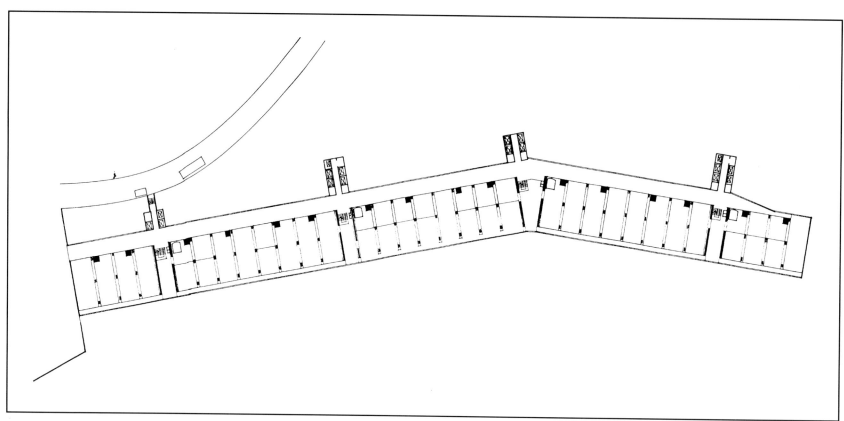

空中庭園
Sky Courtyards

英國倫敦建築協會建築學院(AA學院)畢業作品 (1968)
Final Thesis at the Architectural Association School of Architecture

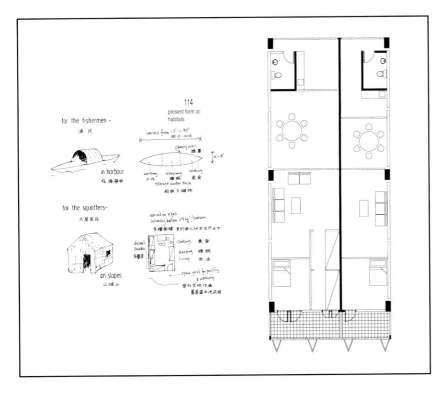

這是在香港南面鴨巴甸區徙置漁民及木屋區居民的大型公共屋村。項目利用了高層的設計盡量提供大面積的綠地及容納高密度的人口(每公頃1000人),24公頃的地盤容納的人口為24,000人。為了適應人口的增長及生活方式的變遷,設計構思採用了一靈活的結構架及預制間牆和可移動的間隔板牆。單元的設計是根據現漁民及木屋居民的生活方式,以一過渡性的佈局,使居民能適應高層居住的環境。

This is a public housing project to rehouse the fisherman and squatter population near the fishing port of Aberdeen in the south of Hong Kong. The high rise solution maximizes the amount of open space and achieved the high density required (1000 persons per acre). The total population of the 24 acres site amount to 24,000. In order to cater for the ever changing population growth and living habits, a flexible structural framework has been used with precast in-fill wall panels and movable partitions. The unit layout is a transitional arrangement sympathetic to both the fisherman and the squatters' existing living pattern to minimize the taumatic experience of living in a high rise environment.

在潘衍壽土木工程師事務所的作品
Projects at Peter Y S Pun & Associates

邵氏製片廠警衛屋
"Guard house", Shaw's Studio

香港清水灣道邵氏製片廠員工宿舍 (1968)
Shaw's Studio Staff Quarters, Clear Water Bay Road, Hong Kong.

這個項目是在香港潘衍壽土木工程師事務所任職時以主任建築師身份所設計及實現的第一項工程。"警衛屋"位於整個員工宿舍的主要入口，是整個宿舍項目的焦點。屋的二層是一理髮店，首層是一間小賣店。設計採用了脫模混凝土外墻及外露預制混凝土旋轉梯，是極受"獸性"主義的設計構思所影響。這個設計構思形成了以後在設計手法上，多採用永久性外墻選料及外露樓梯的個人建築風格。

This is the first project constructed as the designing architect at Peter Y.S. Pun & Associates. The "Guard House" is the focal point of the staff quarters and stands at the main entrance of the complex. It contains a hairdresser on the first floor and a tuck shop on the ground. The design is heavily influenced by "brutalism" with off form concrete finish and precast exposed spiral staircase. This design approach developed into a preference for permanent external finishes and exposed staircases in later designs.

邵氏大屋
Shaw's residence

香港清水灣道邵氏製片廠內（1968～1970）
Shaw's Studio, Clearwater Bay Road, Hong Kong

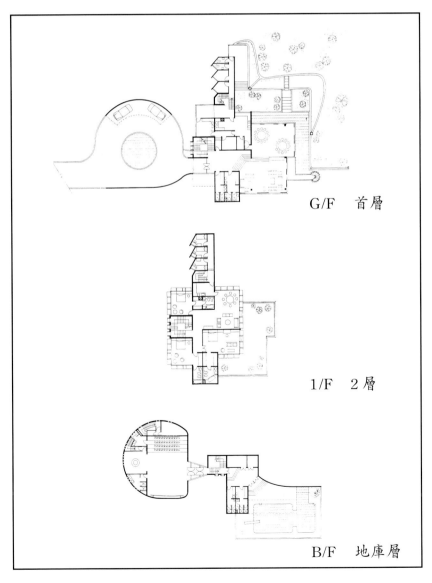

G/F 首層

1/F 2層

B/F 地庫層

大屋功能不單止提供私人住所，更提供了大型宴客場地。首層有大面積的起居室及飯廳，2層有私人住所，包括獨立起居室及飯廳和廚房。地庫有一裝設有水底拍電影設施的游泳池及一個60人的電影院。服務性的功能例如員工宿舍、浴室及樓梯分為3組垂直筒形的組合，起居室及睡房夾於筒形組合中，外形清澈地表露了垂直筒形組合的強烈性而主要的生活空間位於其中，做成一雄壯堡壘的外姿。

The residence is both for mass entertaining and private dwelling. The complex consists of large living and dining areas on the ground floor with a private residence on the lst floor fully equipped with living & dining area with kitchen and a separate basement wing containing a swimming pool designed for under water filming and a private cinema for 60 people. The service functions including servants' quarters, bathrooms and stairs are grouped in three vertical cores with the living and bedroom areas fully glazed to benefit from the surrounding landscape. The design purposely expressed the strong service core elements with the living function suspended in between to create a castle like majestic quality in the extenal treatment.

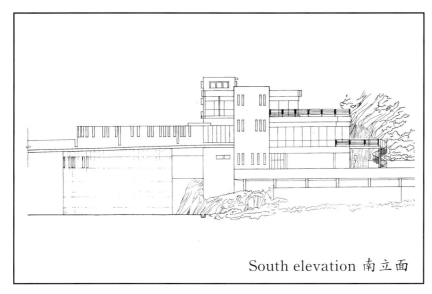

South elevation 南立面

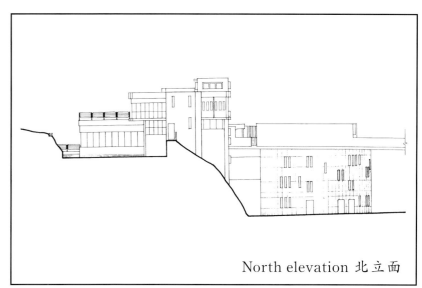

North elevation 北立面

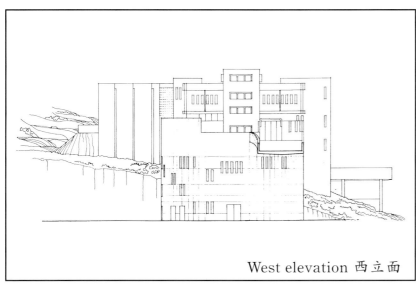

West elevation 西立面

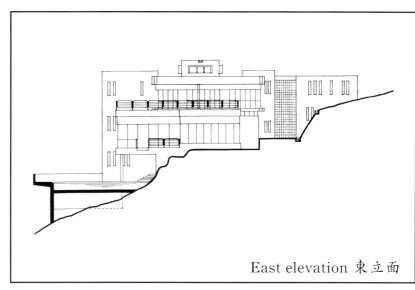

East elevation 東立面

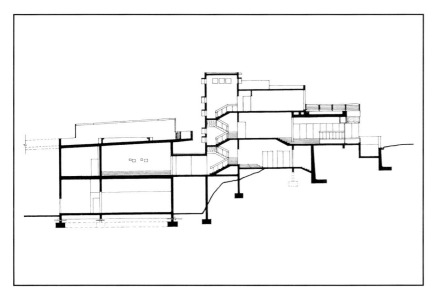

翡翠明珠戲院
Jade and Pearl Cinemas

香港銅鑼灣百德新街室內設計
Interior fitting out at Paterson Street, Causeway Bay, Hong Kong.

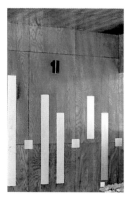

蜆殼油站
Shell's Petrol Stations

香港各地 (1968~1972)
At various locations in Hong Kong.

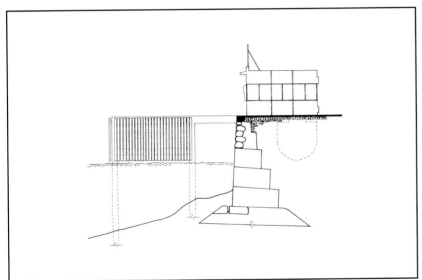

這是一系列的電油站設計，包括2個預制項目。
This is a series of petrol filing station design including two with precast construction.

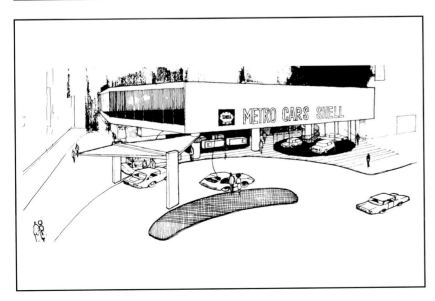

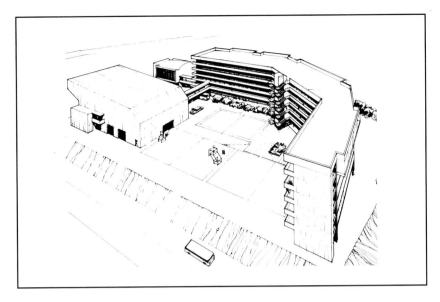

蕭明天主教中學
Siu Ming catholic secondary school

香港九龍葵涌10B區 (1969～1972)
Kwai Chung Area 10B, Kowloon, Hong Kong.

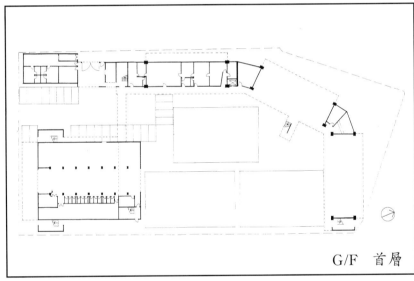

G/F 首層

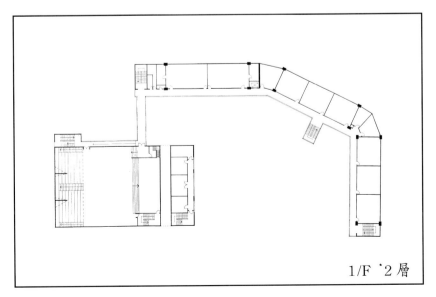

1/F 2層

一處25個課室的政府資助中學。設計的構思很清徹地把"被服務"性的功能，例如課室及實驗室及"服務"性的功能，例如廁所、樓梯等，而多功能的禮堂及主要樓梯則與主樓隔離，作為整間學校的視覺焦點。主樓的定位是為了向籃球場作出遮擋西面太陽的作用，整個項目的設計對本地氣候提供了適當的設施，例如西面立面的混凝土遮陽板及東面立面的走廊，課室及實驗室更有自然通風的功能，外牆是採用了脫模混凝土，提供了經濟的外牆選料。

A government subsidized secondary school with 25 classrooms. This design makes a clear separation between the services function and the served function by grouping the classrooms and laboratories between the lavatories and stairs. The multi purpose school hall and the main staircase are separately located to make them as visual points for the whole school. The classroom block is sited in a position to give the basketball courts protection against the west sun. The whole project is very much concerned with climatic protection with cast in-site sun shades in the west elevation and over hanging corridor in the east elevation. It is also designed for natural cross ventilation in all the classrooms and laboratories. The external finishes are off-form concrete finish to provide an economic permanent finish to the building.

培敦中學
Pooi Tun Secondary School

香港九龍斧山道 (1969-1972)
Hammer Hill Road, Kowloon, Hong Kong.

This 24 classroom school was designed shortly after the Siu Ming catholic secondary school with a much more difficult site condition having hill slopes on two sides of the site. The plan is very compact and is much dictated by the terrain of the site and the amount of accommodation required. This is the first project where the hexagonal shape is used extensively as the best form to fit the function, for example, in the laboratories where the teacher's end required less space, the staircase where the corners are usually waste spaces and the shape of the assembly hall where the hexagonal shape enhances the acoustic effect and permits the flexibility of a centre stage and side stage layout. The siting of the assembly hall allows one end of the hall to rest against the slope while the space gained under the hall is utilized as a covered play space and staff quarters. The roof of the classroom block is utilized as play space while the roof of the laboratory block houses the staff quarters. The service areas such as lavatories, lift and stairs are singled out as separate elements framing the classrooms and laboratories at both ends. The central core acts not only as the visual focal point but also as the centre of the school where access to all facilities begins. The external finish is off-form concrete with highlights of painted window wall panels. The overall design presented a concise solution to the urban school situation.

這24個課室的中學是在設計簫明天主教中學後不久設計的。但工地兩側有山坡，情況比較複雜。項目的佈局緊湊，完全受到工地標高及業主要求所牽制。這是在平面設計上使用六角形的第一個項目，由於六角形是最適應功能上的要求，例如實驗室內教師執教的範圍是比較小的面積，樓梯的角落是多餘的面積，及禮堂用六角形不單可改進擴散及反射聲的效果，而且可靈活地把舞台放在中央或一旁。禮堂的定位使禮堂一方面坐落山坡上，另一方面騰出禮堂下層的空間來作半戶外活動及員工宿舍之用。課室的頂部有室外運動場，而實驗室的頂部則加設員工宿舍，服務性的功能如廁所、電梯及樓梯等分佈在課室及實驗室的兩旁，中部的塔樓形成整個項目的視覺核心及在功能上作為中心區，由此伸展到其他部分。外牆採用了脫模混凝土，只有窗板塗上色彩作為外貌的點綴，整個設計表達了對市區學校的一個恰當的構思。

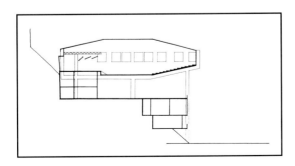

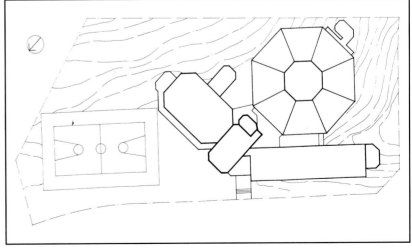

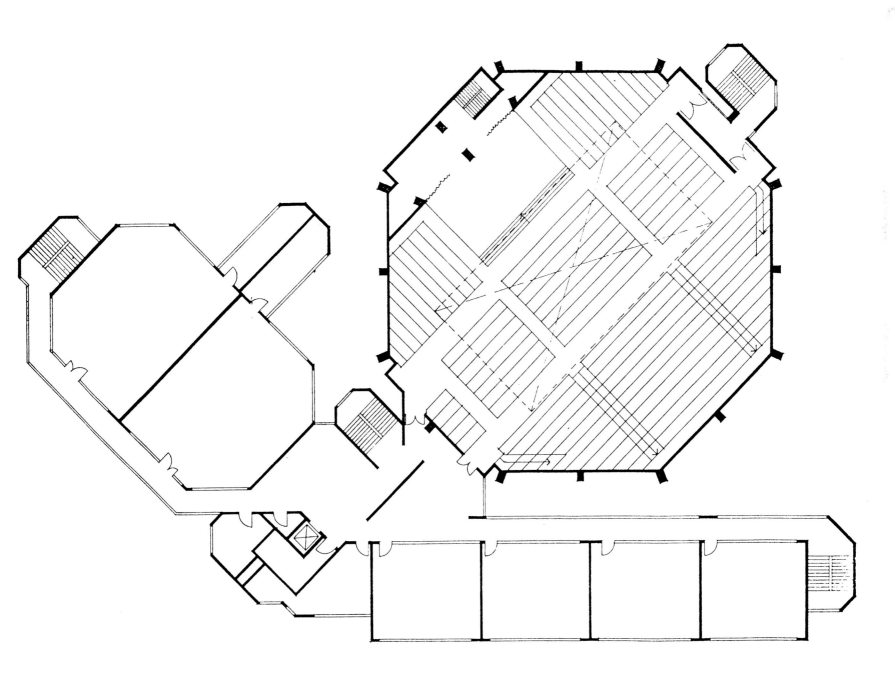

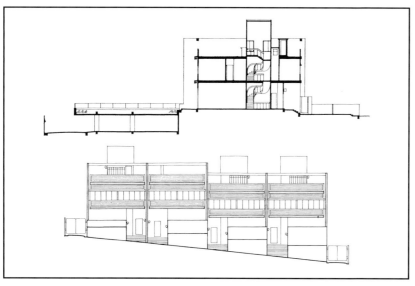

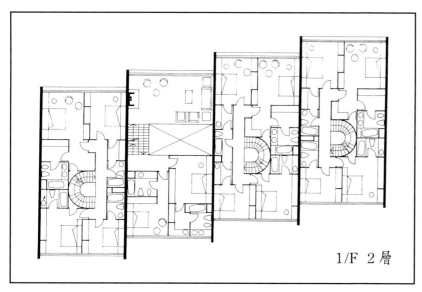

1/F 2層

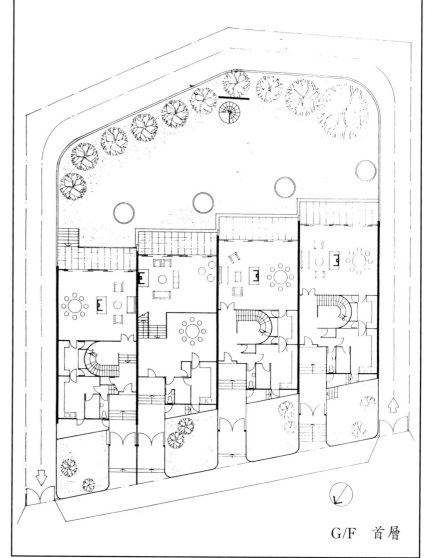

G/F 首層

筆架山道四間排屋
4 Terrace houses, Beacon Hill Road

香港九龍塘筆架山道 (1969～1972)
Beacon Hill Road, Kowloon Tong, Hong Kong.

4間排屋各有業主。3個業主的要求相同,而另個業主要求與眾有別。工地位於筆架山道比較幽靜的一段,所有排屋的入口有小花園,3間屋有入口樓梯大堂,從大堂進入客廳,客廳中有火爐,把起居範圍與飯廳分隔。客廳外有平台花園及山谷遠景,2層有4間睡房及背對背的浴室。其他一間屋是以半層式佈局,入口處一邊是飯廳而飯廳對下是客廳,一層另有一客廳及2個睡房。

排屋的構思不單提供了每間屋同樣的外向,並且從兩屋分享結構牆,得到經濟的建造效果。每個起居室有南向的戶外平台。平台與平台雙連的花園,可提供各戶彼此交際的場所。

The four houses are designed for four individual clients and their families. Three for the clients have similar requirements and one client has a different brief. The site is situated in the quiet end of Beacon Hill Road and all houses are approached through a small garden in the front.

Three of the houses each have an entrance and stair lobby which leads into the main living room. The living room has a centrally located fire place which separate the living and dining areas and overlooks into the podium garden and the valley below. All four bedrooms are located on the first floor with back to back bathroom layout. The odd house has a split level layout and the entrance opens onto a half landing with the dining area on one side overlooking onto the living area below. The first floor has an addition living area and two bedrooms.

The choice of using the terrace layout is mainly to provide all houses with the same aspect of orientation and the economic aspect of having party walls. The private living area each has an outdoor space facing south and leads directly onto the communal garden where social intercourse of the families can take place.

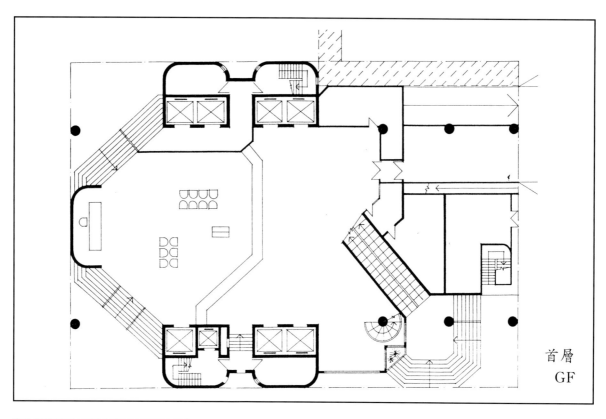

首層
GF

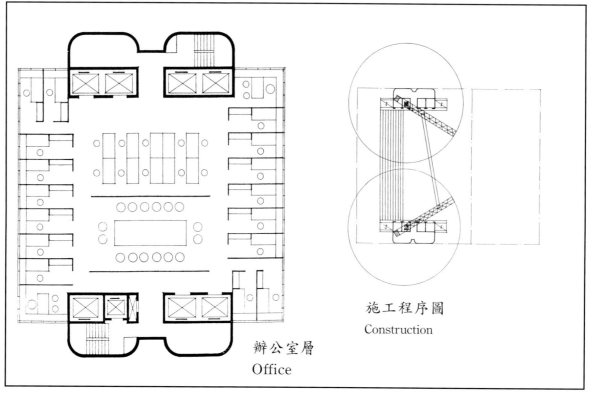

辦公室層
Office

施工程序圖
Construction

海外信托銀行大廈
OTB Bank building

香港灣仔告士打道160號 (1971)
160 Gloucester Road, Wanchai, Hong Kong.

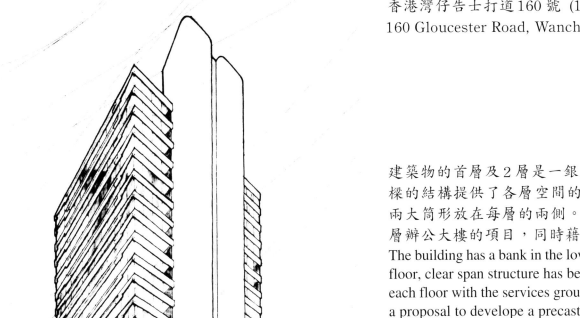

建築物的首層及2層是一銀行，3層至27層是辦公層。無樑的結構提供了各層空間的靈活性。服務性的設施分組為兩大筒形放在每層的兩側。這是嘗試以預制方法去建造高層辦公大樓的項目，同時藉此縮短建造工期。

The building has a bank in the lower two floors and offices to the 28th floor, clear span structure has been used to create a flexible space on each floor with the services grouped in two cores on the side. This is a proposal to develope a precast system for the construction of high rise office building and to provide an alternate construction method which will result in shorter construction time.

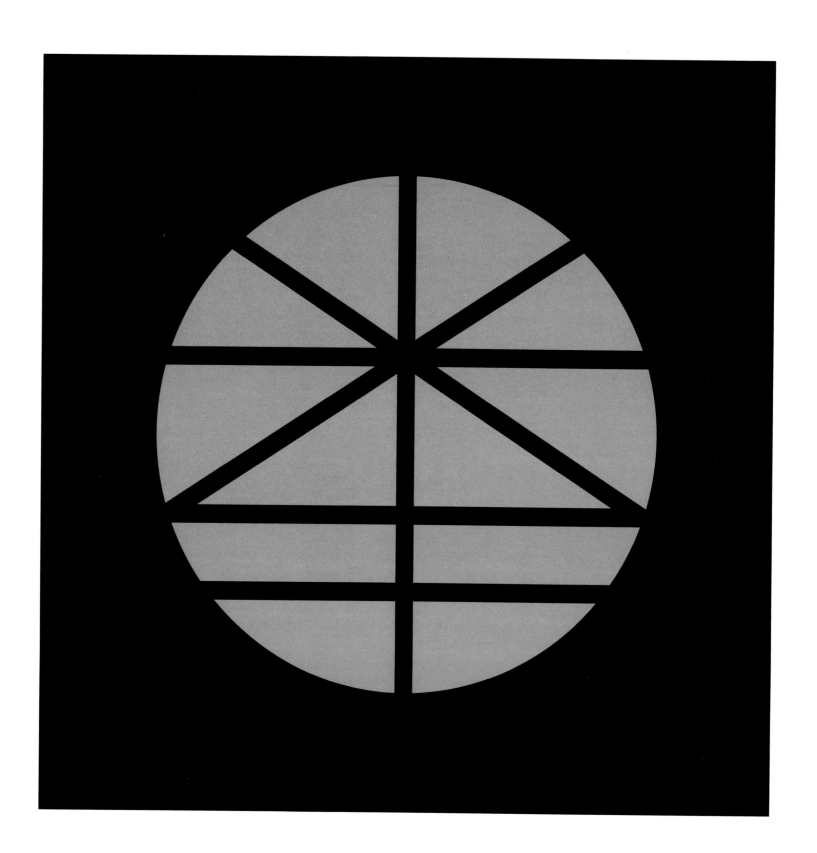

潘祖堯則師事務所 (1973～1986)

Ronald Poon Associates

羅素/潘建築師集團 (1976～1986)

Russell/Poon Group Partnership

潘祖堯顧問有限公司 (1986～　)

Ronald Poon Consultants Ltd.

石澳村屋
Village shop/house, Shek-O

香港石澳 (1973~1974)
Shek-O, Hong Kong.

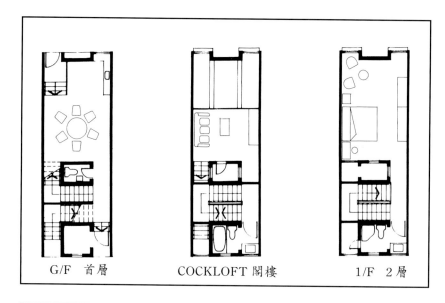

| G/F 首層 | COCKLOFT 閣樓 | 1/F 2層 |

這間簡單的村屋設計，表露了後期設計中常用的設計構思。浴室、樓梯及化糞池組成垂直管道，樓版兩面伸展的結構解決了在兩旁有樓的情況下，做地基的困難及遮陽板和通風效果對地方氣候的交代，都是對業主要求作出一適當的設計構思。

This simple shop house has expressed a design philosophy which appears time and again in later design work. The grouping together of the bathrooms, staircase and septic tank in a vertical stack, the simple cantilevering structural solution which fits the space requirement and resolve the foundation difficulty of having to build between two adjacent buildings and the sun shades and cross ventilation effect which expressed a concern about the climatic conditions, all contribute to an appropriate solution for the brief.

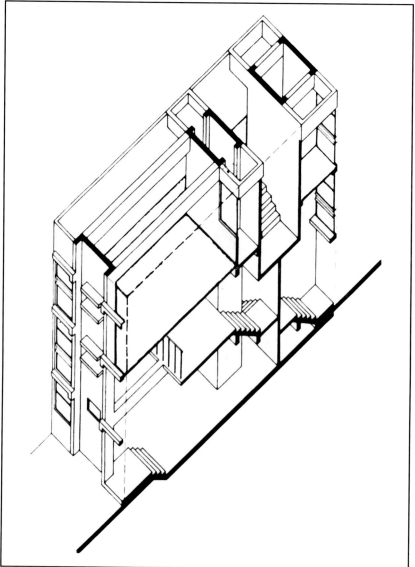

又一村公寓樓

Apartment Complex, Yau Yat Chuen

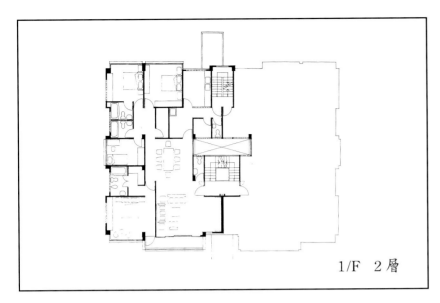

1/F 2層

香港九龍又一村 (1973)
NKIL4613, Cassia Road, Yau Yat Chuen, Kowloon, Hong Kong.

一個市區低層住宅的項目，有6戶4個睡房的公寓，以中間光井的手法，解決了每層中部的通風及日照的問題。

This is a low rise urban housing project with six 4 bed room apartments. The design introduced the central light well to facilitate cross ventilation and day lighting to the central area of the plan.

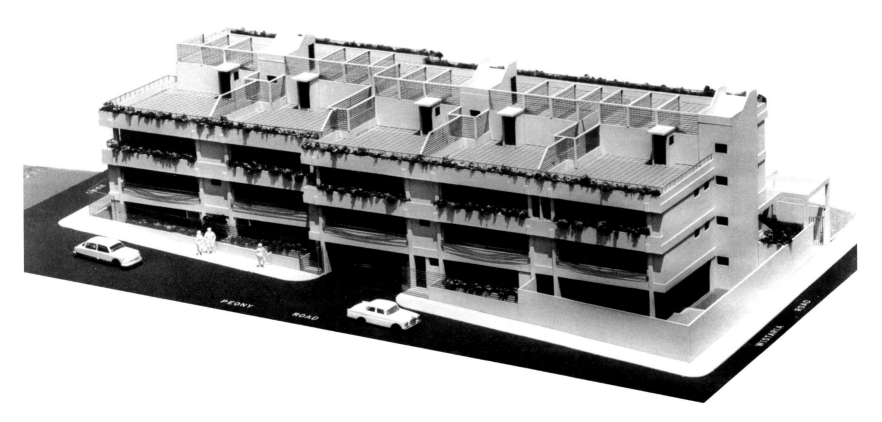

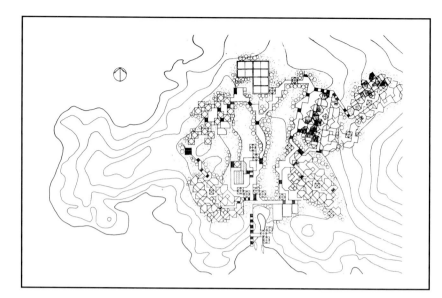

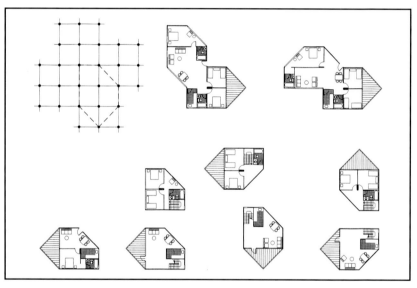

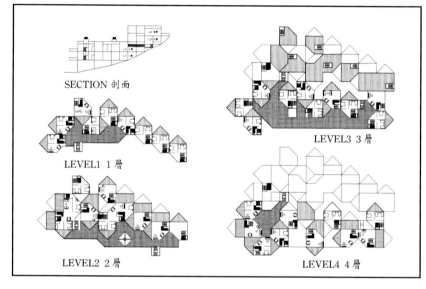

| Playground | Housing Cluster | Walkway | Riding Ring + Pool | Clubhouse | Housing Cluster |
| 遊樂場 | 住宅區 | 步廊 | 馬場＋游泳池 | 會所 | 住宅區 |

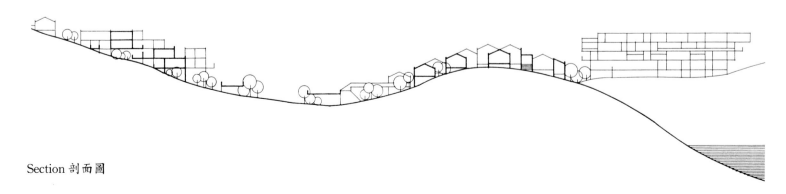

Section 剖面圖

蒲台島度假村
Po Toi island resort

香港島南面　(1973)
South of Hong Kong Island.

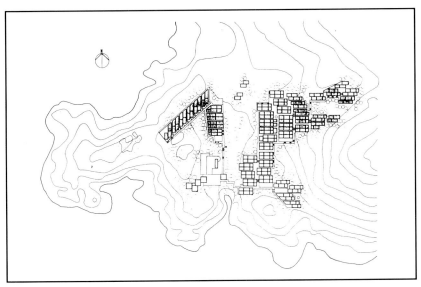

蒲台島位於香港島南面，約半小時的海程，便可到達。有兩方案：一為鄉村屋群；二為包羅萬有的度假村。鄉村屋群是為了本地人周未度假用，設施有限。村屋沿山坡而建在山谷的周圍，創造出一鄉村的佈局。度假村的方案有4組居住單元，單元以半六角形的平面佈局做出比較有度假風味的空間。單元重叠組成一組，每組依山而建，面向遠景。

The island is situated south of Hong Kong and can be reached by a half hour boat ride. There are two alternate proposals, one being a simple village housing and the other a comprehensive holiday resort. The village housing caters for local week end usage and has limited facilities. The houses are sited along the terrain overlooking the valley in the centre of the site creating a village environment. The holiday resort proposal has 4 clusters of residential units, all designed with a part hexagonal plan to create a holiday like space. Units are staggered one above another and the cluster rests along the contour to give views to all units.

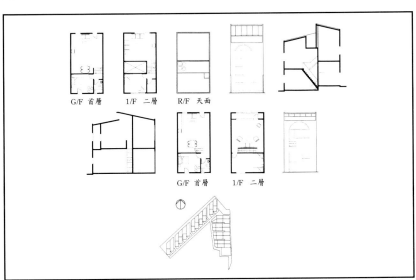

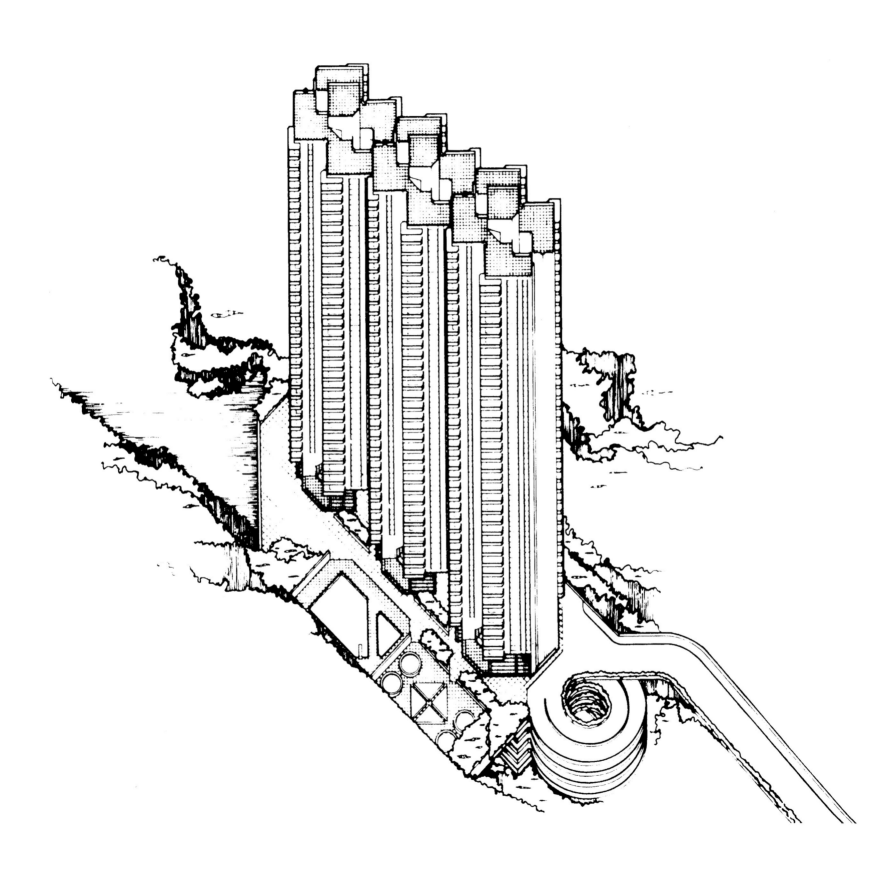

南豐路公寓大樓
Apartment complex, Nan Fung Road

香港北角南豐路　　(1979)
Nan Fung Road, North Point, Hong Kong.

與羅素建築師合作設計
Jointly designed with Mr. David Russell

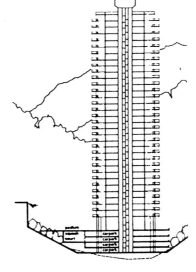

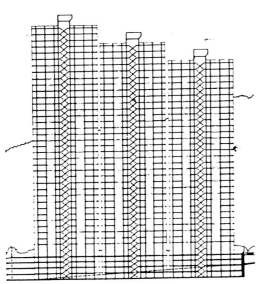

工地坐落於一山谷底，周圍有山溪、步行山徑及滿佈樹林的山坡。項目有4層停車場，車路直達停車場頂部的平台。平台有多項公共設施，例如游泳池、花園、幼稚園、居民會所、游樂場、壁球室、溫習室、管理處和維修設施。平台的標高與附近的山坡小徑相連，原在工地的水塘加以保留更成為平台的一部分，增加休息的景觀。項目中有3憧不同高度大廈，劃分為30、33及36層高，每層有4個佈局相同的單元。電梯大堂及樓梯皆有天然日照，單元面積為150平方米，內有3個睡房。由於工地坐落大自然的郊野公園，所有的居住單元都有4面外牆，盡量享受到日照、通風及周圍景色的好處。

The site is set at the bottom of a valley on the edge of natural wooded slopes with running mountain streams and hillside walks. The lower four storeys are car parks with vehicular access to the podium deck. The podium level contains communal facilities such as swimming pool, garden, kindergarten, resident's club, covered play spaces, squash courts, study rooms, estate offices and service areas and is at a level where easy links are provided to the hillside walks. The oringinal reservoir on the site was retained to provide an added dimension of recreational facility and formed part of the podium landscape. The development consists of three separate towers of varying heights 3033 and 36 storeys. There are four identical flats per floor. Natural lighting is provided to all lift lobbies and staircases. The flats are 150m2 and has three bedroom each. Taking into account the unique environment of being sited almost in the middle of a country park, the flats are designed with almost four external sides to provide maximum light, air and view to the dwelling.

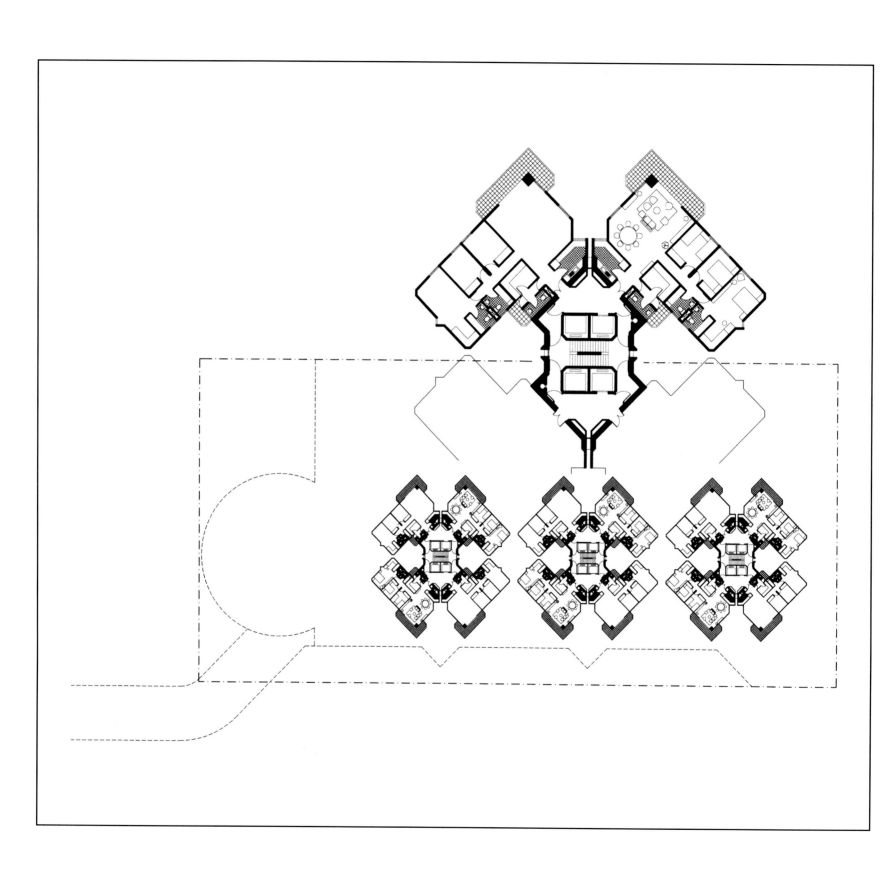

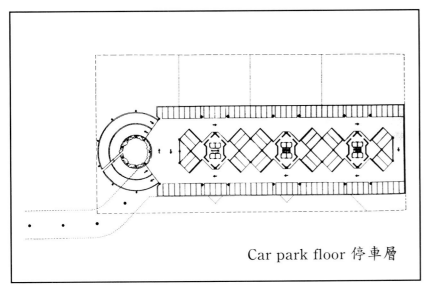
Car park floor 停車層

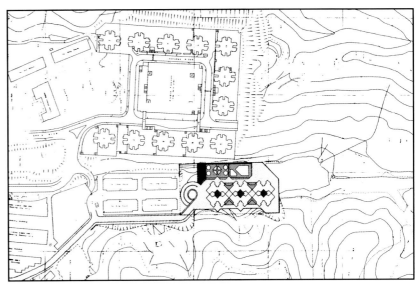

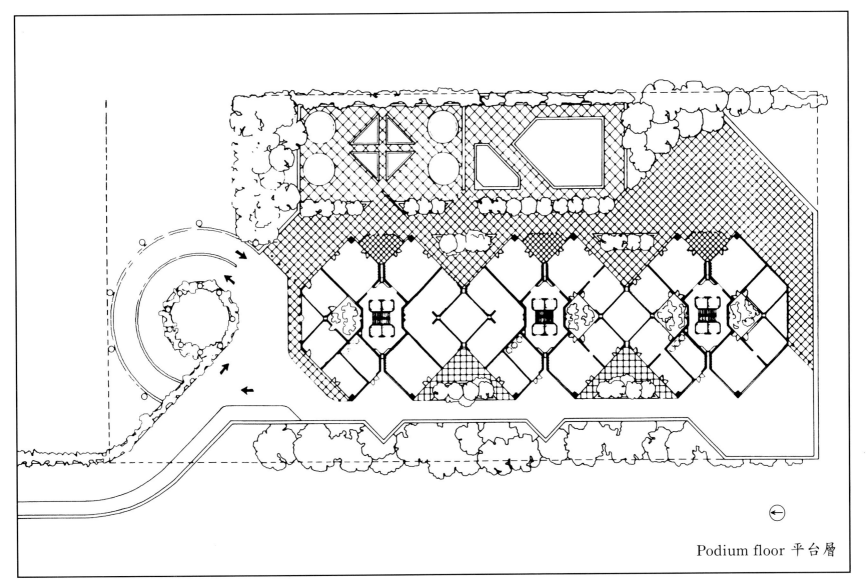
Podium floor 平台層

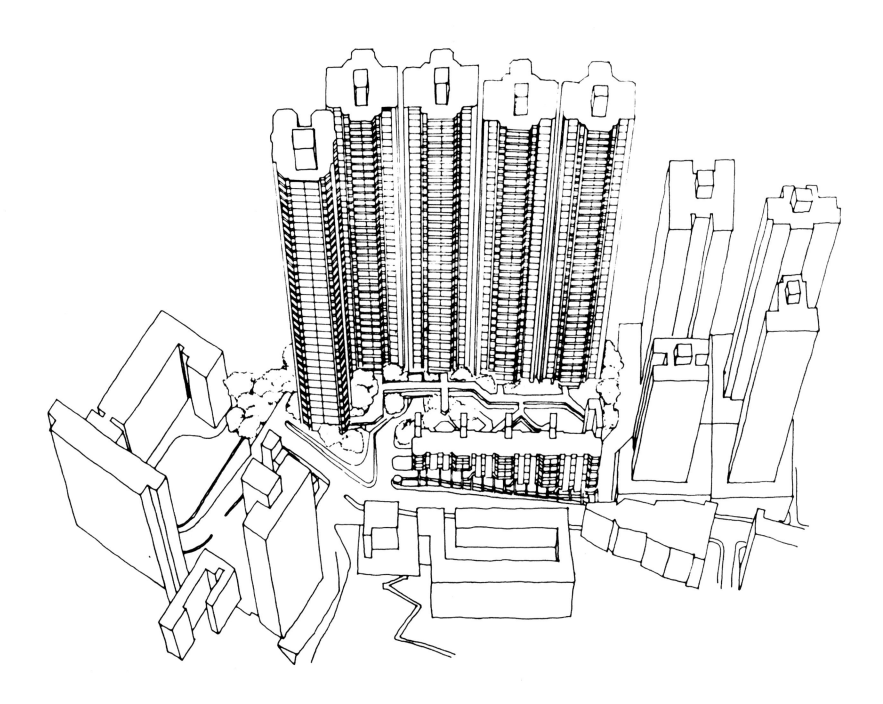

山光道公寓大樓

Apartment complex, Shan Kwong Road

香港跑馬地山光道(1979)
Shan Kwong Road, Happy Valley, Hong Kong.
與羅素建築師合作設計
Jointly designed with Mr. David Russell

這是另一高密度的市區住宅。項目工地背山位於山坡上，有4層停車場。停車場頂部為設有公共設施的平台花園。項目有3種不同的大廈：一種有36層高，"飛機型"佈局，3個睡房的公寓；一種有4個睡房的公寓；另一種有低層公寓。高層的公寓有三面外牆及通風設施，項目在市區高密度的情況下，提供了不同大小的居所。

This is another high density urban housing project. The site is situated on a slope with the mountain behind. It has 4 storey car park with a podium deck containing all community facilities. There are three types of blocks, one 36 storey high with the typical "aeroplane" shape layout, with 3 bedroom units, one 36 storey high with 4 bedroom units and one low rise block along the road. The units in the high blocks have 3 external sides and through ventilation. The design here tries to provide comfortable environment to different family sizes in a high density urban situation.

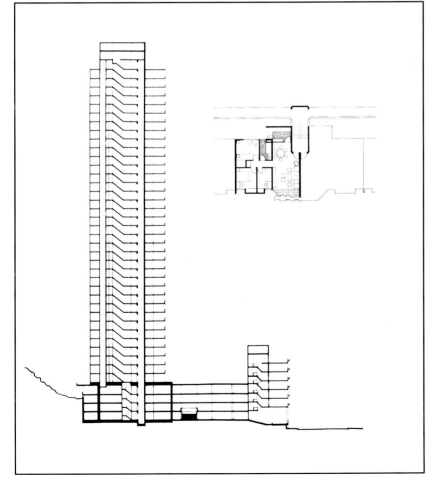

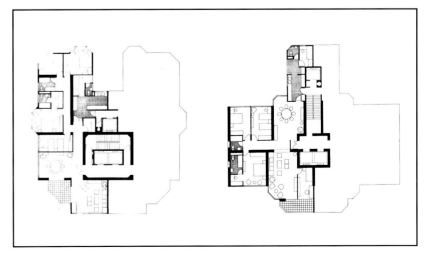

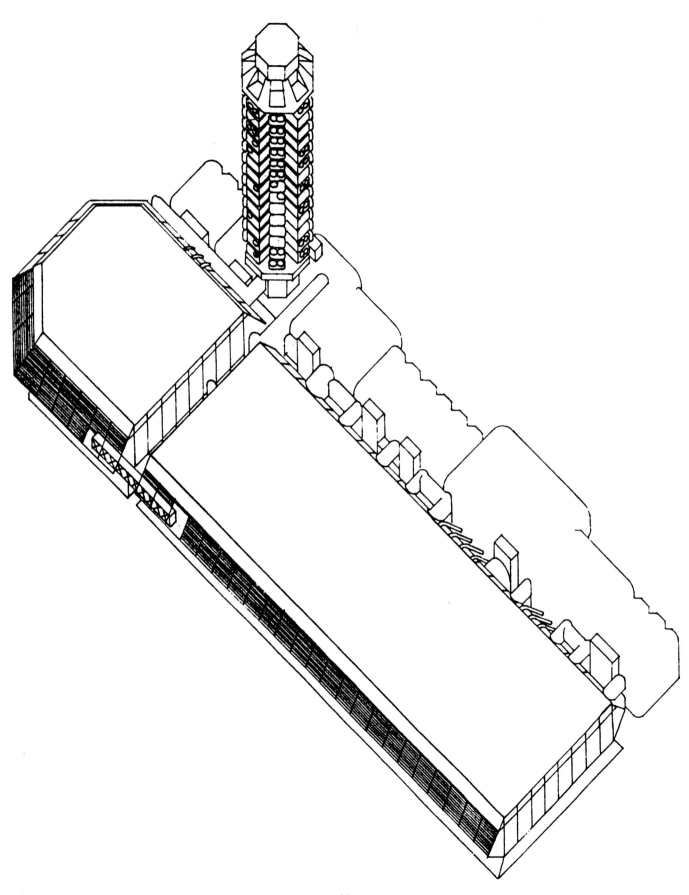

銀禧體育中心
Silver Jubilee Sports Centre

香港沙田　(1978～1979)
Shatin, Hong Kong.

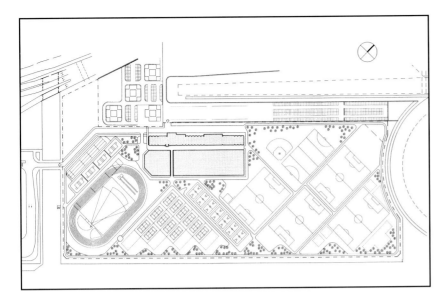

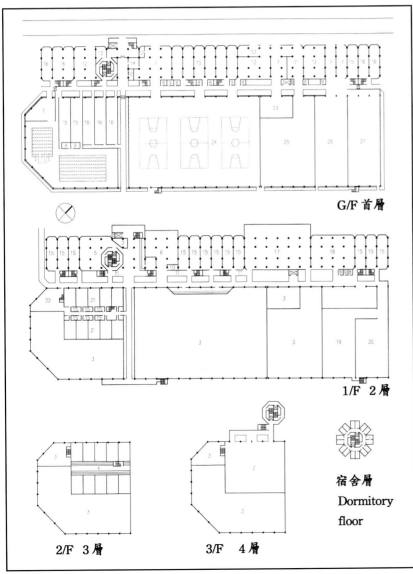

中心的構思與廠房生產綫相同，因為兩者的功能都是與出產產品有關。體育中心出產優秀的運動員，輔助性的功能如實驗室、更衣室、管理辦公室及用品商店等在大樓的一旁組成單行組合，通過相連的走廊與另一旁的訓練設施如室內球場、體操禮堂、訓練游水池等相連。受訓的運動員宿舍是一高層塔樓，有240個單元，位於中心主要入口的上層，作為整個中心的視覺焦點。

The centre is designed like a factory production layout as there is a similarity between the two types of building both are for the manufacturing of products. The centre products top quality sportsmen. The supporting elements like the science laboratory, changing rooms, administrative office, sports shops etc. are grouped in a line on one side of the complex while the training facilities such as games courts, gymnastic hall, training pool etc. are sited on the other side linked by a circulation corridor in between. The trainees are housed in a high rise residential tower with 240 units rising above the main entrance area of the centre acting as a visual focal point to the complex.

1. 1-2人房間　One to two bed units
2. 餐館　Restaurant
3. 挑空　Void
4. 看台　View gallery
5. 商店　Sport shop
6. 行政中心　Administration
7. 貯物室　Storage
8. 教練室　Coach room
9. 急救室　First aid
10. 清潔室　Cleaner store
11. 自助售賣機　Vending machine area
12. 機房　Plant
13. 洗衣房　Laundry
14. 工作室　Work shop
15. 更衣室,貯物櫃,衛生間,浴室　Changing, lockers, toilets, showers
16. 戶外運動更衣室　Dirty changing for outdoor sports
17. 教練學院中心　Coaching academy & education centre
18. 運動科學實驗室　Sports science laboratories
19. 武術場　Martial arts hall
20. 排舞室　Dance studio
21. 壁球場　Squash courts
22. 酒廊　Bar
23. 舉重練習場　Weight & strength training hall
24. 主要運動場　Main sport hall
25. 體操場　Gymnastic hall
26. 多功能廳　Multi-purpose hall
27. 射擊場　Projectile hall

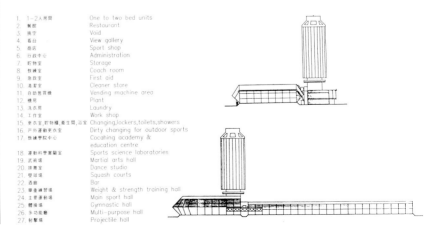

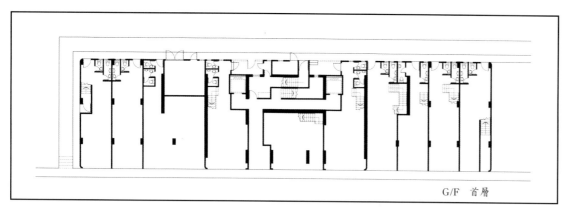

G/F 首層

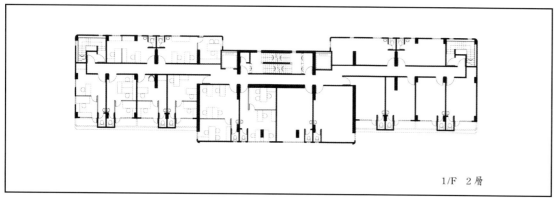

1/F 2層

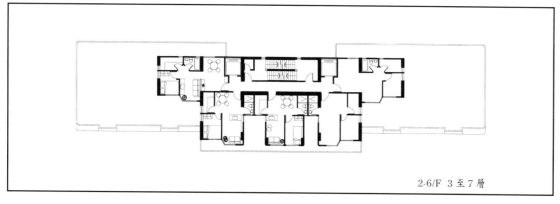

2-6/F 3至7層

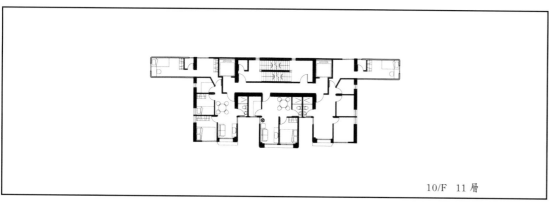

10/F 11層

加達樓
Cactus Mansion

香港灣仔麥加力歌街1至19號 (1973～1979)
1-19 Mcgregor Street, Wanchai, Hong Kong.

這個工地是位於皇后大道東及軒尼詩道兩條主要交通幹線之間比較清靜的小巷中，寬11.5米長45米，週圍有戰前低層的舊樓及戰後多層的大廈。這個項目雖然規模比較小，不能對週圍市區環境有太大的影響，但希望能對一般在香港已過時的建築設計作出新意。項目樓高50米，首層有16間商舖，2層有16間辦公室，3層至17層有57個住宅單元。單元面積從42平方米至64平方米。高層住宅的面積受到嚴格的建築條例影響，但巧絕的住宅佈局，提供了合理的居住環境，包括在一間睡房的單元都提供了小平台花園。外墻的設計是這個項目的精點，墻版的選料以鮮明顏色的小磁磚為主，用不規則的色彩分隔，務求給個別單元提供不同顏色的外墻及對整個外貌提供一強烈的色彩，避免以後居民加建蓬帳、乾衣架及籠屋時對大廈的外貌產生不良的效果。

The site is situated in the quiet backwater between two main city arteries of Queen's Road East and Hennessy Road with a narrow width of 11.5m and a 45m frontage. The urban fabric in the area is a mixture of pre-war low rise and post war medium rise buildings and though the scale of this project in itself is unlikely to transform the overall environment of the area, it is hoped that at least it will set an alternate example to some of the dated run-of-the-mill designs one mostly encounters in Hong Kong. The building rising 50m above the street level, comprises a shopping arcade on the ground level, a commercial first floor and 14 levels of domestic units above. The building provides 57 nos of flats varying from 42m2 to 64m2, 10 nos of shops and 16 nos of offices. Though the very stringent building regulations on narrow street have produced very restricted floor areas on the upper floors, the internal layout of the flats manages to provide a reasonable living environment including one bedroom flats with roof gardens. The elevation design is a special feature of this project. The in fill wall panels have random prime colour mosaic finish. This random colouring of the elevation is designed to give identity to individual flats and to create a dominating feature of the facade to counteract the visual disfiguring effect of additional awing, clothes drying racks and cages which the owners may construct later.

弱能兒童院

Home for the severely handicapped childred

香港大埔　(1979～1990)
Taipo, Hong Kong.
與羅素建築師合作設計
Jointly designed with Mr. David Russell

弱能兒童院是一兩層高庭院式的建築群，充分反映出地方的風格，加上屋頂上採用了太陽能斜板的設施，配合環保的要求。院內首層共有8間能容納10個小童的睡房及4間大型起居室，2層有48間員工宿舍。為適應地方氣候，每間房都具備自然通風的條件。應業主要求，場地上保留了原有的小屋，屋內設有小教室，形成了整個項目的主樓。整體設計的佈局就是根據小屋的外形擴展成三組庭院的組合。弱能兒童的起居條件中，特別要嚴格保持冬暖夏涼的環境。為此，兒童起居室頂部設有透光玻璃，便於引入冬天的陽光；又設有活動的反射太陽的保護罩，來對抗夏天的烈日。院內設計充滿了大家庭的氣氛，可使兒童增強歸屬感。員工的宿舍位於2層，能遠望週圍的田園風光，使工作了整天的員工能有鬆弛心身的幽美生活環境。此項目獲得1994年香港復康聯會的設計優異獎。

The Home is a two storey courtyard complex set in a countryside village. The slanting solar panel on the roof reflected a concern on local identity and environment. The Home consists of 8 ten children bedrooms and four large lounges. There are 48 staff units on the first floor. Every room is designed with cross ventilation. The brief called for the preservation of an old building in the middle of the site containing a chapel which forms the focus of the complex. The shape of the existing building helped to determine the three courtyards in the complex. Severely handicapped children require strict environmental conditions and the glass roofs above all lounges admit the winter sun and movable sun reflectors in the roof help to reflect the summer sun. Overall the Home is designed as a big family dwelling to create a homely atmosphere for the children. The staff quarters on the first floor has views over the surrounding countryside which provide a relaxing environment after a day's work. This project received the Joint council for the Physically and Mentally disabled 1994 Design Award.

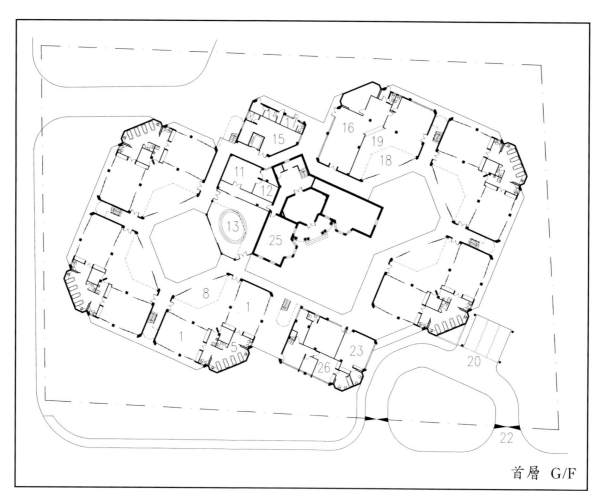

首層 G/F

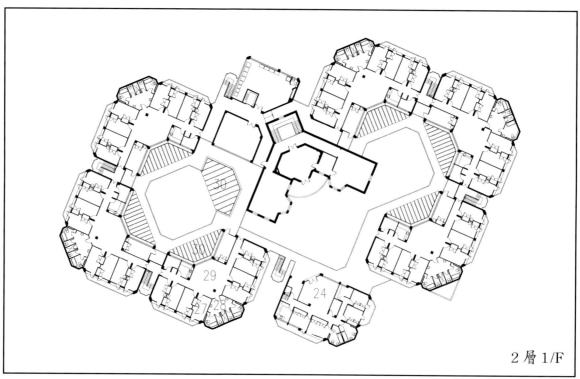

2層 1/F

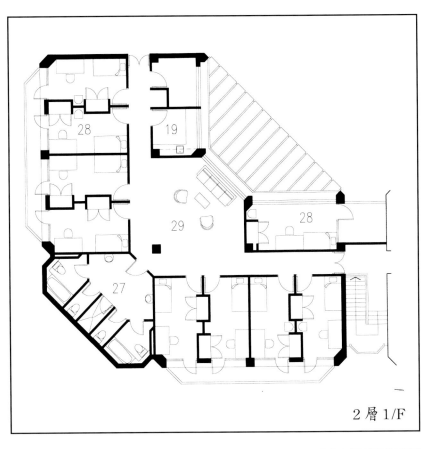

2層 1/F

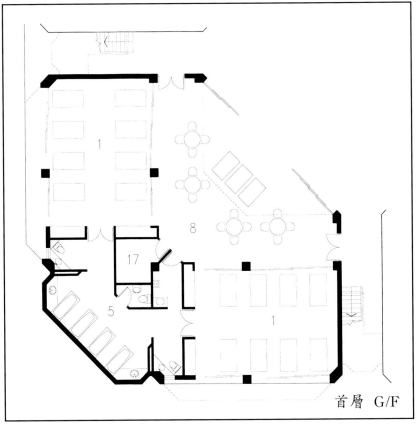

首層 G/F

	説明	KEY
1.	兒童房	Children's room
5.	兒童浴室	Children's bathroom
8.	遊戲室	Playroom
11.	醫療室	Clinic
12.	醫生房	Doctor's room
13.	泳池	Pool
14.	狗房	Dog kennel
16.	廚房	Kitchen
17.	儲物室	Store
18.	員工飯堂	Staff canteen
19.	輔助廚房	Kitchen
20.	停車場	Carpark
22.	入口	Entrance
23.	辦事處	Office
24.	女主任寢室	Matron's room
25.	教堂	Chapel
26.	客房	Guest room
27.	員工浴室	Staff bathroom
28.	員工睡房	Staff bedroom
29.	員工休憩室	Staff lounge
30.	天窗	Skylight

豐樂閣

Albron Court

香港堅道99號 (1981～1985)
99 Caine Road, Hong Kong.

Albron Court is situated in the mid-level of Hong Kong with the harbour in front and mountain at the back. The design is developed based on the restrictive building regulation on street light angle and ventilation to create a new design approach which reflects a characteristic of the time with essence of the traditional design and local identity. The service core in the middle of the building has been painted bright red and the surrounding 4 blocks are alternatively finished with grey and white tiles. The external treatment of the fire escape stairs together with the staggering effect of the various blocks created a spiral effect at the top part of the building like a flower in full bloom. The special colour effect and the characteristic of the building is such that it can be identified very easily even from as. far as Kowloon reflecting the description depicted in a Tang Dynasty poem " a red dot amongst ten of thousands of green ones ". The floor layout is based on market demand and each floor layout is designed with the flexibility of having 8 nos 65m2 units or 4 nos 130m2 units. The shape of the building is developed with the concern for light and ventilation and each unit has three external sides. This way the units not only has good natural ventilation but also good views all round. The upper part of all windows each have an aluminum sun shade and each unit has a balcony. All in all the design not only meets the local climatic condition but with the particular characteristic and configuration of the building, it also fully reflects the colourful life style of Hong Kong and the local traditional identity. Albron Court has 92 apartments. The second floor is a large podium with swimming pool, coffee shop etc.. There are shops on the ground floor and the lower ground floor.
The 3 basements are car parks containing 130 cars.

豐樂閣建於香港半山區，背山面海。建築設計在滿足香港建築條例對住宅建築採光通風的嚴格要求下，以創新的演譯手法創造出既有時代特徵又有傳統內涵及地方風格的一幢建築物。大樓中部的電梯組及防火樓梯，採用外塗鮮紅色的中筒旋轉其間而直插雲端，週圍四幢樓身外塗灰、白兩色，高低錯落有緻，蔚為壯觀。從而，創造出樓身頂部螺旋形的外貌，酷似盛開的花朵。由於其特殊的造型，由九龍隔海相望，千百座高樓大廈中豐樂閣頗有唐詩"萬綠叢中一點紅"的意境。

在平面佈局上，根據香港特殊的市場需求，以靈活的住宅設計去適應市場對大小單位需求的變化。每層單位分佈可由8個65平方米的改為4個130平方米的單位。樓身的造型是根據通風和採光的要求設計的，使每個單位都有三面外牆。這樣，不僅起居室有良好的自然通風條件，且從每一個單位都既能遠眺海景又能近觀山色。外窗頂部裝有鋁質擋太陽遮頁，每單位有露台一個。總括來說，整個設計是提供了多項完全符合香港氣候的設施，而整幢大廈的色彩及造型都充分反映了香港多姿多彩的生活方式及地方傳統風格。

豐樂閣共有92個公寓，2層設有大型平台花園、游泳池、咖啡廳等公共設施，首層及地下層設有商場，地下三層車庫共設130個車位。

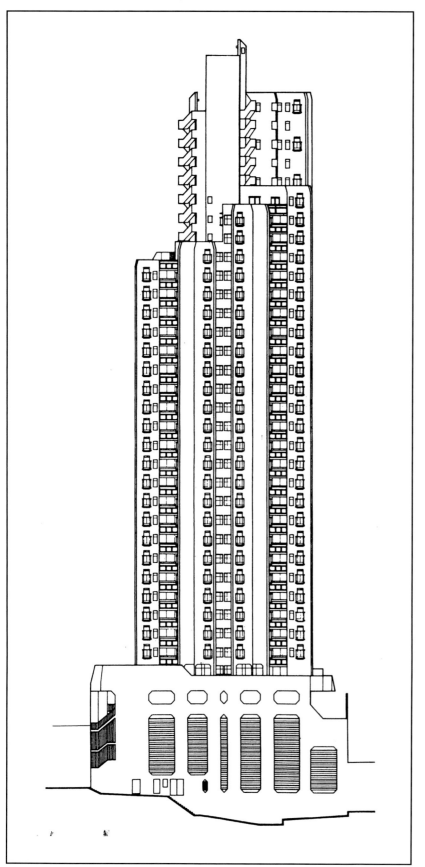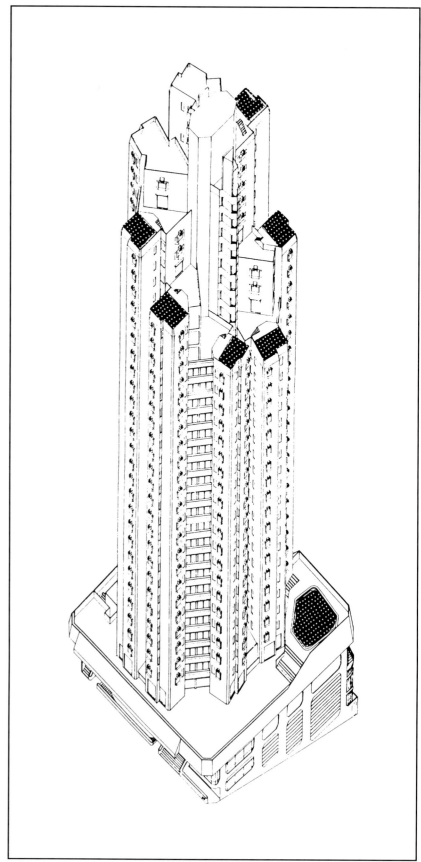

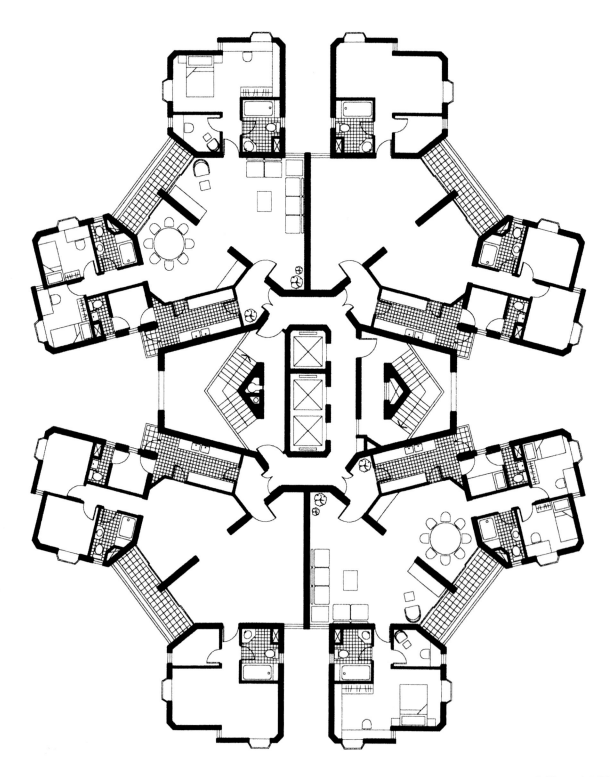

3/F to 19/F 4層至20層

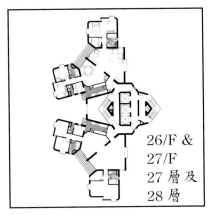

26/F & 27/F
27層及28層

25/F
26層

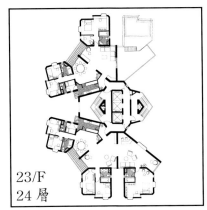

23/F
24層

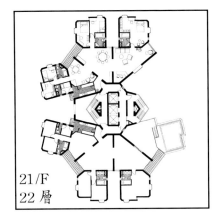

21/F
22層

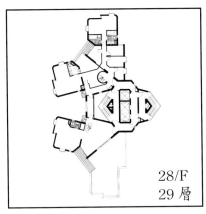

28/F
29層

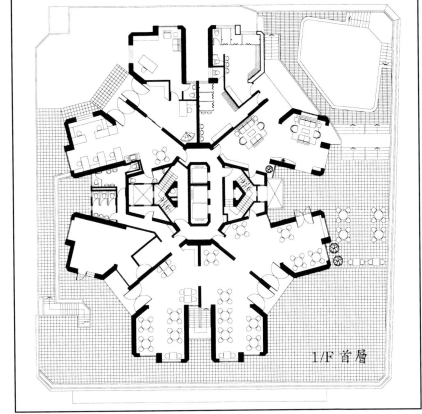

1/F 首層

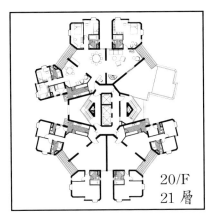

20/F
21層

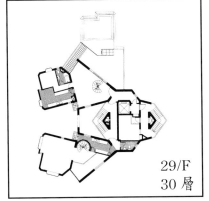

29/F
30層

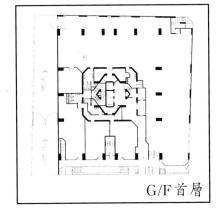

G/F 首層

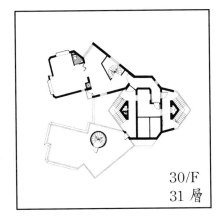

30/F
31層

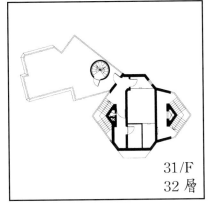

31/F
32層

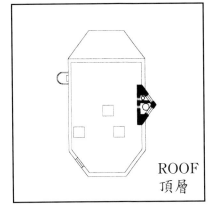

ROOF
頂層

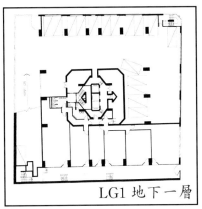

LG1 地下一層

紙幣廠

Printing Factory for Thomas De la Rue Ltd.

香港大埔工業村 (1982~1985)
Taipo Industrial Estate, Hong Kong
與羅素建築師合作設計
Jointly designed with Mr. David Russell

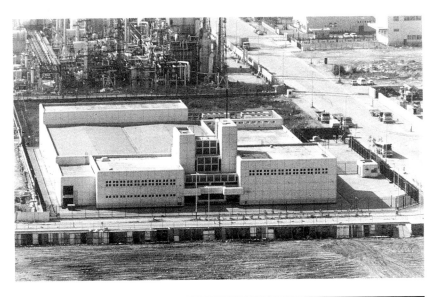

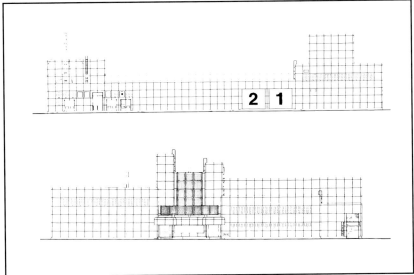

工地位於新填土區，業主的要求與眾不同。作為印制紙幣及其他保密文件的印刷廠，建築物要有溫度及濕度的控制，提供良好的印刷工作以及儲藏紙和保安的條件。整個建築物的設計佈局是採用了1600 x 1600的規格，外墙以6個規格為一組，其中有伸縮縫。外墙磁磚以1個規格來組成，外墙用規格的方式可以方便以後加建門窗或設備時避免干擾整個外墙的外貌。為符合高度保安的要求，盡量減少外窗，而有窗的地方，多加橫直墙架。墙身的結構採用了雙層加固式的水泥空心磚而印刷廠部分用鋼架屋頂。這一種結構在香港不常採用，是為了經濟和保溫的效果。墙身內橫直的加固鋼筋提供了額外的保安設施。這項設計工程獲得香港建築師學會1986年優異設計獎。

The site is located on recently reclaimed land. Client's requirements were very specific and rather unusual. As security printers (printers of bank notes and other security documents) the building has to provide a temperature and humidity controlled environment for the printing processes and for the storage of paper as well as for a very high degree of security. The building is planned on a horizontal and vertical module of 1600 x 1600. The external walls are built in 6 modules wide with expansion joints between each panel. The external wall tiling is arranged in panels of one module square. The grid like expression of the joints facilitates the insertion of any opening that may be required. Such openings can be made at any time without spoiling the appearance of the building. For a high security building, window openings have been kept to a minimum and where these do occur, they have been recessed behind thick mullions and transoms. The building is of reinforced load bearing double skin hollow concrete block wall construction and with steel stanchions and roof trusses over the main printing area. This method of construction is unusual in Hong Kong and is chosen for its economy and good insulation quantities. Horizontal and vertical reinforcement within the hollow block coincidentally provides a secure skin that would be difficult to breach. This project received the HKIA Design Merit Award in 1986.

城市理工大學
City Polytechnic University

香港九龍塘 (1983)
Kowloon Tong, Hong Kong.
羅素／潘建築師集團與英國拿士敦建築師合作設計
Russell/Poon Group Partnership in association with Denys Lasdun Redhouse Softley.

It is considered that the natural landscape of the site has sufficient merit to warrant preservation and renewal as far as possible, particularly in respect of the wooded hill areas on the on the west side rising above the road and bordering the park. The bulk of the building have therefore been sited along the opposite ridge and its slopes roughly parallel with Tat Chee Avenue and extend down into the valley in a terraced formation.

The complex composes of two major zones. A split teaching and lecture zone spine arranged along Tat Chee Avenue with its main horizontal and vertical circulation simply and clearly defined. The other zone has three separate building complexes, comprising the sports and amenities centre, the administration and the learning resources centre in the valley beyond the spine. Two zones are linked by a central plaza which forms the focal point of the campus and centre of all main pedestrian routes.

Architecturally, the intention is to create a campus that not only fulfills all academic needs but also offers to students, staff and the general public (who should be brought into the life of the campus where possible) a harmonious but stimulating environment with areas of quiet and repose, that cater for all the academic and social needs of its inhabitants. In this respect the series of linked spaces and terraces, radiating out from and terracing down to the central plaza are most important and enable the buildings to extend into the landscape. It is believed that the enjoyment of a complex such as this is mainly conditioned by the connections and spaces between buildings.

工地內有山谷及富有天然條件的山坡，是應盡量保存，尤其是西面路旁及公園側的林木山坡。因此把大部分建築物放在對面的山坡上，與德智路平排及一層層的向山谷伸展。大學主要分為兩大部份，一部分包括教學及課室設施，一連串的沿德智路平排，建築物內的縱橫人流網絡以很簡單及清晰的手法處理；第二部分有3幢位於山谷中的建築物包括體育及輔助設施中心、管理辦公樓及學習資料中心。兩個部分由一中心廣場相連，廣場為整間大學的核心，所有主要行人通道由此伸展。

設計構思務求不單是通過清靜及休閒的空間創造出一個能滿足教學的需求的校園，更要向學生、教師及公眾人士提供和諧及富有意義的環境去滿足所有大學內的人們的學術及社會上的需要。因此從中心廣場伸展的連貫空間及平台是非常重要而且使到築物能伸展到綠附近的林木中。建築物與建築物的聯繫及它們之間的空間對享受一個優美的環境是能創造條件。

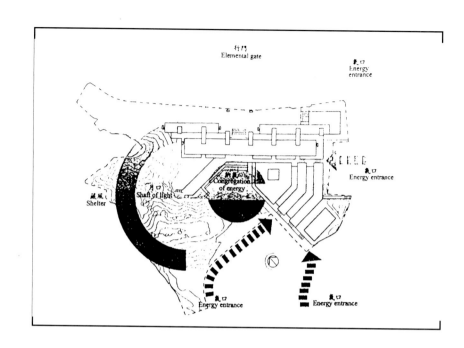
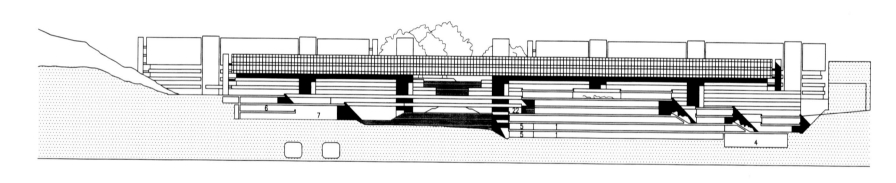
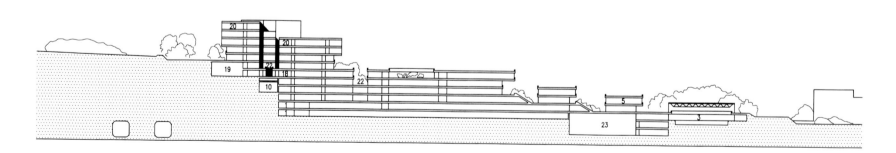

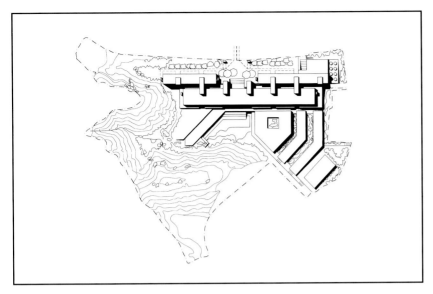

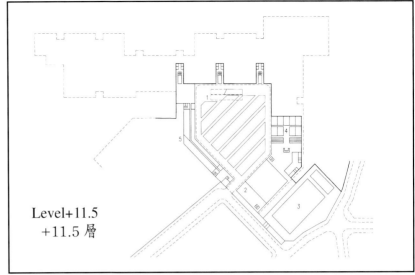

Level+11.5
+11.5 層

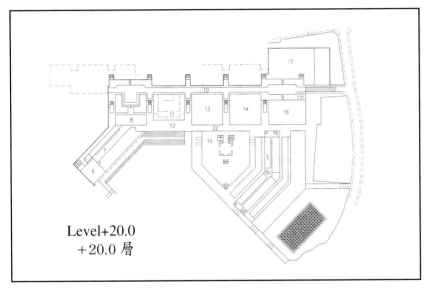

Level+20.0
+20.0 層

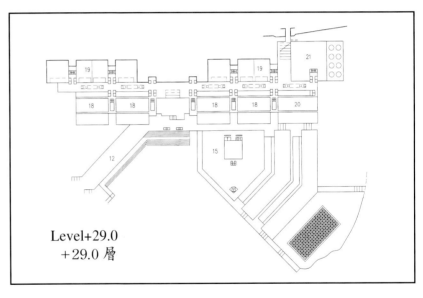

Level+29.0
+29.0 層

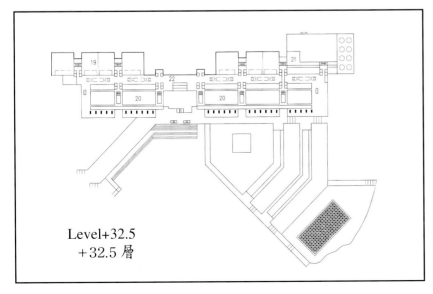

Level+32.5
+32.5 層

說明	Keys		
		12. 平台	Terrace
1. 停車場	Car park	13. 語言中心	Language centre
2. 網球場	Tennis courts	14. 電腦中心	Computer centre
3. 游泳池	Swimming pool	15. 圖書館	Library
4. 壁球場	Squash courts	16. 維修工場	Maintenance workshop
5. 行政	Administration	17. 機械室	Plant room
6. 員工飯堂	Staff dining	18. 實驗室	Laboratory
7. 學生飯堂	Student dining	19. 課室	Lecture rooms
8. 廚房	Kitchen	20. 行人步廊	School of study
9. 貨物起卸區	Loading bay	21. 工場	Workshop
10. 服務通道	Service road	22. 學生通道	Students street
11. 禮堂	Assembly hall	23. 體育館	Sports hall

牡丹山莊
Xian Mountain Inn

中國西安市郊，臨潼縣 (1985)
Lin Tung County near Xian, China.

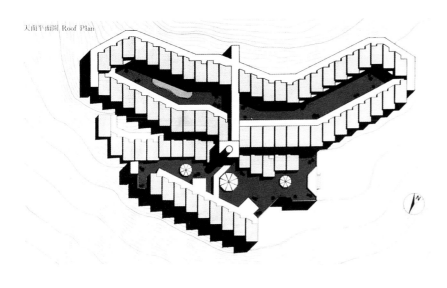
天面平面圖 Roof Plan

牡丹山莊的工地位西安市以東臨潼縣著名的華清池後面山坡上。這個方案是盡量利用本地材料，來嘗試建造一個低層的飯店，它反映了本地獨特的風味與富有傳統色彩的格調，這個表現手法同大城市常見的玻璃幕墻、旋轉餐廳等是大相異趣的，並有明顯的對比做法。

這個飯店包括飯堂、劇場、會議廳與商業服務中心等全面性設施。房間共有187間，大部分房間一方面是面對中國式的庭園，另一方面又能面對週圍開闊的田園風景。

The site is situated on a mountain slope overlooking the Huaqing Hot Springs in the outskirts of Xian, off the road leading to the site of the Museum of Terra-Cotta Warriors. The project is an attempt to design a low rise hotel complex, which is to be built with the maximum use of local materials, to provide an appropriate solution for the locality with an atmosphere full of local traditional environment for habitat. This approach contrast sharply with the trend of high rise, curtain wall international style of solutions we see everywhere in the major cities of China.

The project is self contained with restaurant, theater, conference rooms and business centre facilities together with 187 rooms. Most rooms are grouped around landscaped gardens with a commanding view over the countryside.

雕鷹山莊

Lanzhou Mountain Inn

中國蘭州市五一山造林站 (1986)
Lanzhou City, Gansu Province, China.

The project is located in a valley overlooking the city of Lanzhou and the Yellow River. It is surrounded by beautiful landscape full of fruit trees and vegetable farms with many places for horse riding and walks. The first phase consists of 168 rooms including 12 nos. 3 star standard suites, 22 nos standard rooms and 52 nos 2 star standard rooms built into the mountain side. The design of the 2 star standard rooms are based on the modern cave construction system with natural ventilation to preserve energy and to provide an economic room facility. This part of the Inn aims to prove that earth construction is feasible. The design of the Inn is based on the traditional narrow courtyard houses in the north, facilitating the winter sunlight into the building and stopping the summer sun penetration. The fish ponds in the complex reflect the atmosphere of a mountain inn set in a traditional garden environment. The Inn has a 100 person restaurant, coffee shop, swimming pool, shops, management office and senior staff quarters. There is a service centre equipped with maintenance facilities, staff canteen, staff quarters and district staff training school. The building material used includes locally produced bricks and precast concrete elements. All stairs are built of wood and all external walls are constructed by ram earth method to harmonize with locality. This is an attempt to design an appropriate and economical hotel in the developing regions of China using local material and local construction methods. The design is different from the usual glass curtain wall and revolving restaurant solutions one normally comes across.

鵰鷹山莊位於蘭州市北側的一個山谷裏，鳥瞰黃河與市容。此處週圍風景幽美，果林滿山，林場中有豐富的瓜果產品和綠化用的苗圃，其中有很多地方可供騎馬與散步。第一期包括168個房間，其中包括有3星級的12個套房及22個標準房。此外，有52個2星級的房間建在山坡上，這一部分以現代化的窟洞構造為主，有天然通風系統，節省能源。提供一些經濟客房，是為了實現生土建築的新生。山莊的構思，是據北方傳統狹窄的四合院盡量吸收冬天的陽光進入建築物，及阻止夏日陽光的照射。庭園中的魚塘，有聲有色地反映山莊的氣氛，好像置身於傳統的中國古典名園境界之中。飯店有100人的食堂、咖啡廳、泳池、商店、飯店辦公樓及高級職員宿舍，此外還有一個服務中心，包括齊全的輔助設備及員工飯堂、宿舍和地區性飯店員工訓練學校。
建築物用本地出產的磚與預製混凝土構件建成，走廊樓梯用木結構來適應鄉土的風味，四週的圍墻，用黃土建成，這是嘗試以本地風格及盡量利用本地材料，本地現代生土建築的技術來建造適合中國發展中地區用的廉價飯店的一個例子，是與流行的玻璃幕墙、旋轉餐廳之類的方案不同。

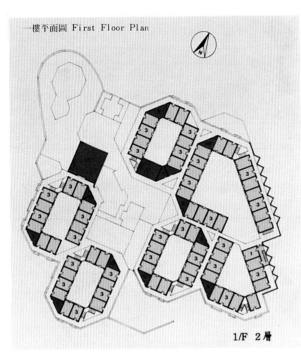

一樓平面圖 First Floor Plan

1/F 2層

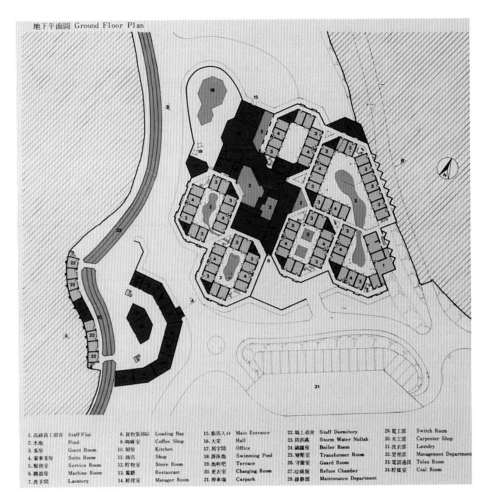

地下平面圖 Ground Floor Plan

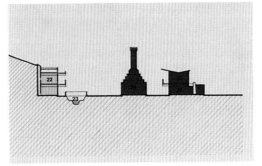

1. 高級員工宿舍	Staff Flat	8. 貨物裝卸區	Loading Bay	15. 飯店入口	Main Entrance	22. 職工宿舍	Staff Dormitory
2. 水池	Pond	9. 咖啡室	Coffee Shop	16. 大堂	Hall	23. 防洪溝	Storm Water Nullah
3. 客房	Guest Room	10. 廚房	Kitchen	17. 寫字間	Office	24. 鍋爐房	Boiler Room
4. 豪華客房	Suite Room	11. 商店	Shop	18. 游泳池	Swimming Pool	25. 變壓室	Transformer Room
5. 服務室	Service Room	12. 貯物室	Store Room	19. 酒吧	Terrace	26. 守衛室	Guard Room
6. 機器房	Machine Room	13. 餐廳	Restaurant	20. 更衣室	Changing Room	27. 垃圾房	Refuse Chamber
7. 洗手間	Lavatory	14. 經理室	Manager Room	21. 停車場	Carpark	28. 維修部	Maintenance Department
29. 電工部	Switch Room						
30. 木工部	Carpenter Shop						
31. 洗衣部	Laundry						
32. 管理部	Management Department						
33. 電話通訊	Telex Room						
34. 貯煤室	Coal Room						

South elevation 南立面

North elevation 北立面

East elevation 東立面

West elevation 西立面

South/north section facing west 向西/南北剖面

North/south section facing east 向東/南北剖面

West/east section 西東剖面

聽葉山房

House, Ilan.

台灣宜蘭市　(1989)
Ilan, Taiwan.

這是一間鄉下度假屋，位於山坡底，面向綠色的田野。房子離地而建，避免水浸及山泥傾瀉。設計上採用了斜頂及木欄杆是反映地方傳統民居的風格。

This is a country retreat house situated at the bottom of a mountain slope facing out onto the fields beyond. The house is raised to avoid flooding and landslide. The design incorporated slopping roofs and wooden external railings to express a spirit of the traditional house design.

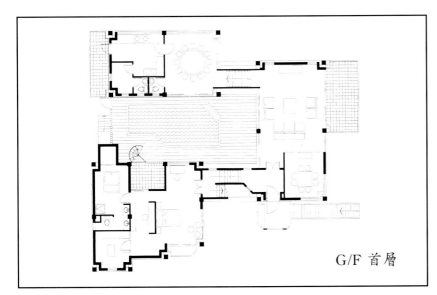

G/F 首層

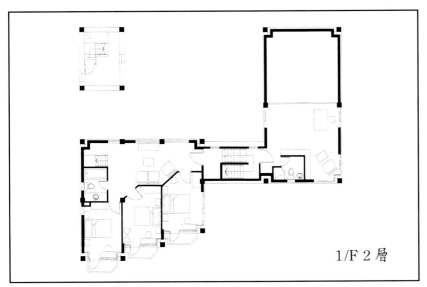

1/F 2層

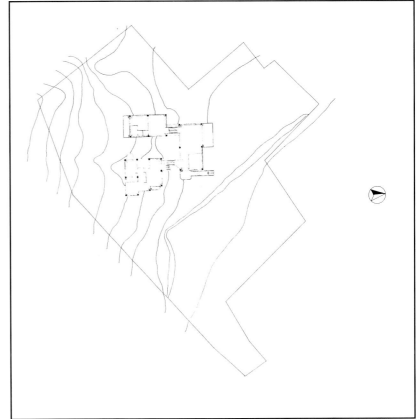

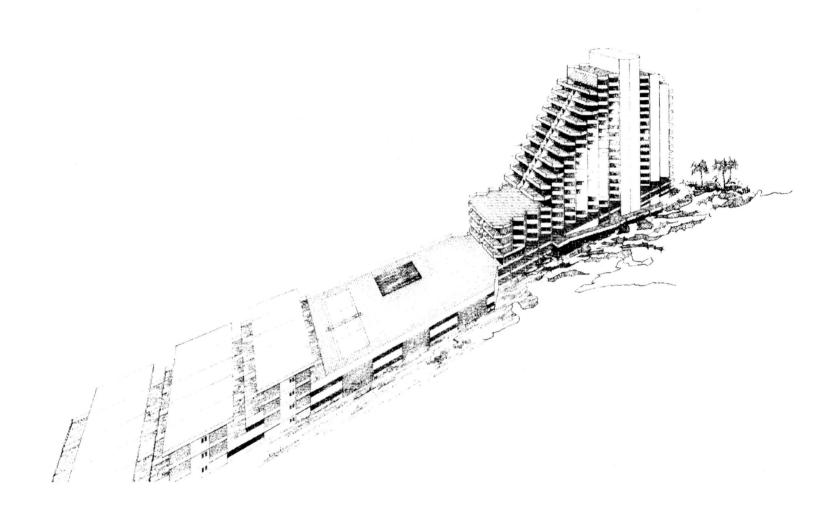

華獻度假公寓大樓

Holiday Condominium, Hua Hin

泰國華獻市高德其海灘 (1989)
Khaotakieb Beach, Hua Hin, Thailand.

這個項目與北芭堤雅的度假公寓大樓是同一類又窄又長的工地。公寓大樓坐落近路的部分，主要是避免車流深入項目。而康樂設施例如泳池及健康中心是在項目的中部，近海灘的部分增設比較豪華的庭園房子。雖然受到又窄又長的工地限制，但高層的單元都能各有海景。

This project has the same long and narrow beach front site like the holiday condominium at North Pattaya. The condominium block is sited nearer the road to minimize car traffic access into the site with the recreational facilities like the pool and health clubs sited in the middle. The beach front area is occupied by rows of courtyard houses to provide a more luxury accommodation. The design provides sea view to all condominium units despite the narrow site restriction.

G/F 首層

Typical floor 標準層

北芭堤雅度假公寓大樓

Holiday Condominium, North Pattaya

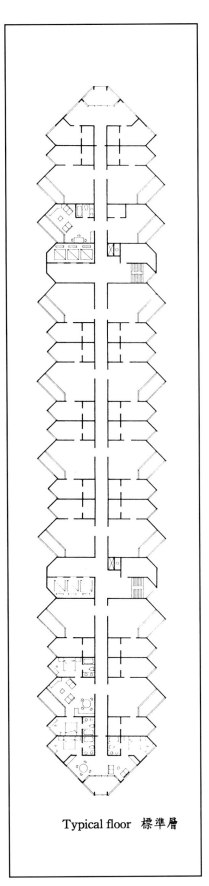

泰國北芭堤雅　(1989)
North Pattaya, Thailand.

工地又長又窄的造型是屬於芭堤雅一種常見的情況。公寓大廈是提供度假居所給曼谷的居民，整體佈局務求每個單元都能有前面的海景及後面的山景，公寓分為一個房間及兩個房間的單元，首層有商舖康樂及飯食設施，包括一游泳池。

The site is a typical narrow and long beach front site in Pattaya. The condominium is designed as second home for city dwellers from Bangkok. The unit layout aims to provide every unit with a view of the sea in the front and mountain view at the back. The sizes of apartment varies from a single room unit to two bedrooms. The ground floor contains shops and recreational & restaurant facilities including a fun pool.

G/F 首層　　Typical floor 標準層

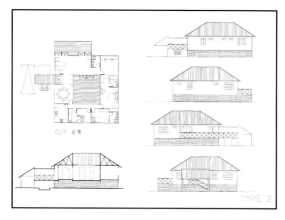
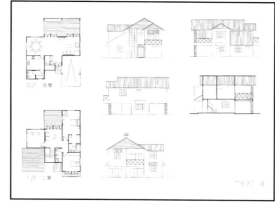
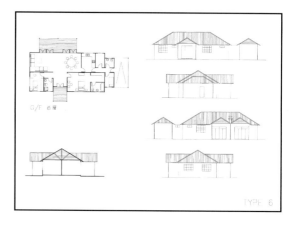

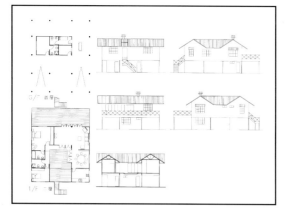
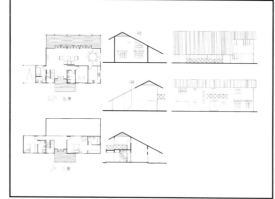
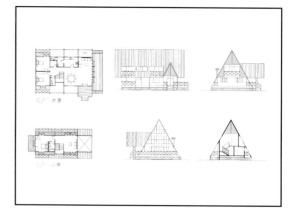

清邁山莊度假村

Misty Hill Holiday Homes, Chiang Mai

泰國北部近清邁市郊米特區 (1989)
Mae Taeng near Chiang Mai, Thailand.

山莊是提供給曼谷市本地及外國居民度假之用，位於郊外大自然的景色中。工地總面積115,200平方米，分為43個別墅地段。為了適應不同生活方式的泰國人、中國人及外國人，個別別墅的設計有10種不同的風格，提供給買家選購，包括有2層入口的泰式、A字結構的瑞士山居、中式四合院、美式農莊、英式鄉舍等。大部分別墅用木建造及有柚木瓦的屋頂，務求保持鄉村的氣氛並與週圍環境配合。

The homes are designed for both local and expatriate families from Bangkok. It is situated in the middle of the countryside with wonderful natural scenery all around. The total site area is 115,200m2 and is divided into 43 no. of individual house lots. To meet the various style of living of the potential purchasers which include thai, chinese and westerners, the proposal developed ten house types for selection by potential purchasers. It includes thai house with raised first floor entrance court, swiss chalet with A frame, chinese courtyard house, american ranch house, english country cottage etc. All houses are mainly built in wood with teak shingle roofs to maintain a uniformity of the village formation and to harmonize with the surrounding.

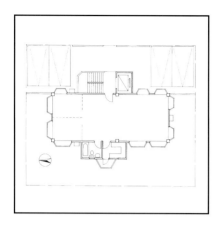 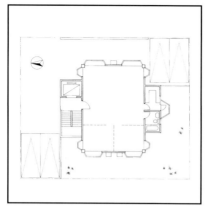 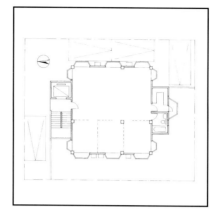 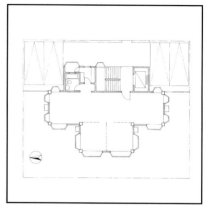
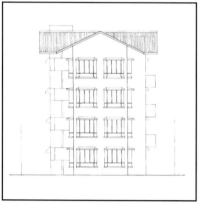   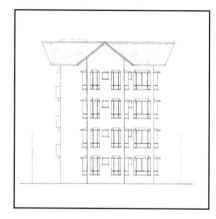
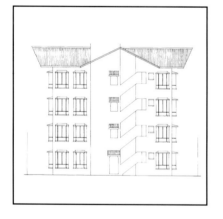   
 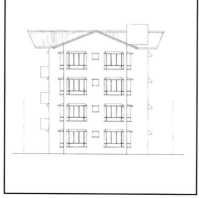 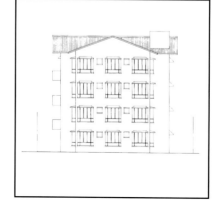 

蘇榮特皇路住宅小區

Housing estate, Survintawongs Road

泰國曼谷市 (1990)
Survintawongs Road, Bangkok.

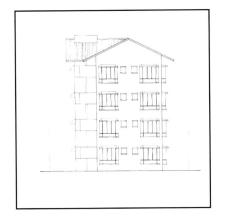
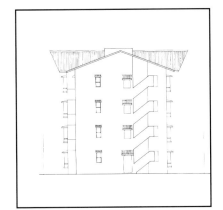
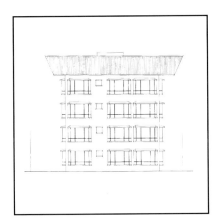
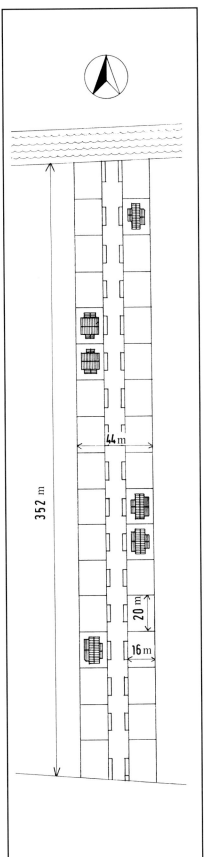

這是曼谷市內標準的又長又窄的工地，一面對路，一面對河。如果要有一個有經濟效率的開發方案，只有把公寓樓放在中間道路的兩旁，樓是每層102平方米至125平方米面積。由於公寓不單只提供居所，更要提供家庭式工場的功能，所以每層的主要空間可以靈活地由戶主自行作出房間的分布間格。

This is another typical long and narrow site in Bangkok between the road and the canal. To provide economical development, there is no alternative but to site the apartments blocks on the side with a central access road. There are 5 block types of various sizes ranging from 102m2 to 125m2 per floor. The basic layout is flexible leaving individual families to freely partition the main space in each apartment as the apartments not only serves as dwelling units but also as family workshops.

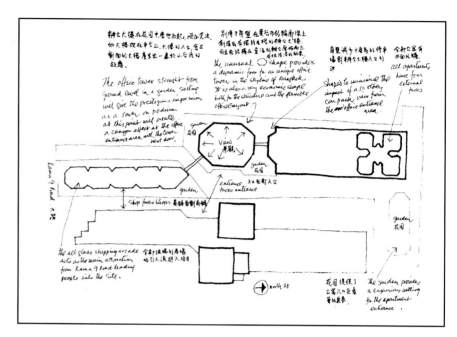

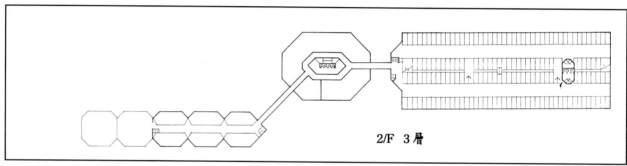

2/F 3層

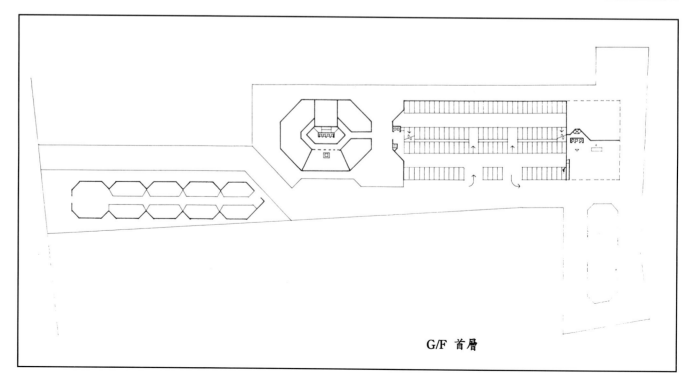

G/F 首層

拉曼9街商住大樓
Commercial/residential complex, Rama 9 Road

泰國曼谷市華馬9路 (1991)
Rama 9 Road, Bangkok, Thailand.

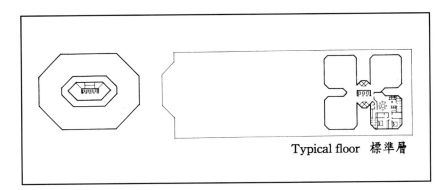

Typical floor 標準層

在一個又長又窄的工地上，主樓被迫定位於深入地盤比較寬闊的地方，但又能維持辦公大樓的壯觀是一難題，其中一個解決辦法是以一全玻璃覆蓋的商場，引行人進入中心地帶。項目包括一座八角型35層高辦公大樓，一多層停車場，一座商場及一座26層高104個2間睡房的公寓大樓。有蓋的空中步廊把商場、辦公樓及停車場相連，北面的道路通過花園提供了公寓大樓的入口。

The problem of a long and narrow site is the difficulty of having to site the office building deep into the site and away from the road and still maintain the prestigious aspect of an office block. One obvious solution is to use a podium design from the road side but it will provide excessive commercial space which was not in demand. The best solution is to provide an unusual attractive glass cladded shopping arcade, both physically and visually directing the pedestrian into the site. The project consists of a 35 storey hexagonal shaped office block, a multi level car park, a shopping arcade and a 26 storey apartment block with 104, 2 bedroom units. Covered elevated walkways are used to link the shopping arcade with the office tower and the car park. The road at the northern end of the site serves as access to the apartment block facing a garden.

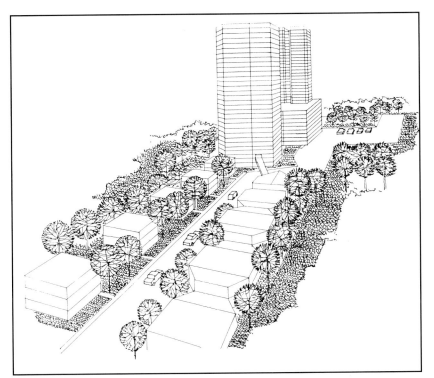

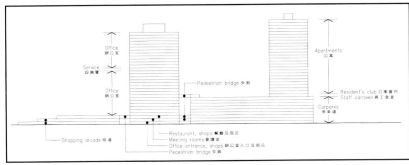

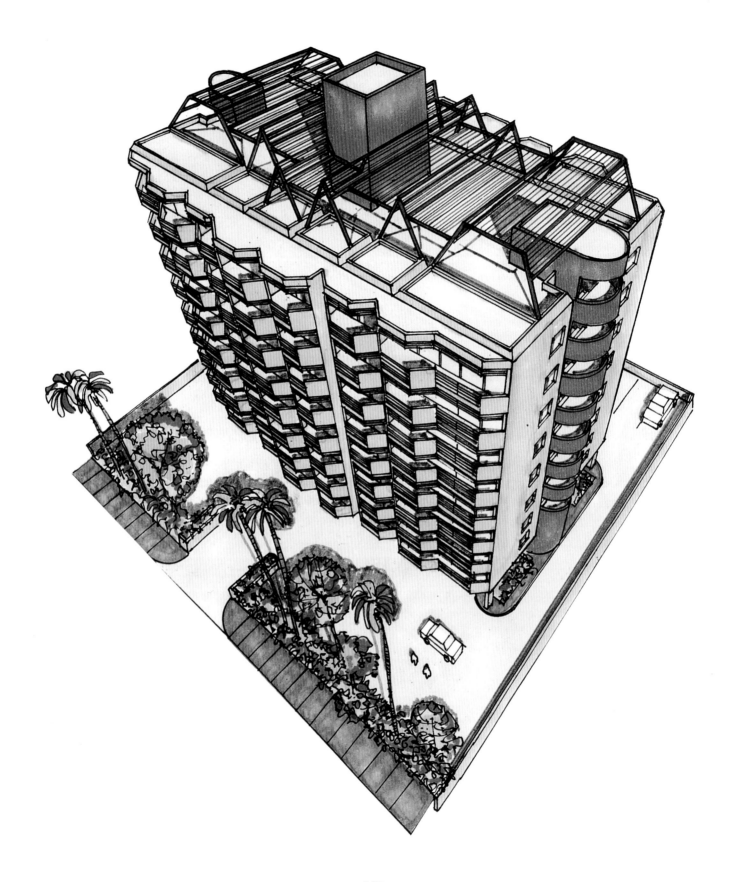

曼谷市42街公寓大樓

Residential condominium, Soi 42

泰國曼谷市華金坑道42街(1991)

Soi 42, Ramkamhaeng Road, Bangkok. Thailand.

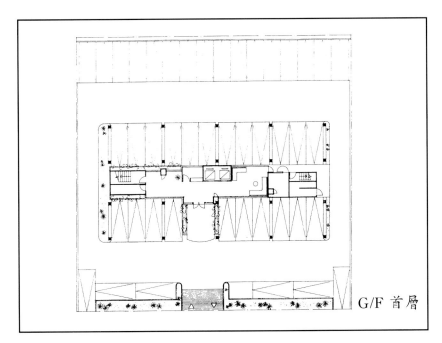

G/F 首層

公寓大樓是向中等收入的曼谷居民提供住所。樓高9層，每層有12個單元，4個有一間睡房，另外8個是單房的單元。公寓的露台用管制的欄杆便於通風及提供戶外空間和遮太陽的作用。屋頂的設計是嘗試創造出一些本地風格。

This is a residential project for the middle income city dwellers of Bangkok. The building is 9 storey in height and has 12 apartments per floor, 4 of which are one bedroom type and 8 are one room apartments. The balconies have open railing to facilitate air penetration and provide an outdoor space and sun shading for the apartment below. The roof treatment is an attempt to create some form of local identity to the building.

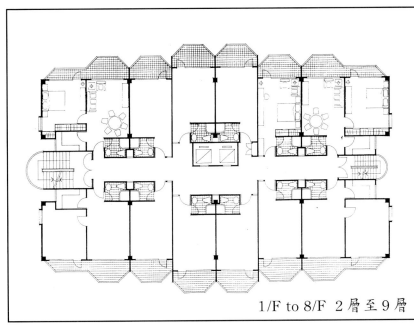

1/F to 8/F 2層至9層

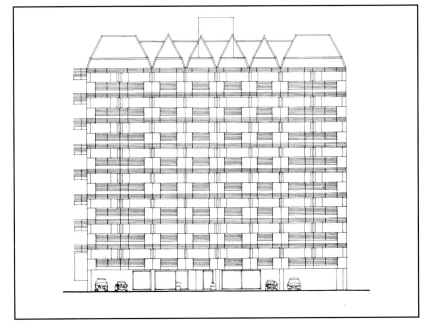

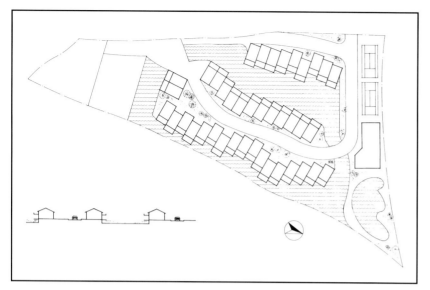
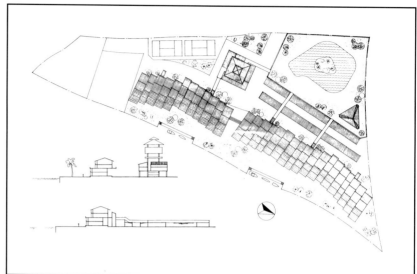

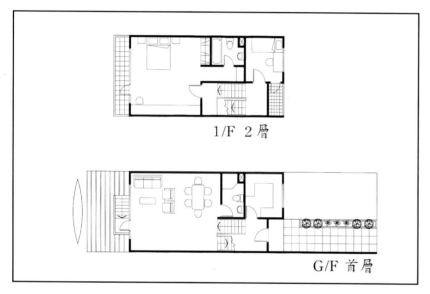
1/F 2層
G/F 首層

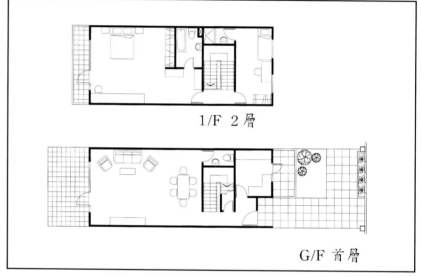
1/F 2層
G/F 首層

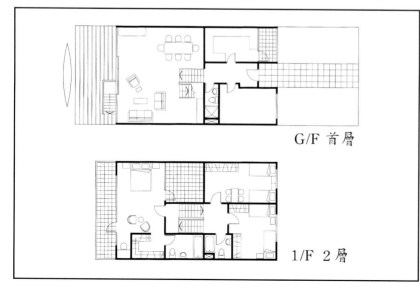
G/F 首層
1/F 2層

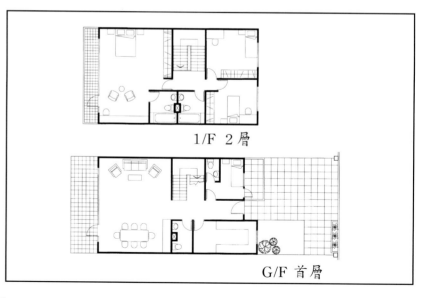
1/F 2層
G/F 首層

曼谷市郊河畔小築
Riverside housing, near Bangkok

泰國曼谷市郊河畔　(1992)
Ban Khlong Om Yai near Bangkok, Thailand.

這個項目有兩個方案；第一個方案有雙層的2層高單元及較大的沿河花園和停車場頂部花園，第二方案有2至3層沿河住宅及各住宅有停泊小船的碼頭。構思務求創造一個鄉村的氣氛，會所是整個鄉村的中心。中心有娛樂設施，並且形成為項目的視覺標誌。整個項目以斜屋頂及木結構的造型來反映地方風格及適應氣候的要求。

This project has two proposals. One consists of double stack of 2 storey units with communal garden along the river and on top of the communal car park. The other has two to three storey houses each sited on the water edge with mooring facilities. The idea is to create a small village atmosphere. The club house, being the centre of the village, is designed with recreational facilities and forms a visual focal n ode to the complex. The design attempts to use sloping roofs and wooden construction forms to reflect local identity and provide appropriate protection for the climate.

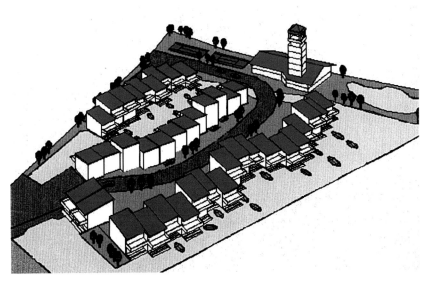

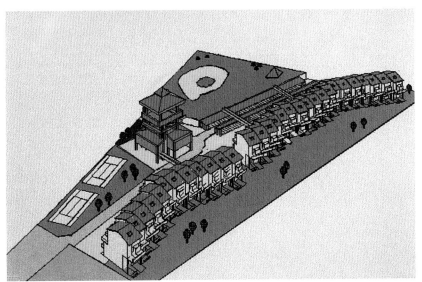

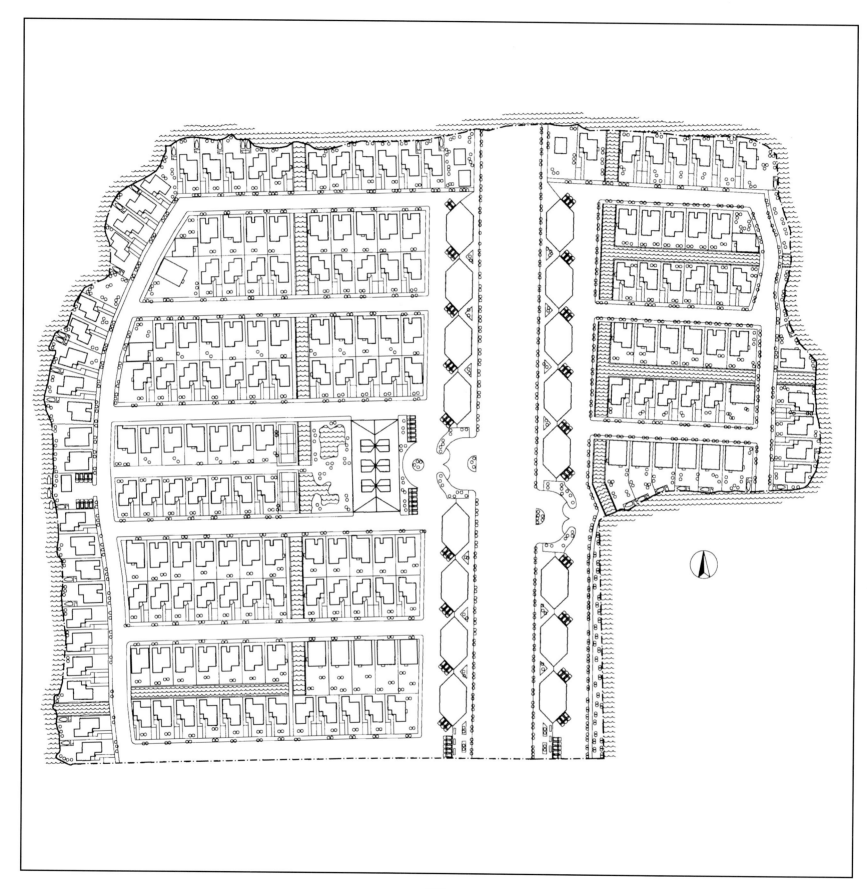

蕭山市金家浜花園
Riverside Garden, Xiao Shan

中國杭州南面蕭山市(1993)
Xiao Shan City, South of Hangzhou, China.

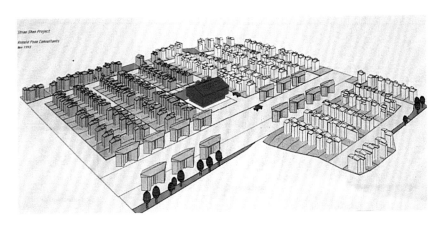

項目位於杭州南面海畔的蕭山市，工地內有一公路橫跨。有180間花園別墅及沿路旁的低層公寓，部分別墅位於河畔有私人船泊位，多條人工運河穿插其中，提供了水鄉的氣氛。區中心的會所成為整個金家浜的核心，別墅的設計分為5個種類，3種4～5間睡房的有南向的後園，2種5間睡房的有南向前園。有2種公寓的設計，3間睡房單層式及4間睡房雙層式。所有別墅的頂部有簡單的斜頂，務求與週圍現有的民房協調。

The project is situated along the river in the city of Shao Shan south of Hangzhou. The site is divided into two parts by a road. The development consists of 180 villas with individual gardens and low rise apartment blocks along the road. Some of the villas are sited on the water edge with private landing docks. Canals have been introduced throughout the project to create a river community atmosphere. The community centre is the focus of the development. Five types of house plans have been developed, three for south facing rear garden with 4 to 5 bedrooms and two for south facing front garden with 5 bedrooms. There are two types of low rise apartments one storey with 3 bedroom and two storey with 4 bedrooms. All houses have simple sloping tile roofs to blend in with the traditional houses in the neighbourhood.

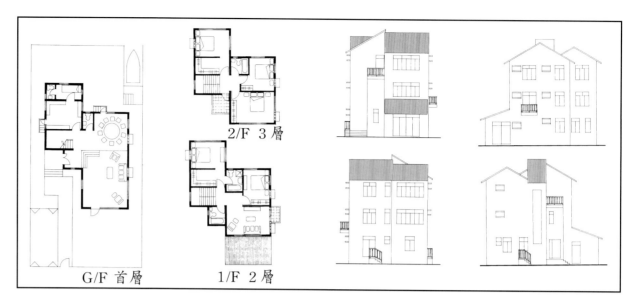

G/F 首層　　1/F 2層　　2/F 3層

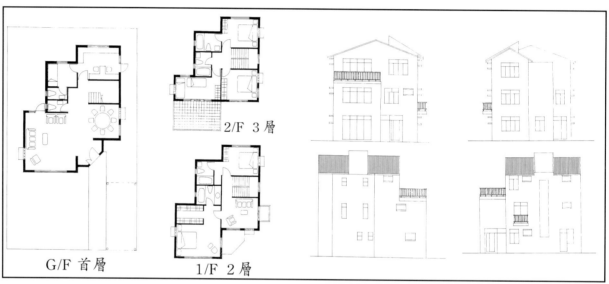

G/F 首層　　1/F 2層　　2/F 3層

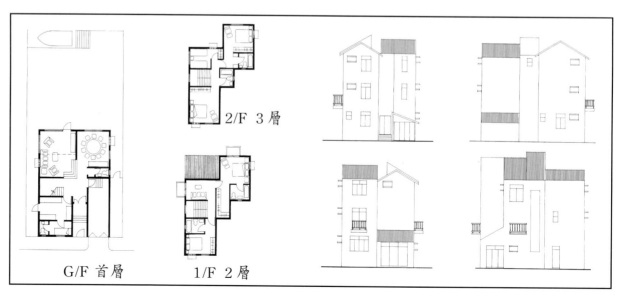

G/F 首層　　1/F 2層　　2/F 3層

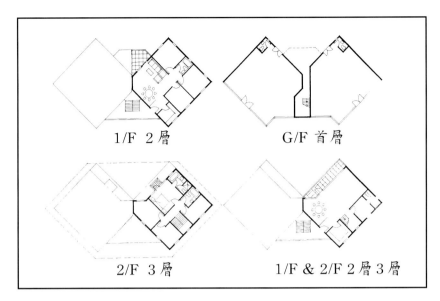

1/F 2層　　　　　G/F 首層

2/F 3層　　　　　1/F & 2/F 2層3層

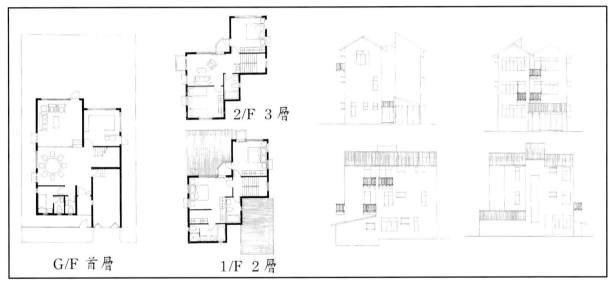

G/F 首層　　　1/F 2層　　2/F 3層

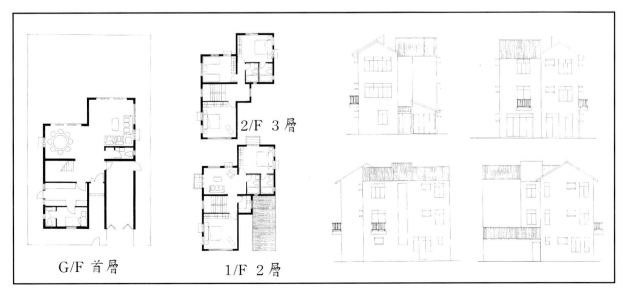

G/F 首層　　　1/F 2層　　2/F 3層

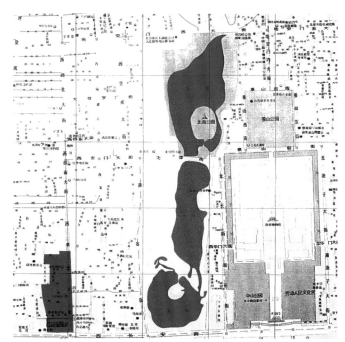

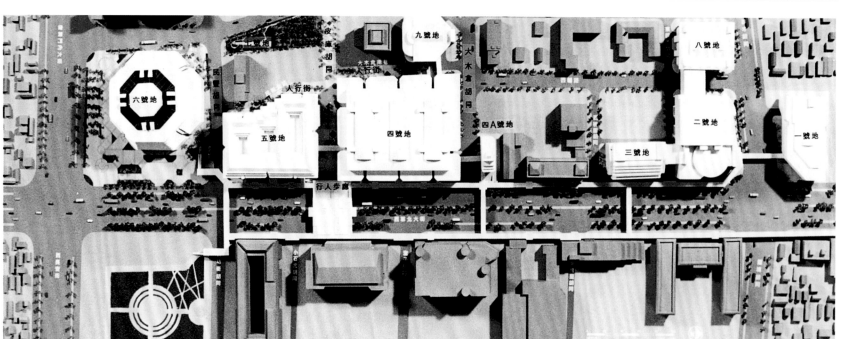

總顧問:香港潘祖堯顧問有限公司　北京西單北大街西側商業區1-9號地塊總規劃

"西西工程"總規劃
Master Plan for "Xi - Xi Project"

中國北京西單北大街西側商業區 (1994)
Xidan North Street, Western Commercial District, Beijing.

The Company was appointed by the Beijing Jing Yuan Property Development Co. Ltd. in June, 1994 to be the lead consultant for the master plan work which included nine sites in total. The work included current assessment and in-depth studies to co-ordinate development work in progress.

The project is situated at the western side of Xidan North Street with a total area of 20.85 hectares. The total gross floor area is 600,000m2 to be completed in 6-7 years' time. Xidan North street will be widened to 70m width with a new parallel road of 30m width and underground parking facilities for 3000 cars. The site is surrounded in the west by the west city cultural district, south by west Chang-an Street and east by Xidan North Street. The site measured 130-132m x 860m and surrounded on three sides by roads.

The aim for the re-development is to rebuilt the district into a modern, multi functional commercial area with inter-connected development framework. The master plan proposed district wide pedestrian system with first floor inter building walkway which linked up all buildings along both sides of Xidan North Street together with cross street high level pedestrian links, one of which has a shopping arcade providing continuity of the shopping activities between the shopping malls on either side of Xidan North Street. It also proposes under road shopping malls which link up basement shopping malls along the main road and underground bicycle routes partially lid by daylight throughout the length of Xidan North Street freeing the main road of bicycle traffic. All sites are linked in the second basement level by an underground network of roads which helps to reduce surface car traffic and provide alternate fire escape routes for all basements.

北京敬遠房地產開發有限公司於1994年6月24日委托，以香港潘祖堯顧問有限公司為首的顧問團對北京規劃管理局所制定的北京西單北大街西側商業區"西西工程"1～9號地塊土地使用規劃條件進行評估及深入研究。因有幾個工地已開工進行個体設計，所以原則上評估及深入研究的工作是以配合開發的現況進行，盡量把現有的總規劃做到更完善的境界。

項目地區位於西單北大街西部，佔地約20.85公頃。建築樓面佔600,000平方米，分6~7年建成。西單北大街將擴寬至70米，並在新發展區內加建一條寬30米與西單北大街並行的道路。整個發展計劃還包括停車大樓及地下停車庫，共提供約3,000個停車位。項目地區北面是西城文化中心；南臨西長安街；東瀕西單北大街的東邊；西面與現有的幾個地塊相連，由西單北大街向外伸展130～320米，地塊由北至南距離860米，三面均臨現有街道。

此總規劃的主要目的是為該地區重建成為現代化的、多用途的綜合商業体而提供各連貫性的發展架構。總規劃提議一地區性行人網絡，以2層高的步廊把面對西單北大街的兩旁建築物相連，加上過街的步廊，包括有一個商場的過街步廊，將街兩旁的商場相連，使商業活動能延續跨過大街。此外增設大街地下商場，把大街兩旁的地下商場相連。更提議一地下有部分日照的自行車網絡沿西單北大街全程，解決了自行車干擾地面交通的問題。全部工地建築的地下2層以地下車道相連，減少地面道路的交通量，並且提供額外地下防火通道。

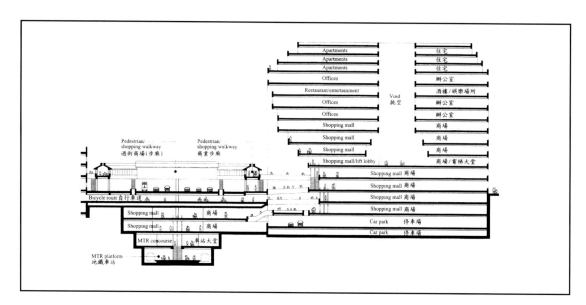
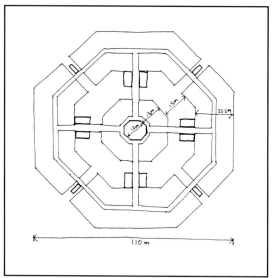
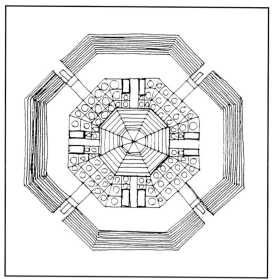
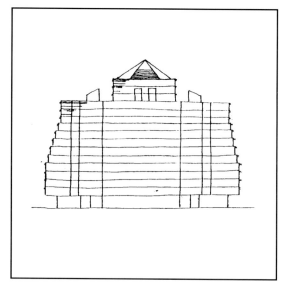
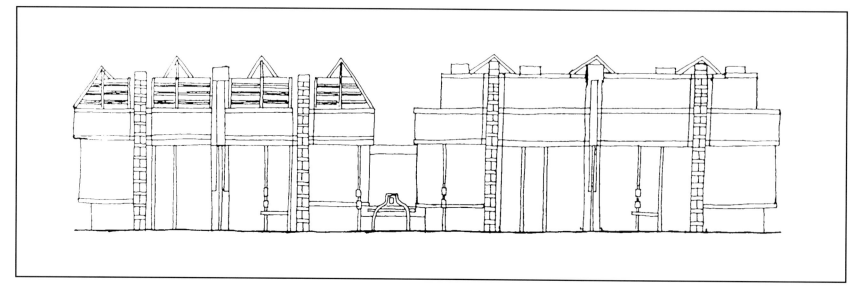

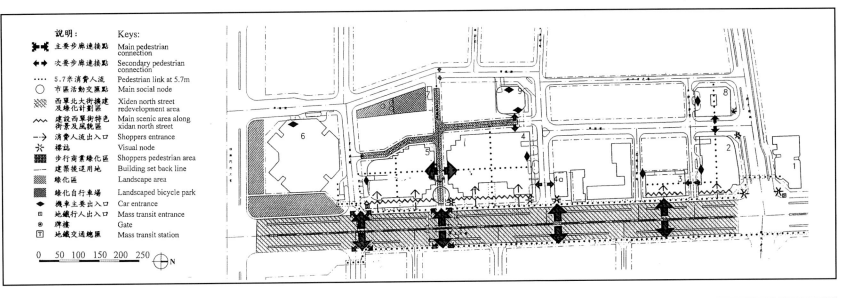

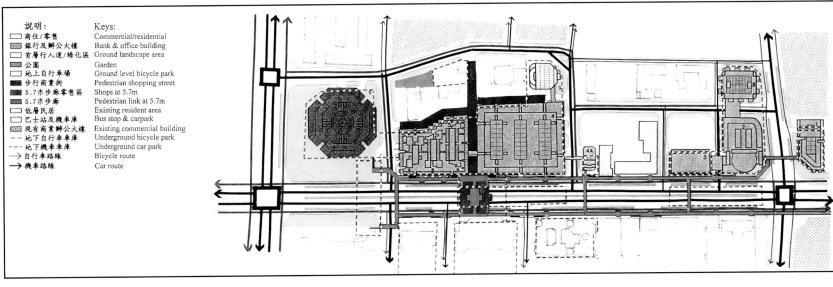

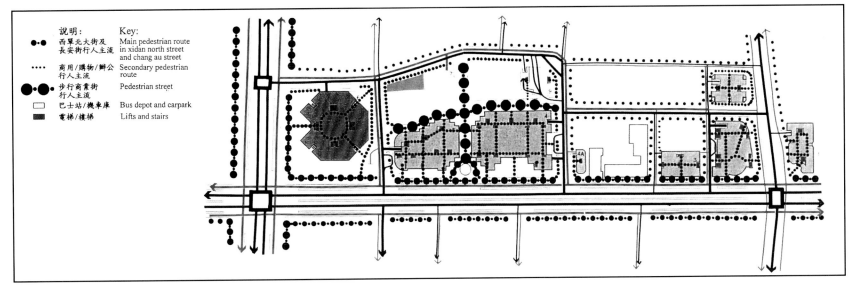

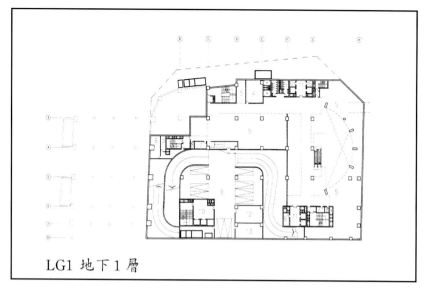

LG1 地下1層

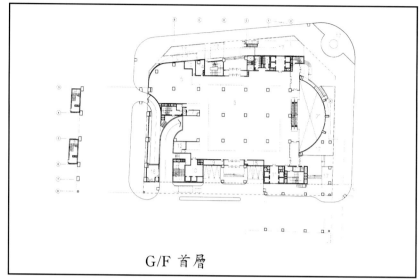

G/F 首層

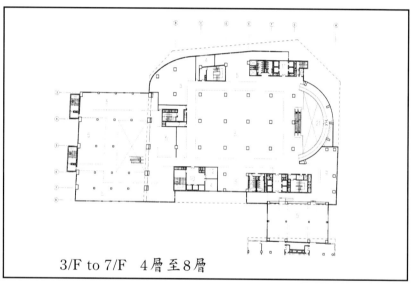

3/F to 7/F 4層至8層

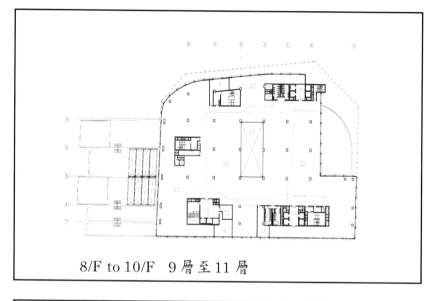

8/F to 10/F 9層至11層

East elevation 東立面

說明	Key		
1. 男廁生間	Male lavatory	18. 煤氣錶房	Gas meter room
2. 女廁生間	Female lavatory	19. 廣播室	Boardcast room
3. 電話上線井	Telephone shaft	20. 行人步廊	Pedestrian bridge
4. 新風機房	Fresh air intake room	21. 挑空	Void
5. 商舖	Shop	22. 辦公室	Office
6. 貯物室	Storage	23. 茶水間	Pantry
7. 配電房	Electrical room	24. 衛星電視控制室	Satellite tv control room
8. 弱電房	Low voltage switch room	25. 通風機房	Fan room
9. 貨物起卸區	Loading bay	26. 電梯機房	Lift machine room
10. 貨梯大堂	Cargo lift lobby	27. 加壓送風機房	Pressurization fan room
11. 客梯大堂	Guest lift lobby	28. 消防及生活用水水箱	Water tank
12. 垃圾房	Refuse chamber	29. 沖洗用水水箱	Flush water tank
13. 電纜分界小室	Electrical meter room	30. 消防及生活用水泵房	Pump room
14. 噴淋控制室	Sprinkler control room	31. 冷卻水塔	Chiller
15. 發電機散熱器房	Generator room	32. 排氣機房	Exhaust fan room
16. 消防控制室及保安室	Fire control & security room	33. 排煙機房	Smoke exhaust fan room
17. 花槽	Planter		

2號地商辦大樓
Office/commercial complex, no. 2 site

"西西工程"2號地, 北京西單北大街
No. 2 site, "Xi-Xi project", Xidan North Street, Beijing.
與吳享洪建築師合作設計
Jointly designed with Mr. Anthony Ng

The building is designed to accommodate the existing commercial and office premises displaced by the development of the "Xi-Xi Project". Most service areas are located along the north and south edge of the building to facilitate a large open area in the middle for the shopping mall. The building is situated at the northern end of the Xidan North Street forming a "gate" building for the Xidan Commercial District. Therefore the east elevation is particularly designed with a curvature to lead the eye into Xidan North Street. The curve shape enveloped a ramp & a void which runs from the basement to the second floor providing a pedestrian link with the shoppers being visible from the street. There are three basements, 2 for car parking and machine rooms with the lowest basement double up as an air-raid shelter. The first basement is partly a loading bay and partly shopping mall. The shopping mall occupies eight floors above ground with three storey of offices on top. The building is linked in the south and west to the adjacent buildings with over the road buildings. It is also linked by underground roads in the second basement to adjacent basements providing vehicular accesses to other buildings and minimizing car traffic on ground. The elevation treatment aims to reflect the function within the building in a simplistic manner with the office floor, containing openings sympathetic to the design grid below and with sun shading to counteract climatic conditions.

這是一個回遷項目，服務性的設施大多數放在樓面的南北側，使樓身的中部成為一個空闊的商場。大樓位於西單北大街的北面，是西單商業區北面的門樓，因此樓面的立面特以圓形的外形，把行人的視綫帶引到西單北大街的方向。圓型立面的背後有一連接地下首層到3樓的斜徑，街上的行人可以看到商場人流的影像。大樓有3層地下層，兩層為車庫及機房，地下3層兼做人防。地下1層為部分商場及部分貨物起卸區，地上8層為商場，而頂部3層為辦公層。大樓在西及南面有過街樓與鄰近建築物相連。地下2層有與隔鄰雙連的車道，使車輛能通到道路對面的地下車庫，減少路面的交通量。立面的設計是以簡單手法去表達商場的功能，而辦公層的窗格，是配合商場的外形加上遮陽架以適應地方氣候。

4及5號地商辦大樓

Office/Commercial Complex, no.4 & 5 sites

中國北京西單北大街"西西工程"4及5號地 (1995-1997)
Sites 4 and 5 of the "Xi - Xi Project", Xidan North Street, Beijing, China.

4及5號地的設計是以"兩位一體"的原則為基礎，把4及5號地的建築體以過街樓及地下車道相連，兩地之間的步行街之東面為區廣場，形成人群集中的地方。4及5號地的主要人流入口，面對西單北大街以4至5層高的半露天空間構成，空間中有觀光電梯及自動扶梯直達各層商場，有豐富傳統色彩的櫥窗，有連繫其他建築物的二層高的步廊，創造出一個富有地方色彩及商業氣氛的商場入口空間。

4及5號地樓高12層，4號樓地下兩層為車庫、機房和人防，有1002個車位，地下首層為娛樂及餐飲設施，地上首層至4層為商場、5層至12層為辦公；5號樓除地下首層為車庫和機房有222個車位及地上4層為商場之外，其他各層的功能與4號樓一樣，此外4號樓的頂部設有網球場兩個。4及5號地總面積為320,000平方米。

4及5號地樓的外形設計清楚地表達不同的功能，辦公的部分為了符合地方氣候及節省能源，加設了遮太陽的柵格，而商場部分則以雍容華貴的大理石裝飾，整個外形為了減輕高密度構成的建築體積，把外形分段，其中以開放式的防火梯作為重心，頂部以後退式的金字塔形柵格遮蓋了各機房的醜貌，表達了傳統屋頂的輪廓綫，加上辦公部分米黃色的鋁合金蓋板，外牆有凹有凸，色彩調和，使龐大面積的一座建築物，在視覺上層次分明，充分表達出首都風貌。4及5號樓在首都規劃委員會舉辦的"1995年首都建築設計匯報展"被選為"首都十佳建築設計方案"之一。

Sites 4 and 5 are designed as a single complex connected by buildings across the street and underground connecting roads. The eastern end of the pedestrian street between the two sites is developed as the district square to facilitate the concentration of pedestrian in the area. The major entries into the buildings are situated on the eastern side along Xidan North Street with four to five storey high atrium interlaced with glass lifts and escalators. The design of the shop front is inspired by the traditional concept and together with the elevated walkway system which links all the sites created an entry environment full of local identity and commercial atmosphere.

The buildings are twelve storey high. Site no. 4 has two storey of underground car parks with 1002 spaces. The first basement houses restaurants and entertainment centre. Four storeys of shopping mall with seven storeys of offices above. Site no. 5 has three basement car parks with 222 spaces and similar accommodation above as site no. 4. There is two tennis courts on the roof of site no. 4. The total area for the two sites is 320,000m2.

The external treatment of the buildings clearly reflected the different functions of the various parts. The sun shades on all office floors cater for the climatic requirements and the commercial parts are cladded in marble to reflect the luxury commercial activities therein. The bulk of the building is divided into several vertical groups by the external fire stairs in order to lessen the massiveness of the whole complex. The simple slanting roof shape echos the silhouette of the traditional roof shape together with the pale gold coloured aluminum cladding of the office floors gives one the impression of a well articulated building with a sense of belonging to the capital city of Beijing. The designs of these two buildings have been awarded "The Ten Best Design in the Capital" award during the 1995 capital Architectural Design Exhibition organized by the Capital City Planning and Construction Committee of China.

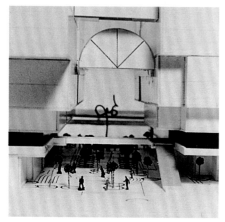

西單北大街西側商業區—四號地與五

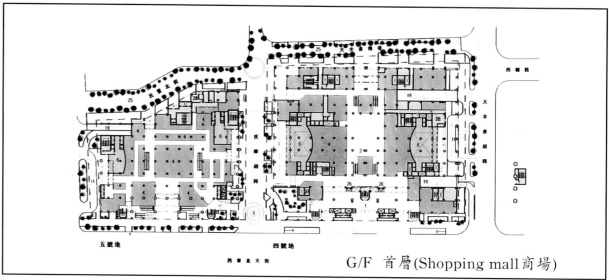

G/F 首層 (Shopping mall 商場)

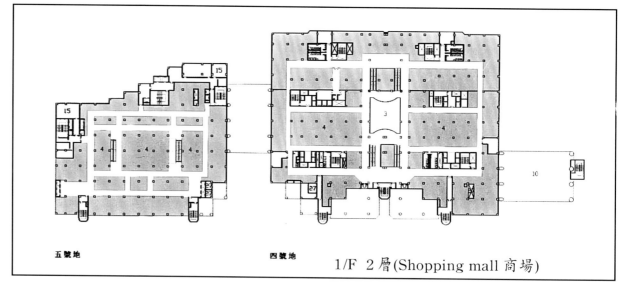

1/F 2層 (Shopping mall 商場)

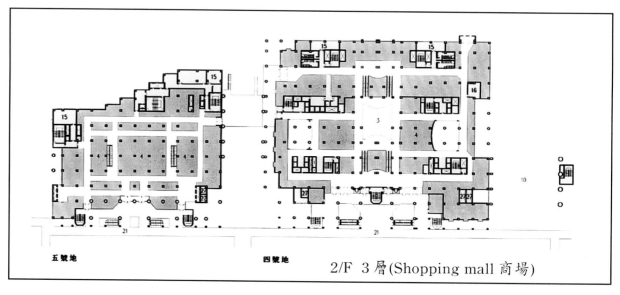

2/F 3層 (Shopping mall 商場)

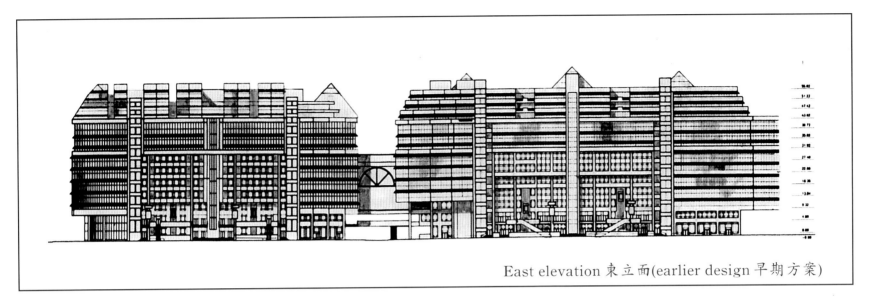

East elevation 東立面(earlier design 早期方案)

East elevation 東立面

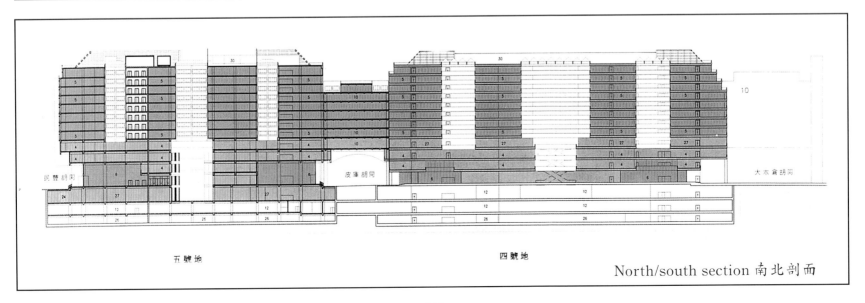

North/south section 南北剖面

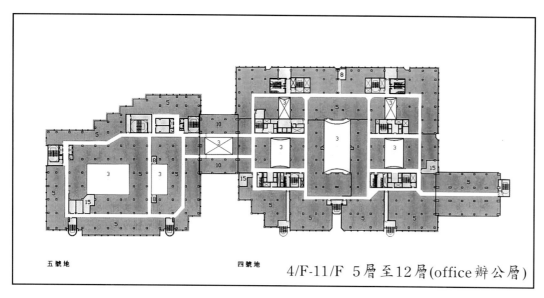
4/F-11/F 5層至12層(office辦公層)

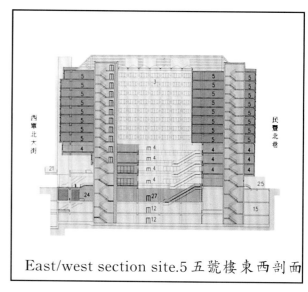
East/west section site.5 五號樓東西剖面

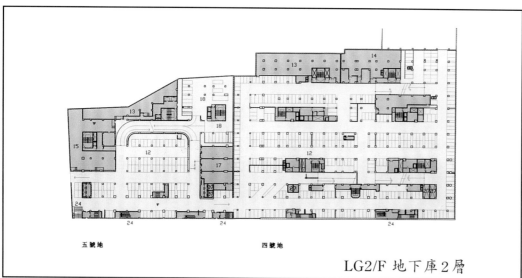
LG2/F 地下庫2層

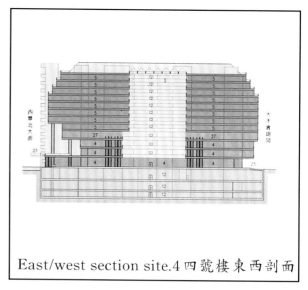
East/west section site.4 四號樓東西剖面

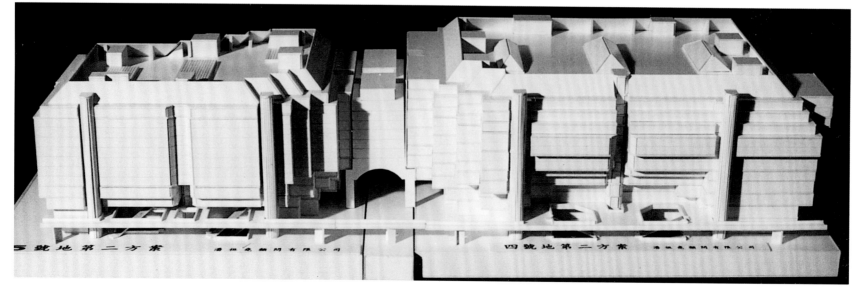

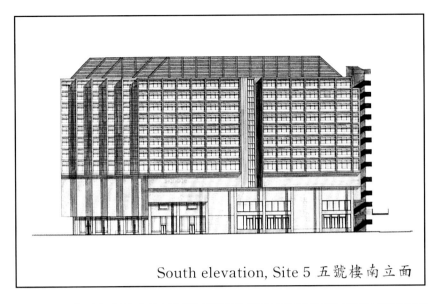

South elevation, Site 5 五號樓南立面

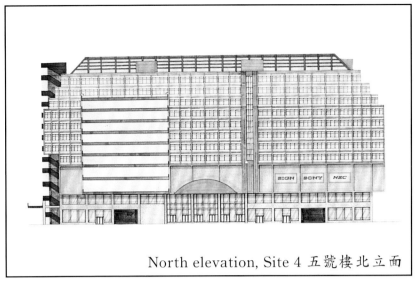

North elevation, Site 4 五號樓北立面

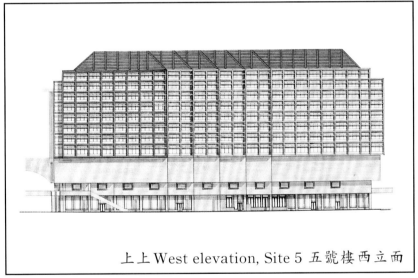

上上 West elevation, Site 5 五號樓西立面

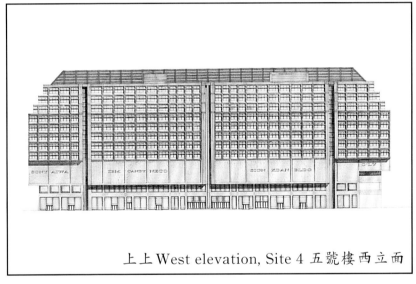

上上 West elevation, Site 4 五號樓西立面

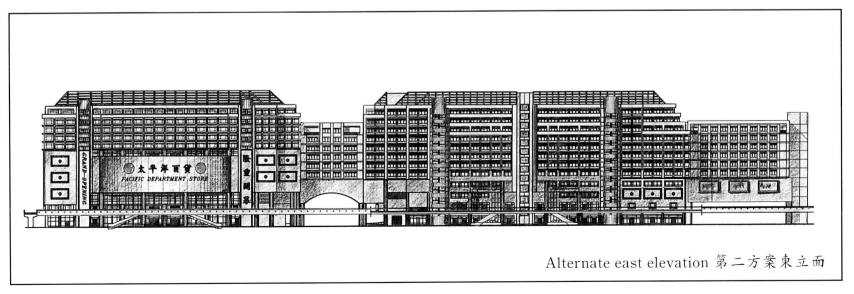

Alternate east elevation 第二方案東立面

武康路公寓大廈

Apartment Complex, Wu Kang Road

中國上海武康路 (1997)
Wu Kang Road, Shanghai, China.

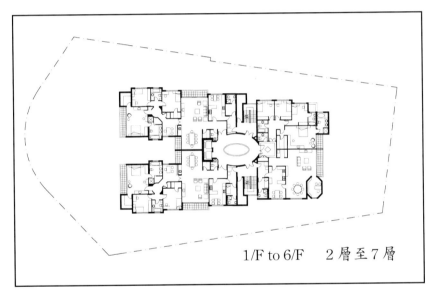

1/F to 6/F　2層至7層

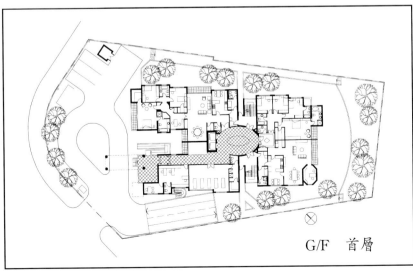

G/F　首層

公寓大廈位於上海一商住區內兩條街的一角。大廈有20間出租公寓,是提供給外國家庭租用。有地下停車場及私家健身室和大廈管理辦公室。公寓的設計為了適應外國人的生活習慣,增設了特大的公寓入口大堂、廚房及書房。橢圓型的電梯大堂,增添了豪華的氣氛。大堂中有一貫通各層的光井,使天然光能達各層的大堂。外牆的造型是與紅綫斜度呼應,加上外牆磁磚的選色,嘗試反映上海傳統民居的色彩及風格。

This is a corner site situated in a district in Shanghai with mixed development of commercial and residential buildings. The project provides 20 rental apartments to expatriate families. Parking facilities is provided in the basement and the complex has its own health club and management office on the ground floor. The apartments are designed with large entrance lobby, kitchen and study to meet the tenants particular way of life. The oval shape lift lobby aims to create an air of luxury and has a light well in the middle to allow penetration of sun light through out the height of the building. The configuration of the external wall is sympathetic to the shape of the site and together with the colour scheme of the external tiles form an attempt to recapture the character of the traditional housing in Shanghai.

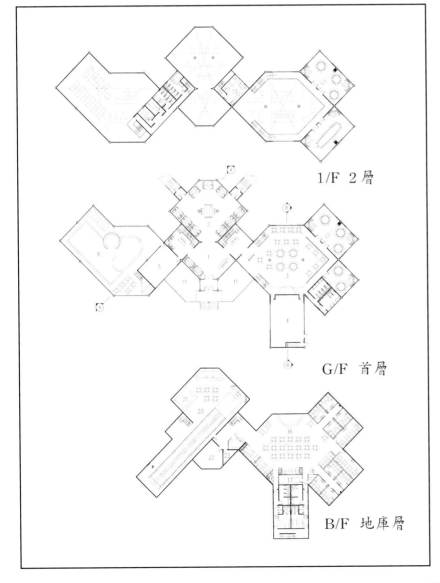

龍潭湖度假村

Holiday villas complex

北京龍潭湖 (1998)
Lun Tam Lake, Beijing.

項目位於公園中的人工湖畔，業主要求一豪華俱樂會所及住所。總規劃是以根據工地視野、日照方向及會所與住所的關系推制出來的中橫幾格局作基礎。會所大堂的中央是位於工地的中心點，會所的其他設施從大堂向北伸展，使全部住所群能有南向的一面。住宅座落於園林中，有傳統蘇州園林的風格。整個項目務求創造出一處具有豐滿對比規律園林氣氛而又有傳統內涵的現代化民居。

The site is situated in a park setting near a man made lake and the brief calls for a high class recreational club with residential accommodation. The master layout is developed from a diagonal and cross grid based on the near square shape site taking into account view, orientation and the relationship between the club house and the residential units. The centre of the lounge of the club house is sited at the centre of the site from where the club house facilities spread northwards. All residential units occupy south facing aspects. The houses are set in a landscaped surrounding like houses in a traditional garden with man made rock formation. The whole idea is to recreate a modern residential complex with the essence of traditional housing, rich in symmetry and garden like environment.

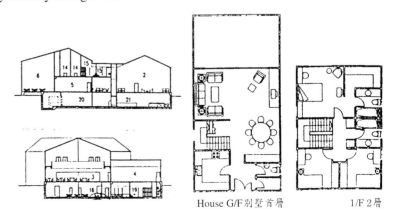

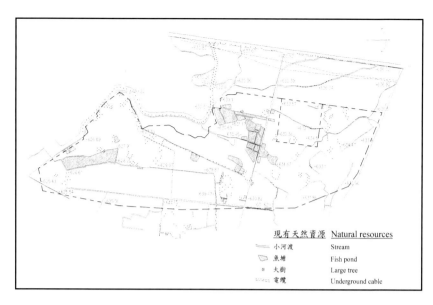
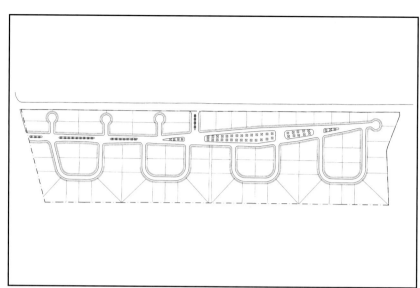
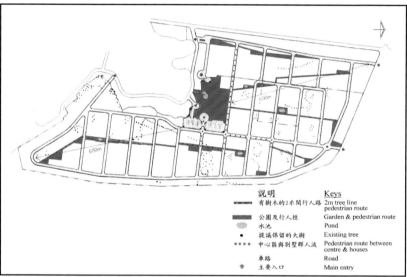
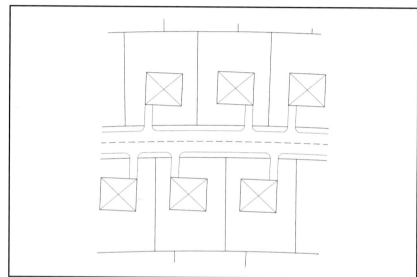
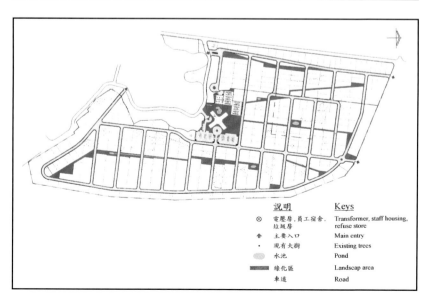
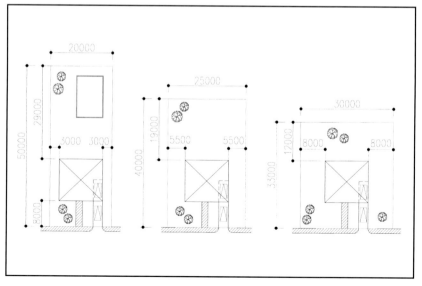

都江堰別墅山莊

Holiday villas, Dujiangyan

中國四川都江堰青城　(1998)
Dujiangyan, Qingcheng, Sichnan Provence, China.

There are three very important elements which a well designed holiday villa project should have, sufficient greenery, a rich variety of recreational facilities and a pleasant surrounding so that the inhabitants feel that they are in a natural setting and full of holiday spirit. The site at Dujiangyan is surrounded by natural setting with mountains and small rivers and has all the natural elements needed. The total area of the site is 840 acres. The widest distance between north and south is 1,300m and from east to west 600m. The level difference between north and south is about 8m. There are six to seven streams across the site with 20 odd large and small fish ponds. From the survey map provided there appears to be at least 350 large trees existing on site.

The planning layout concentrates in providing a direct pedestrian link from the individual holiday villa lot to the centre in order to minimize the use of the car within the site. Inhabitants can reach the centre via tree-lined pavement, rest garden and landscaped walks. There are a total of 412 individual villa lots of 1.5 acre each. The layout of the holiday villa lots are basically north south oriented. More than half of the lots have gardens facing south. For the lots with garden facing north, the villa is set further back from the front providing a garden with suitable depth in the front facing south. In most cases the villa along the road are not sited directly opposite each other so that each villa has a distance front view.

The major streams are preserved using underground piping and in suitable location the stream are preserved on the surface as a landscape feature along the rest gardens and landscaped walks.

In addition as much of the existing mature trees are preserved to maximize the use of existing natural resources and to eleminate part of the regimental effect of the villa layout. The whole development has been designed with the most practical lot sizes and economic road network with maximum use of existing natural resources and suitable landscape areas overcoming the limitation of low cost and providing a high quality holiday villa environment.

一個優良度假別墅區的構成必須有充分的綠化、豐富的娛樂設施及良好的週圍環境，使居民有處身於大自然的懷抱及有濃厚的度假氣氛的感覺。位於都江堰的工地，週圍有大自然的環境，有山、有水是具備優良的天然條件。工地的面積共840畝，由南至北最闊有1300米，由東至西最闊有600米。地形由北到南有8米之高差。有6、7條小河渡縱橫整個工地面積，其中大小魚塘有20多個，成熟的大樹有350棵以上。整個山莊的佈局是注重方便人流從個別別墅直達區中心，避免居民使用汽車代步，減少區內的車流量。居民應能通過兩旁的綠化行人路、漫步小徑及休憩公園到達區中心，山莊共有412個1.5畝的別墅地段。

別墅地段的佈局，基本上大部分是南北向，多於半數的地段的花園是南向，而花園北向地段，提議把別墅的位置從道路北移，使別墅的前面有適當深度的南向前園。沿路兩旁的別墅位置，在盡可能的情況下不會面對，務求使每一幢別墅都有適當的視野。

原有區內主要的小河以地下及地面水道加以保留外，還在南區的別墅區內在一系列的休憩公園及綠化行人徑內保存了一段地面小河及加設小湖，在其他別墅內原小河道中增設小湖，增加區中的水面景觀。此外，現有的大樹加以保留，盡量利用原有資源，同時利用大樹不規劃的佈局去打消別墅佈局的規則性。整個別墅山莊是以最實際的地段面積及最經濟的道路綱組成，盡量利用天然資源及適當的綠化設施，把山莊的格調提高，使到山莊演變成為一個別墅量高而又有充分的度假氣氛的一個度假山莊。

Youth Centre, Shenzhen

深圳少年宮

中國深圳市中心區第28-2地段 (1998)
Shenzhen City Centre Lot 28-2, China.

The site is situated at the north eastern corner of the city centre area on the edge of Lan Hua mountain park and east of the ceremonial square. It looks across to the Cultural Centre designed by Arata Isozaki in the west and high rise residential development to the east. The overall design has been influenced by the international ultra modern theme developed firstly by the conceptual design along the central axis of the city centre by Kisho Kurokawa, secondly by Isozaki's Cultural Centre and John Lee's City Centre building. In an attempt to bring back some form of local identity to buildings in the centre area, the main approach to the building has been designed with a grand stepped entrance area reflecting the essence of the traditional palatial approach and the provision of sun shading devices throughout the building expresses a concern for the local climate.

The complex is divided into several clear areas for the different activities. The science and technology activities are housed in a aircraft hanger type of structure at the north eastern corner of the site mainly as a result of the large and flexible exhibition spaces required and a desire to block off traffic noise from the road intersection. In addition the curved shape serves to act as a visual round off corner effect to the overall city centre and together with the shape of the proposed high rise housing block in the east forms a gateway into the centre area. Its western elevation facing the Ceremonial Square has glass curtain wall which provides a clear view from the street of the activities inside. The multi function auditorium is shaped like a large pebble sculpture sitting on the edge of a large water surface to give added visual attraction to the centre. The cultural activities are housed in a 6 storey curved shaped block with a sloping roof pointing towards the centre of the project. The space auditorium is shaped like a large metallic ball sitting at one side of the main entrance area. The central area is raised one storey up to permit view over the tree top to the Ceremonial Square. It is covered by a reflecting glass roof space frame with a tubular A shaped tower in the centre rising to 60m and contains a globe shaped solar powered clock, solar panels and aerial equipment. The central area has no doors or windows to economize on air-conditioning and to facilitate the meeting point effect of the pedestrian circulation. This tower serves a the focal point of the whole project. The whole project has been designed like a fun fair using variety of building shapes to reflect the multi function activities of the Centre and the vitality of youth. The idea of integrating the youth population into society is expressed by the use of silver grey and white metallic claddings with just the central space frame and tower painted earth yellow echos an attempt to blend in with the cultural area in this part of the city centre.

這個項目位於市中心區的東北角，蓮花山公園側，儀式公園的東面，西面對磯崎新設計的文化中心，東面對高層公寓大樓。黑川紀章設計的中心區中軸綫方案，磯崎新設計的文化中心及李名儀設計的市民廣場，都是富有前衛現代國際建築的風格，少年宮的設計構思不免受他們影響。但為了提倡對本地風土人情有所交代，少年宮的主要入口採用了莊重及大排場的石級，反映了傳統宮庭式的入口，建築物上的遮太陽板反映了對氣候的關注。

項目的各功能劃分清晰，為了要容納大型展品及遮擋工地西北角的交通噪音，科技館採用了一大型飛機庫形狀的建築物。半圓形的科技館與東面的高層住宅造型，形成了中心區東北面入口的排樓。科技館的西面有玻璃幕牆，使館內的活動一覽無遺。多功能大堂造型似一大石舂的雕塑，座落於大水池畔，對少年宮增添色彩。

6層高圓角的大廈是文化活動中心，大廈頂部斜向少年宮的中心區。太空館形似一大金屬球在主要入口的一旁，對少年宮增加了吸引力。少年宮的大堂在2層，從2層可通過工地週圍的樹頂遠望儀式公園。大堂頂部有金屬管金字塔形的反射天窗及一座60米高的A型鐵塔。大堂不設門窗，一方面節省空調設備，另一方面使大堂形成各方人流的自然匯點。塔身中裝置了一球形太陽能推動的大鐘、太陽能收集板及接收器。鐵塔成為少年宮的核心標誌，視覺上把少年宮的建築群聯繫起來。整個設計構思是根據遊樂場的佈局為基礎，利用各不同形狀的建築物，反映少年宮的多功能化及年青人的活力。青少年要融入社會，所以少年宮外牆多用銀灰及白色的金屬板，盡量在色彩上與中心區北面的文化活動區有所協調。

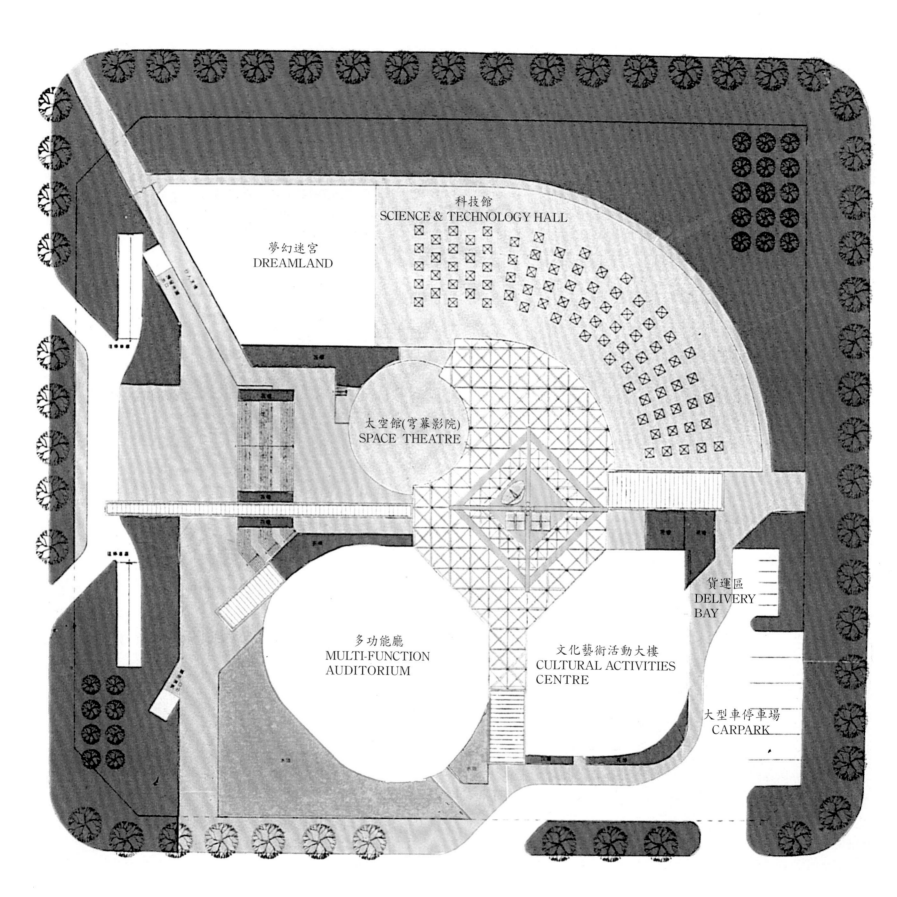

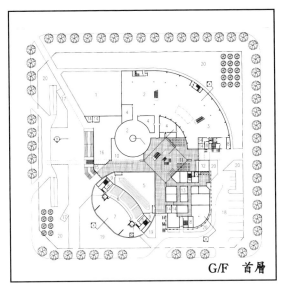

G/F 首層

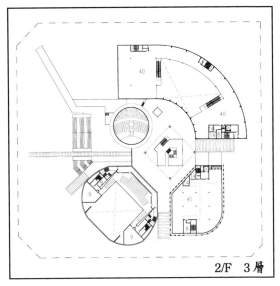

2/F 3層

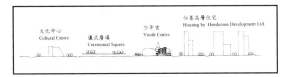

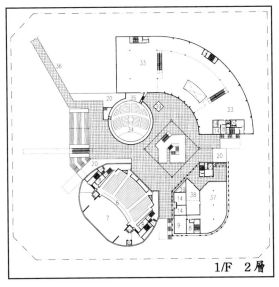

1/F 2層

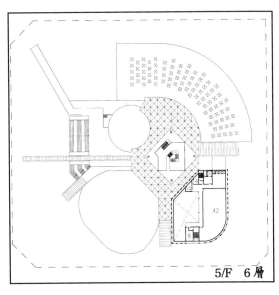

5/F 6層

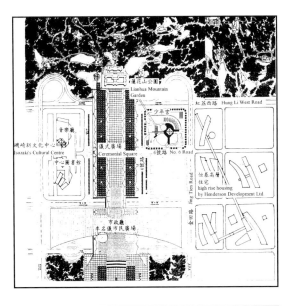

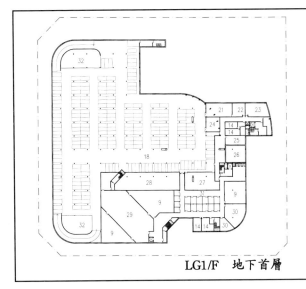

LG1/F 地下首層

1.	夢幻迷宮	Dreamland	24.	垃圾房	Garbage
2.	電腦	Computer	25.	消防泵房	Pump room
3.	通訊	Communication	26.	消防水池	Fire tank
4.	辦公室	Office	27.	中水處理	Waste water treatment
5.	合唱團／大型練習室	Rehearsal room	28.	冷凍機房	Plant room
6.	多功能廳	Multi-function room	29.	舞台下部	Stage under side
7.	舞台	Stage	30.	更衣／淋浴	Changing/shower
8.	機房	Mechanical	31.	化妝室	Make-up
9.	儲物室	Store	32.	預留地鐵	Future subway
10.	禮品店	Gift shop	33.	太空館（穹幕影院）	Space theatre
11.	咖啡廳	Coffee shop	34.	行政轉導	Administration/office
12.	廚房	Kitchen	35.	放映室	Projection
13.	聲樂	Music/vocal	36.	行人天橋	Elevated walkway
14.	洗手間	Wash room	37.	會展廳	Exhibition room
15.	詢問處	Information	38.	課室／會議室	Classroom/conference
16.	自行車停放處	Bicycle parking	39.	售票處	Ticket office
17.	預留地鐵出口	Future subway entrance	40.	能源館	Energy hall
18.	停車場	Parking	41.	心理咨詢／問題研究	Psychological enquiry/problem research
19.	水池	Pond	42.	舞蹈	Dance
20.	花槽	Planter	43.	科技館	Science & technology hall
21.	高壓制房	High voltage switch room	44.	武術	Martial arts
22.	變壓器房	Transformer room	45.	書法輔導	Calligraplry
23.	低壓制房	Low voltage switch room			

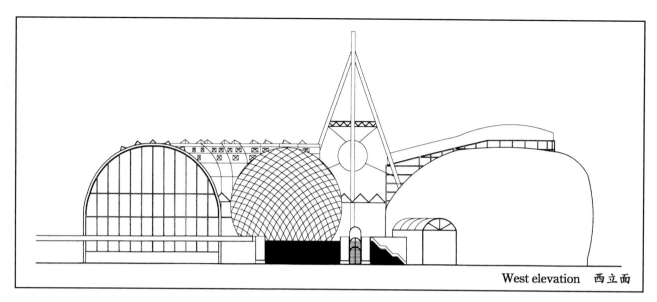

West elevation 西立面

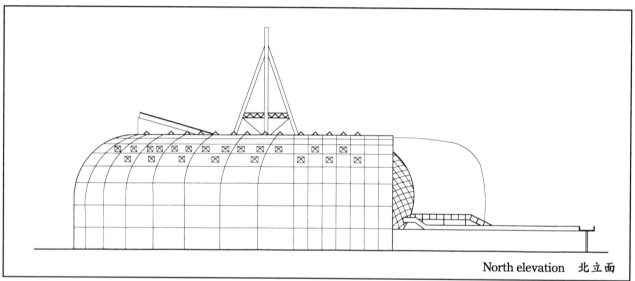

North elevation 北立面

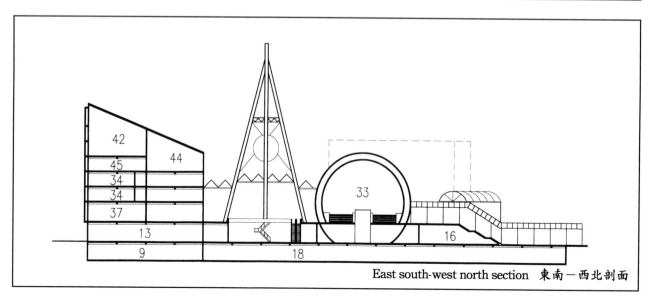

East south-west north section 東南－西北剖面

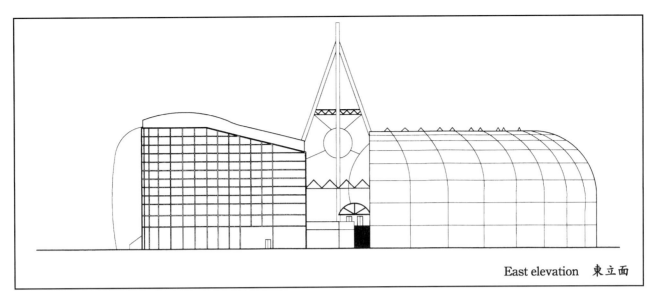
East elevation　東立面

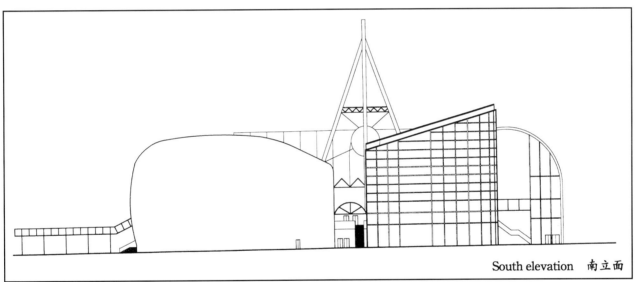
South elevation　南立面

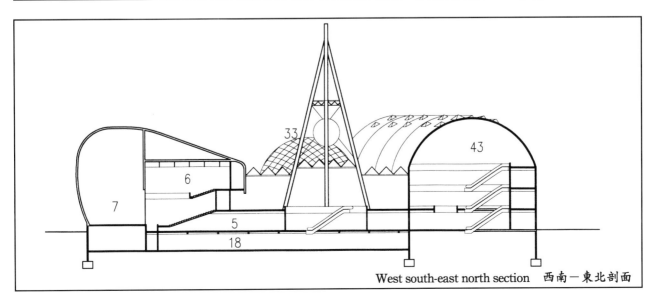
West south-east north section　西南－東北剖面

室內設計
Interior design

室內設計不是我的主要興趣，但在我建築設計的項目內包括的室內設計，我的興趣就不同。因為包括了室內設計，可使建築設計更加完美，同時能提供比較妥善的成果。我早期的室內設計工作是在晚上或週末進行，主要是多賺點錢。1969年至1997年間我做過大約25個工程，但有些設計檔案已失存。最大規模的工程，可算是在潘衍壽事務所工作時做的翡翠及明珠戲院。做室內設計要有細微的心思，徹底了解功能的要求及對設備和材料有深厚的認識，包括最新的產品在內，同時對格調、配色及材料的粗細要有協調的技巧，方能創造出一個好的設計。

Interior design is not my main interest, but I like to undertake interior design work when it forms part of an architectural project as it enables a complete control of the finished work and provide client with well worked out final solution. My earlier interior works were undertaken in the evenings and weekends mainly to earn a few extra dollars. I have done some 25 projects between 1969 and 1997, but unfortunately some of the job records have been lost. The largest interior works undertaken is the Jade and Pearl Cinemas which was done as a project under the employ of Peter Y.S. Pun & Associates. To do interior design well, one must have deep understand of the functional requirements, a good knowledge of equipment and material including new material on the market. At the same time one should have the technical ability to design to the quality required, the colour blending and texture co-ordination.

"起居窩"
"Living Pod"

香港西貢度假屋的室內設計 (1969)
an interior design for a country house in Sai Kung.

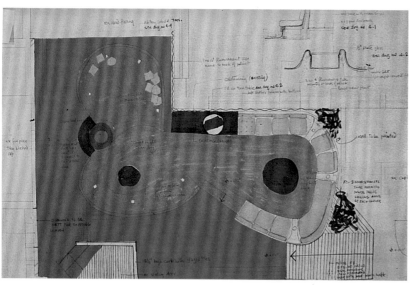

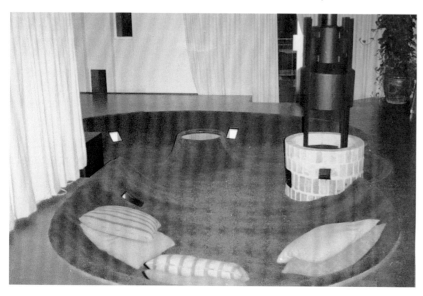

香港節詢問處

Festival of Hong Kong kiosk

香港歷山大廈前　　(1969)
In front of Alexandra House, Hong Kong.

時裝店

Malcolm Starr International Showroom

香港花園道喜爾頓酒店　　(1969)
Hilton Hotel, Garden Road, Hong Kong.

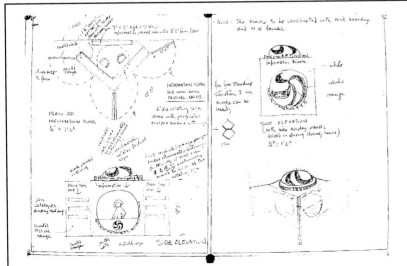
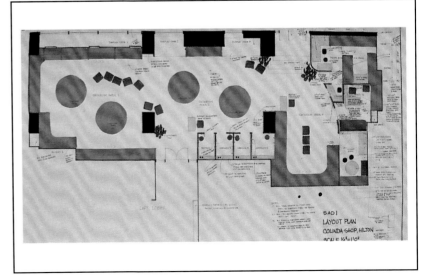

公寓
Apartment

香港僑福道　(1970)
Guildford Road, Hong Kong.

怡和保險有限公司九龍辦事處
Jardine Insurance Co. Ltd., Kowloon Branch Office

九龍彌敦道　(1970)
Nathan Road, Kowloon.

中華文化促進中心
Chinese Cultural Centre

香港德輔道國際大廈　(1984)
International Building, Des Voeux Road, Hong Kong.

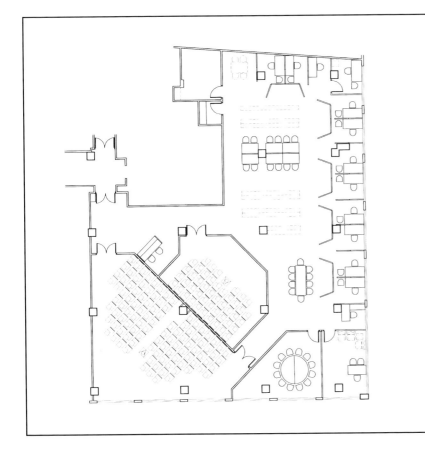

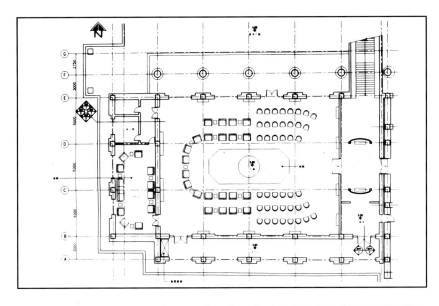

北京人民大會堂香港廳
The "Hong Kong Room", Hall of the People

中國北京天安門 (1997)
Tiananmen, Beijing, China.

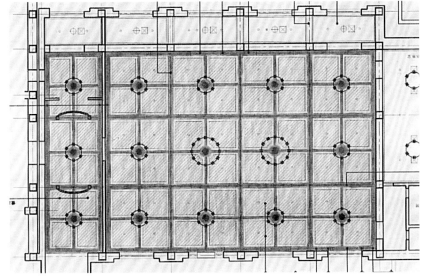

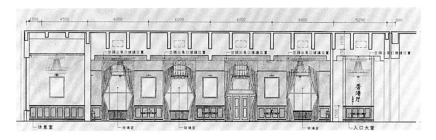

白日依山尽，黄河入海流。欲穷千里目，更上一层楼。

視覺設計

Graphic design

視覺設計能夠通過紙張或其他材料上的設計把創造這個視覺設計的因由充分地表達出來。這是在設計形式上最艱難及最有效的設計方式，所以也是設計者最滿足的成果。我喜愛視覺設計，不只是因為花比較少的時間，而是要有高度的構想能力。我做了少量的視覺設計，開始時做了些公司標志及名片設計，后來更伸展到一些書籍的封面及賀年卡設計。在設計香港科技協進會的會徽時，我採用了一個最簡單及常用的核動力的標誌，提供了一個容易給人馬上能聯想到這個標誌背后組織是什麼。視覺設計要清澈、乾脆及准確。我從1970年起開始設計賀年卡，剛開始的幾年，是採用有趣的圖案或有藝術性設計，但從1984年起，我每年從我拍攝的照片中挑選一張能抽像地反映當時大陸及香港的時勢的照片，加上一打油詩或標以表達我當年的觀點及願望。起初只是一時的興趣，但通過收到我賀年卡的朋友的鼓勵，不知不覺已做了14年，現在的問題是我覺得一年比一年艱難，因為我能拍得比較適當的照片一年比一年少，而我自問才疏學淺，對每年能作出合適的打油詩或標語沒有把握。1987年賀年卡的照片裏一老人正在跨門坎，而標題是"門坎已過"，喻意國家已克服了困難，正是恢愎國泰民安的時代，照片拍於上海玉佛寺內(見158頁)。1988年的照片是一老伯在門坎上坐着休息，標題是"門坎上小休"，是暗示國家有些不安，應當靜坐再作打算(見159頁)。1990年的照片是我坐船下三峽時拍的，所以標題是"山窮水盡疑無路，柳暗花明又一村"，希望能給朋友們一些安慰(見159頁)。1991年的照片是我在應邀參加1989年國慶節在釣魚台國賓館內拍攝，通常從香港到京參加國慶的10多個團體減到有20多人參加，1993年國家又回愎國泰民安的狀況，所以標題是"百花須今再齊放，難能保花永久開"，是希望在普天同慶之時，我們也要居安思危(見160頁)。1994年的照片是在參加政協視察團到漓江所拍，所以標題是"世事複難無靜處，輕舟一片萬重山"(見160頁)。1995年的照片是我在非洲肯尼亞參加英聯邦建築師協會3年一次的大會時所拍，標題是"世事變遷難捉摸，還是靜坐望天明"(見160頁)。1996年的照片是我當年到西藏在布達拉宮前拍的，剛巧當年香港正是受到97回歸的影響，標題是"港事繁複心掛牽，持仗呆坐白牆邊，祈望明天比今好，木口木面又一年"(見161頁)。1997年的照片是在當年參加政協江蘇省考察團時所拍，而精神文明正是熱門話題，所以我以標題表達了我對國家推動精神文明遇到的一些難處，標題是"佛門面貌雖華麗，少僧座姿欠文明"(見156頁)。

Graphic design has the ability to express through the design on paper or other medium a concised message of whatever is behind the creation of such a piece of design work. It is one of the most difficult and effective design means and therefore the end result is most satifactory to a designer . I enjoy graph design as it takes relatively less time to execute, but it involves a great amount of ingenuity. My involvement with graphic design is rather limited. It started with the design of company logos and visiting cards and later to incluode book covers and new year greeting cards. In the design for the logo of the Association of the Advancement of Science and Technology, I used a simple conventional graphic symbol of the nucleur energy to convey the nature of the organization behind the logo. The simplicity of the sign gives an easy and immediate recognition of what the organization is about. Graphic design has to be clear, crisp and precise to be successful.

I started designing my own new year card in 1970. In the first few years I have used interesting pictures and artistic designs, but from 1984 onwards I decided every year to use selected photographs taken by me to illustrate abstractly the state of affairs in either Hong Kong or the mainland with a poem or a title to express my view point and aspiration. It started being an ad hoc effort, but, through the encouragement of friends who received these cards, I have by now done this for 14 years. The problem is the effort is getting more and more difficult, as years passed by I have taken less suitable photographs, but, worse still due to my lack of poetic creativity, it is becoming more and more difficult to produce a suitable poem. In the photo for 1987 card there is an old man crossing the threshold. The title "passed the threshold" illustrated that the mainland has crossed over to a period of prosperity. The photograph was taken in the Jade Budda Tample in Shanghai.(see page 158) In the photograph in 1988 an old man is sitting on the threshold. The title reads "Resting on the threshold". This illustrated that there is an air of uneasiness in the country and the best way is to sit it out. (see page 159) The photograph in 1990 was taken during my trip down the three Gorges in the Zeng Jiang River. The poem reads " When there is no clear road ahead, do not despair, a way out is just round the corner" to give my friends a word of comfort. (see page 159) The photograph in 1993 was taken during my stay at the Daoyutai State Guest house at the invitation of the Government to attend the National Day Celebration in 1989. In that year, the usual number of invited groups from Hong Kong to attend such event had dwindled to only twenty odd persons being invited and we were up graded to stay at the national guest house instead of the usual hotel. In 1993 the Country has returned to a healthy state. Therefore the title " Though hundred flowers are blooming again, it is difficult to blossom forever" is to suggest vigilance in good times. (see page 160) The photograph in 1994 was taken during a CPPCC members, inspection tour of Li Zhiang river in Guangxi Province. I cannot remember the state of affair at the time, but, perhaps it was full of problems. Hence the poem "Life is complicated with no peace, unlike the serenity of a boat meandering down the river". (see page 160) The photograph in 1995 was taken during the Commonwealth Association of Architects Triannual Conference in Nairobe, Kenya in Africa. The situation in 1995 was not very clear and the pocm reads " It is difficult to grasp where changes in our future lies, thus it is best to sit quietly with forward looking eyes". (see page 160). The photograph in 1996 was taken in front of the Potala Palalce in Tibet in that year and Hong Kong was going through a period of uncertainty affected by the 1997 handover. The poem reads "Hong Kong affairs make me cry, it is better to sit with starring eyes, hoping for better times. How time flies". (see page 161) The photograph in 1997 was taken during another CPPCC inspection tour of the Ziangsu Province in that year and the country was very much concerned in improving the quality of life. The poem reads " Beauty behold the Monastery door, ruined by the vulgar display of the monk's posture". I have taken the opportunity to illustrate one of the problem where improvement should be made (see page 156).

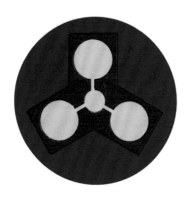
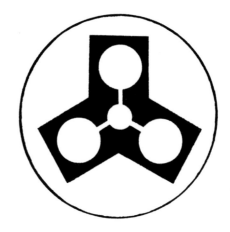

香港科技協進會會徽　(1985)
The logo of the Hong Kong Association
for the Advancement of Science & Technology

PETER Y.S.PUN & ASSOCIATES
潘衍壽土木工程師事務所　(1986)

HONGSTRESS LTD.
康力有限公司　(1970)

SUNDRIDGE ENTERPRISES LTD.
新力工程企業有限公司　(1970)

WAI TAK MARITIME LTD.
懷德航海有限公司　(1970)

邀請卡及公司簡介
Invitation card & company brochures

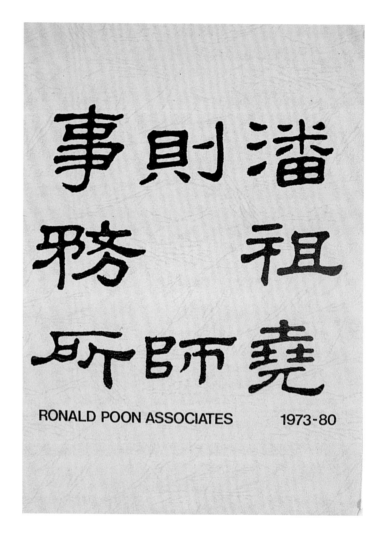

公司簡介
Company brochures

名片
Visiting cards

INSIDE THE SQUARE, OUTSIDE THE CIRCLE

**BY
RONALD
C.Y.POON**

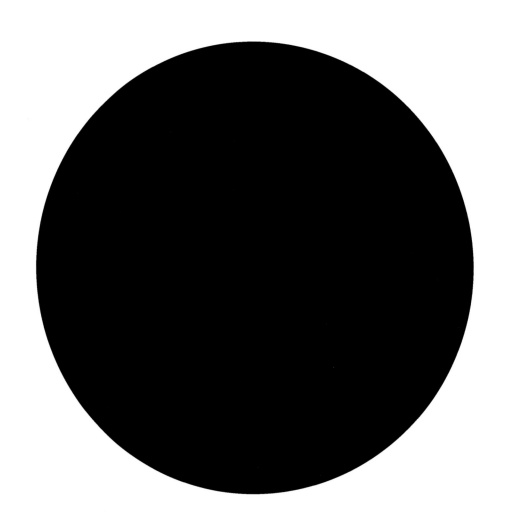

書封面
Book covers

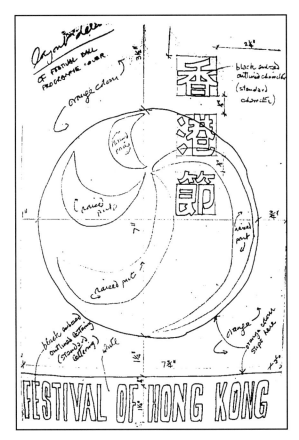
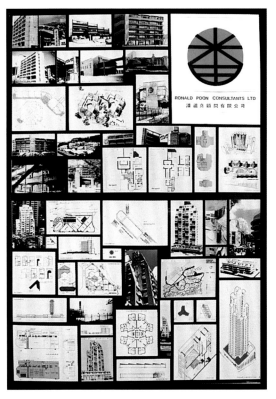

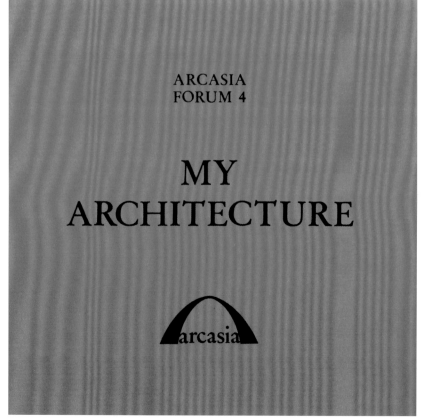

信 箋
Letterheads

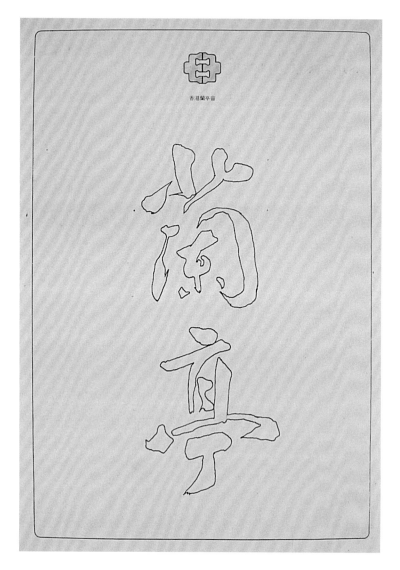

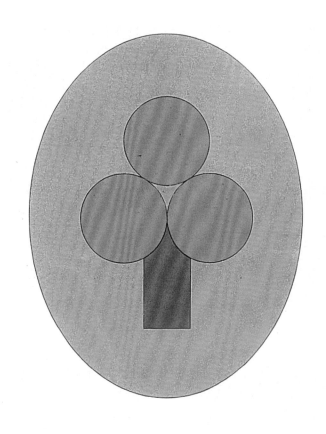

 香 港 植 樹 日

香港植樹日標誌
Hong Kong Tree Planting Day sign

1997
佛門面貌雖華麗
少僧坐姿欠文明
Beauty behold the
Monastery door, ruined by
the vulgar display of the
monk's posture

1998
眼前遍佈荒涼貌
有水石頭也見青
Though destitution is
before us, with water
green comes to the "barren
rock"

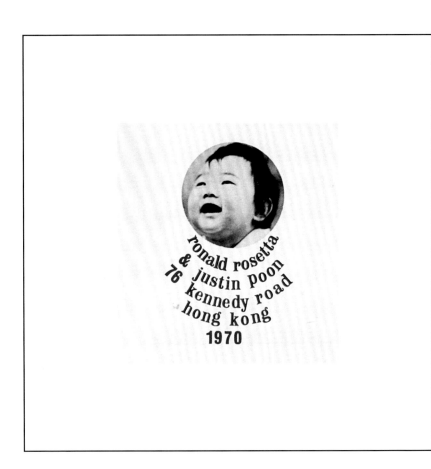

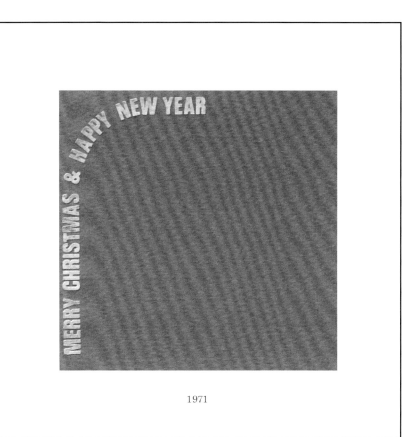

1971

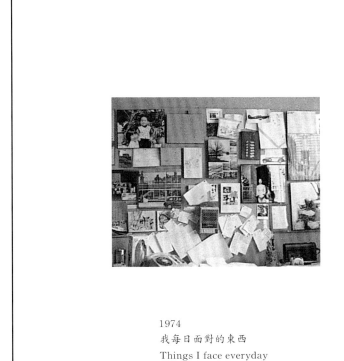

1974
我每日面對的東西
Things I face everyday

1973

1984
清朝"後現代派"古建築樣本位於北京西郊的香山
Post modernism in Qing Dynasty chinese architecture (1644~1911AD) found in the Fragrant Hills, west of Beijing.

1985
鄉土風情
"Indigenous reflection"

1986
中國傳統建築的奧妙
The subtlety in traditional chinese architecture

1987
門坎已過
Passed the threshold

1988
門坎上小休
Resting on the threshold

1989
方圓過後又方圓
閒步踱來認蘇州
The evolution of the circle and the square can only be digested through an unhurried stroll

1990
山窮水盡疑無路
柳暗花明又一村
When there is no clear road ahead, do not dispair, a way out is just round the corner

1991
黑白陰陽在眼前
但願世事均如此
Wishing that everything is as righteous and straight forward as the contrast of light and shadow before us

1992
前路崎嶇又艱巨
但願過後見青天
Though the way forward
is tortuous and difficult,
hopefully a bright future
awaits us.

1993
百花須今再齊放
難能保花永久開
Though hundred flowers
are blooming again, it is
difficult to blossom
forever.

1994
世事複雜無靜處
輕舟一片萬重山
Life is complicated with
no peace, unlike the
serenity of a boat
meandering down the
river

1995
世事變遷難捉摸
還是靜坐望天明
It is difficult to grasp
where changes in our
future lies, thus it is best to
sit quietly with forward
looking eyes

1996
港事繁複心掛牽
持杖呆坐白牆邊
祈望明天比今好
木口木面又一年
Hong Kong affairs make me cry, it is better to sit with starring eyes. Hoping for better times, How time flies!

攝影
Photography

攝影是我在設計工作上第一件工具。從10多歲開始，我就對攝影有濃厚的興趣，但至今對攝影的技術性運作總是有所抗拒，我只是利用攝影機去記錄在一剎間我認為有意義的構圖，包括美麗的景色和豐富表情的人像，所以我的照片只是掠影而不是高度技術性的產品。我的照片是記錄了當時當地的情景，但對我來說，不單是記錄了一時我觸摸到的美麗景色或有趣的人像，而且提供了一個永久性的紀錄給其他人去欣賞。我拍攝過數百張照片，多數是從70年代開始在大陸考察或旅遊時拍的，我挑選了一些來表達我對天然景色及人生的興趣。

Photography is, in fact, my first media in design. I started liking photography in my teens but to-date I have never been able to be interested in the mechanics of photography. I am only interested in using the camera to capture a picture which on a glance appears to have design merits, be it a scenery or a human face. Therefore my photographs are normally snapshots and not highly composed and technical products. My photographs are instant record of the time and place wherever I happened to be there. but it gives me great pleasure because it not only record a moment of joy having encountered a lovely scenery or a face full of expression, it also provides a permanent record for others to enjoy. I have taken hundreds of photographs particularly during my trips in China since the seventies and I have only selected a few to illustrate my love for nature and the humanity.

taken by self timer in Glastonbury, Somerset 1961 自動拍攝於英國森馬石州嘉蘭士頓比利市

繪畫
Painting

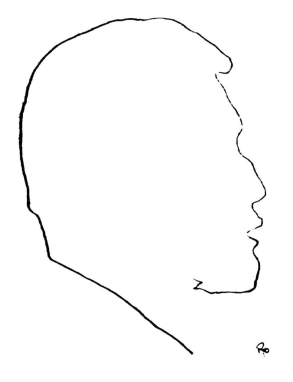

我對繪畫的興趣,始於我在英國邦密夫美術學院就學的時候,多數的畫是我當時在英國就學時回港度假所繪,用的材料大部分是以油漆為主。我的畫是受到大自然的抽像形狀及中國書畫中構圖方式的影響,可惜從此以後,因為沒有太多空閒的時間,只能在與家人度假的時候,才能繪制一些素描。

My interest in painting started during my days in the Bournmouth College of Art in the early sixties, Most of the paintings shown here were done during my holidays in Hong King in my student days mostly using household paints. My paintings are very much influenced by abstract shapes in nature and the form of composition in chinese paintings. Since then owing to lack of time I can only manage to do some sketch work during my family holidays.

出發點
The point of no return

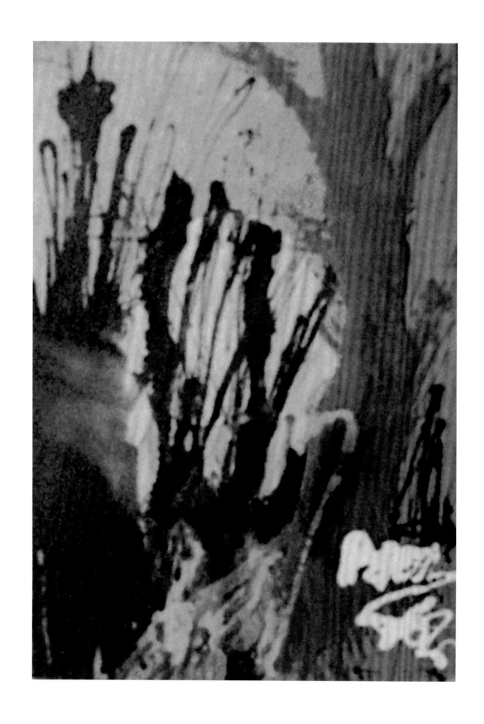

太陽下山了
The setting sun

少女
Young girl

月升
Moon rise

日落
Sunset

一籃生果
Basketful of fruit

播種
Seed sewing

抽像 1
Abstract no. 1

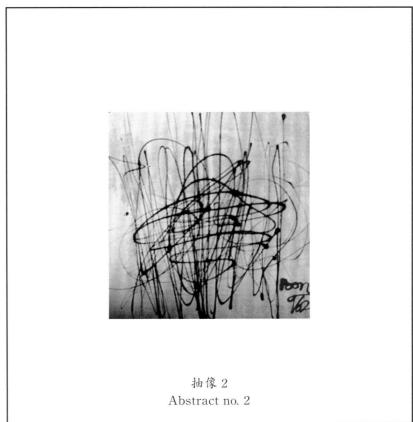

抽像 2
Abstract no. 2

日落林中
Sunset in the forest

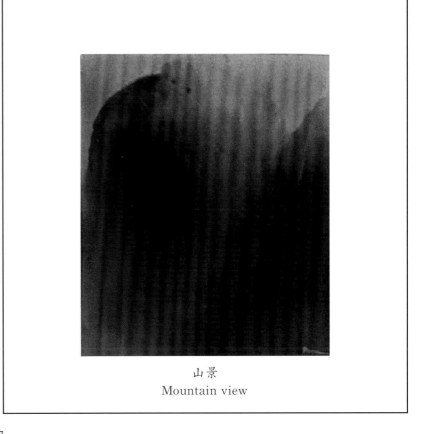

山景
Mountain view

太陽之舞
Dancing sun

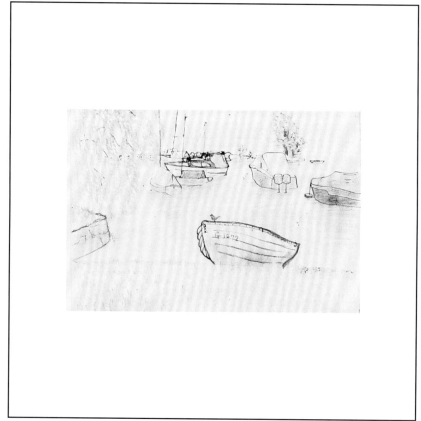

古物收藏

Antique collecting

古物收藏可能是我祖傳的嗜好，先父潘恩愷收藏煙壺及郵票，先祖父潘佩如收藏瓷器，而先曾祖父潘正煒以收藏書畫出名，在清道光時取堂名聽颿樓。因我仰慕先曾祖父的收藏，所以取堂名為"小聽颿樓"。我收藏古物的習慣，由1968年開始，當時正從英國學成回港，任職月薪港幣1,800元，而我可以奢侈地用了港幣500元去買我在一古物店集古齋尋獲的先曾祖父潘正煒(季彤)的書法手卷（見215頁），可見得我對收藏的熱愛度。初期我對書法特別感興趣，70年代得到先祖父部分的收藏后，開始對瓷器產生興趣，其后木器、竹器、文房四寶、古玉、陶器也成為收藏的目標。80年代中期，更開始收集明式及紫檀的家具。雖工作繁忙，我認為古物收藏是生活中不可或缺的一項活動。它能使我索求一種超脫物質享受之外的理想，尋求精神寄托和追求豐盛的人生。古物有它的靈氣和精華，古物收藏使我能理解博大精深的傳統文化，提高作為一名建築師不可或缺的文化素質，探索傳統文化與現代建築的關係，這是收藏中沉澱的無形財富。在收藏古物的過程中，我是不斷的伸展，改變或收窄我的收藏目標，這可能是通過時間的考驗及個人人生觀的改變而影響到收藏的範圍。早期我收藏的書法包括各種字體，例如行書、行草、楷書、草書等，但最近10年我已集中收藏狂草，主要是我覺得狂草富有藝術意味，是各種書法中最耐人尋味的。結果我把以前收藏的行書、楷書等藏品出讓，給別人擁有的機會。我認為，收藏的原則只要曾經擁有，不在乎天長地久或留給子孫永享。因為一方面這是太自私了，另一方面多數的子孫不一定喜歡收藏，就算喜歡也不一定要收藏與您同一樣的東西，只要有其他藏家接手，那保存古物的使命就得到落實。

I probably inherited the love for collecting antiques from my ancestors. My late father My Poon Yan Hoi collected snuff bottles and stamps. My late grandfather Mr. Poon Pui Yu collected porcelain and my late great grandfather Pan Cheng Wei collected painting and calligraphy in the Dao Kuang period of the Qing Dynasty and named his collection "Ting Fan Lau". I named my collection "Little Ting Fan Lau" in admiration of my late grandfather's collection. My habit of collecting started in 1968 when I returned from England after my studies. At the time my monthly salary was HK$1,800 and I was so involved with collecting that on one occasion I spent HK$500 to purchase a hand scroll of calligraphy by my late great grandfather found in an antique shop called Tse Ku Tsai (see page 215). I started being interested in collecting calligraphy. In the 70's after having part of my late grand father's collection I extended my collecting to include porcelain. Later on I also started to collect wood & bamboo pieces, stationery objects, antique jade and earthenware. In the mid 80, I started collecting Ming and zitan furniture. Even with a busy work schedule, I consider antique collecting an essential activity in one's life because it can help you in the search for a spiritual accomplishment and enrichment in life. Antique has it own spirit and essence. By collecting I start to appreciate the greatness of our traditional culture and through the exploration of the relationship between traditional culture and modern architecture I am able to improve the necessary quality of my cultural aspiration as an architect. This is the invisible treasure of antique collecting. In the process of collecting I very often extend the range of my collection, change direction and narrow the choice. This is because through time the area of collecting is effected by one's outlook in life. In the early days I collected all different style of calligraphy such as Xing Shu, Xing Cao, Kai Shu, running script, etc. and in the last ten years I have concentrated in collecting running script only. This is because running script is rich in artistic expression and has a more long lasting attraction. As a result I sold most of my pieces on the other styles of calligraphy so that other collectors can have a chance to own them. The principle of collecting should be one of transitional ownership and not permanent ownership and particularly not "to save it for your descendants to enjoy". Otherwise it will be too selfish and besides your descendants most likely are not interested in collecting. Even though they may be interested, what they like to collect may not be the same as yours. As long as there are other collectors to continue the possession of the pieces, the preservation of antiquity will be accomplished.

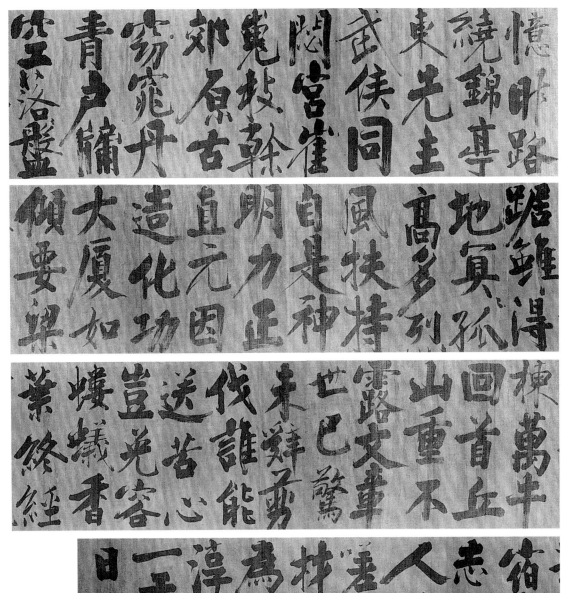

（宋）張樗寮大書古柏行
手卷，1252年作
Hand scroll by Zhang Ji Zhi, Song.

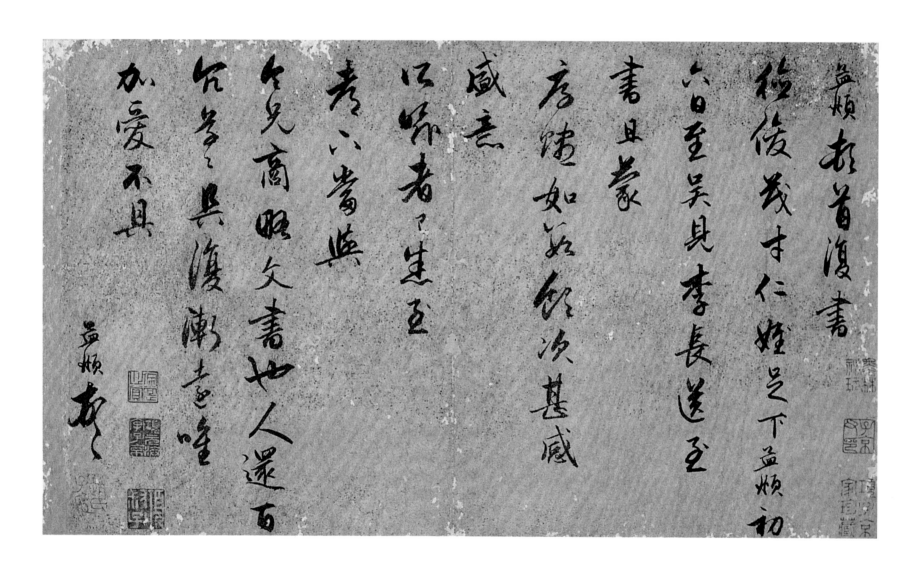

(元)趙孟頫冊頁　Album leaf by Zhao Mengfu, Yuan

(明)文徵明明妃曲冊頁 Album by Wen Zhengming, Ming

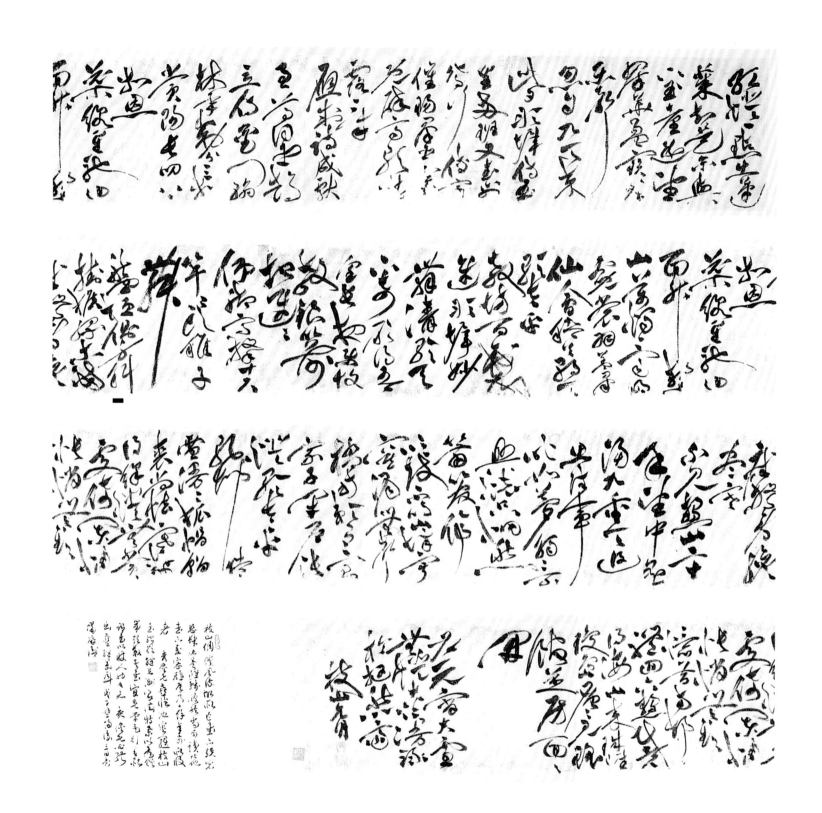

(明)祝枝山狂草詩卷
Hand scroll by Zhu Yun Ming, Ming.

(明)董其昌行書軸
Hanging scroll by Dong Qi Chang, Ming.

(明)元庭行草軸
Hand scroll by Yuan Ting, Ming.

(明)王鐸狂草軸
Hanging scroll by Wang Duo, Ming.

(清)鐵保行書軸
Hanging scroll by Tie Bao, Qing

(清)何紹基行書七言對
Couplet by He Shaoji, Qing.

(清)高劍父草五言對
Couplet by Gao Jianfu, Qing.

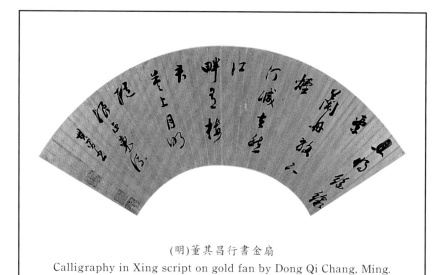

(明)董其昌行書金扇
Calligraphy in Xing script on gold fan by Dong Qi Chang, Ming.

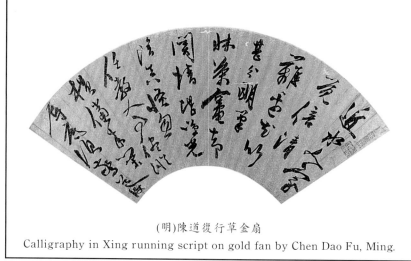

(明)陳道復行草金扇
Calligraphy in Xing running script on gold fan by Chen Dao Fu, Ming.

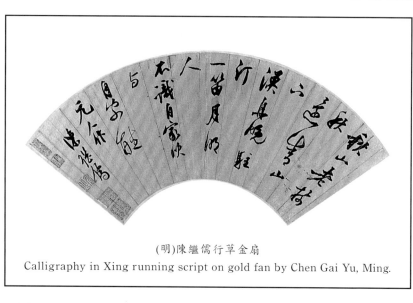

(明)陳繼儒行草金扇
Calligraphy in Xing running script on gold fan by Chen Gai Yu, Ming.

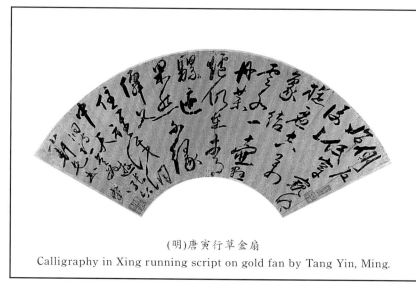

(明)唐寅行草金扇
Calligraphy in Xing running script on gold fan by Tang Yin, Ming.

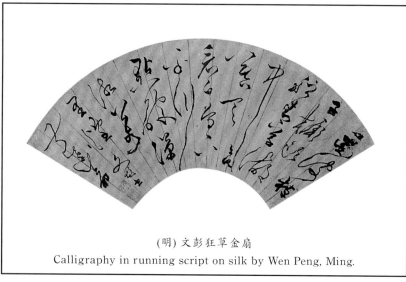

(明) 文彭狂草金扇
Calligraphy in running script on silk by Wen Peng, Ming.

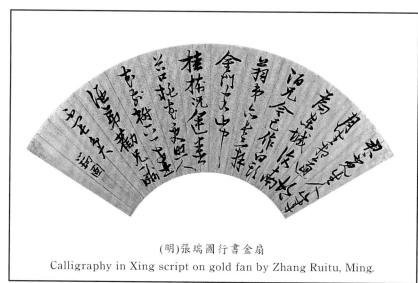

(明)張瑞圖行書金扇
Calligraphy in Xing script on gold fan by Zhang Ruitu, Ming.

(清)潘季彤行書中山松醪賦卷

Hand scroll in Xing script on paper by Pan Cheng Wei, Qing.

(新石器)馬廠形馬家窰文化彩陶兩耳小罐
Small Neolithic pot with two handles

(唐)三彩小酒壺
又 Small ewer in sancai glaze, Tang

(宋)鈞窰天藍地紫斑碟
Dish with flattened rim, lavender blue glaze with purples splashes, Jun ware, Song

(元)青花釉裡紅瓜形小罐二件
Two small jarlets in Qingbai glaze with under glaze brown and blue decorations, Yuan

(明)哥窯小水丞
Small water dropper, Ge-type, Ming

(明)石灣小水牛丞
Small buffalo water dropper, Shiwan glaze, Ming

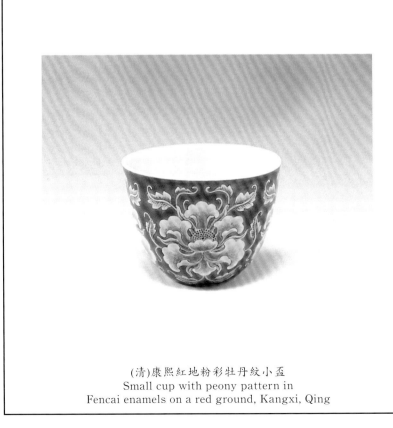

(清)康熙紅地粉彩牡丹紋小盃
Small cup with peony pattern in
Fencai enamels on a red ground, Kangxi, Qing

(清)豇豆紅小瓶
Small vase in Peach Bloom glaze, Kangxi, Qing

(清)瓶形端硯
Duan inkstone in vase shape with brown stripes, Qing

(清)蠟石連座
An amber colour stataemite rock with stand, Qing

(明)田黃小山羊章
A Tianhuang stone rectangular seal with a baby goat knob, Ming

(清)雲石畫
A white variegated marble with frame, Qing

(清)宜興小茶壺
Small Yixing teapot, Qing

(清)黃楊樹根
Small boxwood root, Qing

(清)竹雕東方塑偷桃
Bamboo carving of Dong Fanshuo with stand, Qing

(明)雞翼木文件盒
Chi-Chih Mu document case, Ming

(清)影木根筆筒
Burl wood trunk brushpot, Qing

(明)黃花梨刻螭虎花卉紋三層文件盒
Huanghuali document box. three tiers
with dragon and flower decorations, Ming

(明)黃花梨夾頭榫長案
Huanghuali long an table, 17th Century

(明)黃花梨品字欄杆書架
Huanghuali bookcase, 17th Century

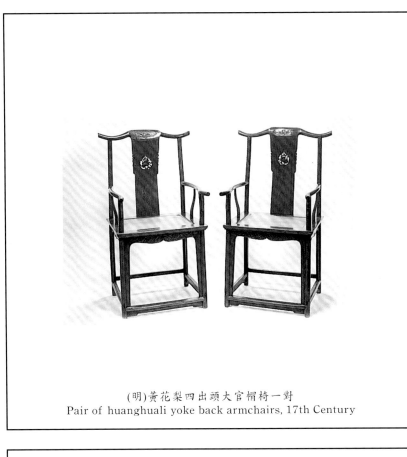

(明)黃花梨四出頭大官帽椅一對
Pair of huanghuali yoke back armchairs, 17th Century

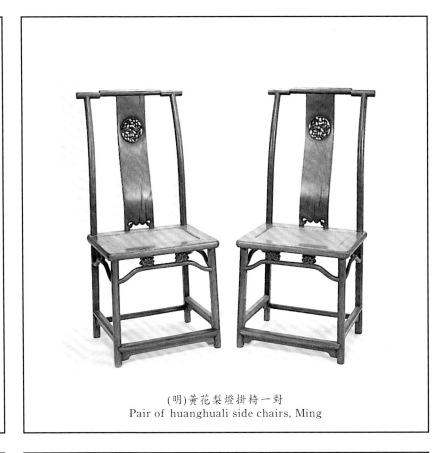

(明)黃花梨燈掛椅一對
Pair of huanghuali side chairs, Ming

(明)榆木圓腳交椅一對
Pair of Yu Mu folding chairs with round legs, Ming

(清)紫檀官帽椅一對
Pair of Zitan spindle back armchairs, Qing

學會工作

Professional institution work

香港建築師學會 (1972~   )
Hong Kong Institute of Architects (HKIA)

我是在1972年加入香港建築師學會的，從此開始了我26年來與學會工作的不解之緣。1973年被推選為理事會成員，任職至1984年。從1974年至1982年曾兼任23個委員會成員及多個委會的主席(見附錄1)。參加學會的工作對一個建築師的成長有很大的幫助，不單可以通過學會的工作去深入了解面對建築師的各種問題，而且可以從中推廣學會的目標，提倡優良的設計及高水平的執業道德，更能推動建築教育改革，受益社會及年青一代的建築師。香港建築師學會是一民間專業組織，除學會內全職的員工外，所有參與學會委員會工作的建築師，都是出於自願，並且沒有薪酬，所以一般投入學會工作的建築師，多是出自一腔熱血，希望能為本行做點事。我對學會工作每一方面都感興趣，所以在1974年至1982年間參與過差不多整個學會內每一個委員會的工作，其中有多項頗有意義的工作，如在1976年編制第一份建築師執業法案初稿、1974年及1981年參與學會銀牌獎評判工作、1979年代表學會會長參加亞洲建築師協會雅加達創會會議、1977年及1981參與學會驗察香港大學建築系的工作及於1978年第一次擔當大型國際會議英聯邦建築師協會3年一次的大會的籌委會主席。

1981年當選為學會會長(1,2,3)，任期為兩年。我是於1973年創辦自己的設計公司，到1981年公司已有8年的運作，可算是上了軌道，可以將每日工作時間的百分之三十投入學會的工作。上任後馬上忙於籌辦學會銀禧會員設計展覽並於6月在香港置地廣場展出及舉行了慶祝晚會(4)，隨後到京拜訪中國建築學會的領導(5)及商量開展籌辦中、港第一次設計交流展覽，首次開通了中國建築學會與香港建築師學會之間的航道，一方面把香港建築師的作品於1981年底在北

1

4

2

5

3

6

北京展出，展覽包括800張展品，代表了香港20多間傑出建築師事務所的作品。北京的展覽開幕儀式，中國建築學會理事長楊廷寶大師及城鄉建設環境保護部部長李錫銘也參加了(6)。這個展覽在北京結束後，便安排到南京、上海、鄭州、西安、成都、昆明及廣州等20多個城市巡迴展出。另一方面，中國建築學會籌辦了一個以"中國古代建築"為題的展覽，於1982年8月在香港新世界中心展出(7)，展出了400多張圖片，包括從未公開展覽過的原裝戰國建築圖案拓本、宋代城市圖案拓本、北京宮殿內景、古代四合院建築圖案等，中國建築學會副理事長王華彬、中國建築學會副秘書長曾堅、香港房屋處處長廖本懷先生參加了開幕儀式(8)。1981年世界著名建築大師富勒到訪，我以會長名義約見並趁機會向他取經(9)。

1982年8月香港建築師學會負責籌辦亞洲建築師協會的第2次年會及委員會會議，我趁機會邀請了大陸及台灣的建築師與會，這是自1949年以後兩岸建築師首次促膝相談，此後於1984年10月，嘗試發起一次大陸、港、台三地建築師在香港的聚會，取名為"蘭亭會"。清華大學吳良鏞教授也題了字"群賢會"(10)，我更邀請了香港著名雕刻家唐積聖先生刻了一枚"蘭亭會"的石章(11)，但因當時條件不成熟，結果會議告吹。

1982年間學會舉辦了有史以來香港第一個國際比賽，是一個住宅群的項目，命名"山頂比賽"。有570個參賽方案，由日本的磯崎新、澳大利亞的約翰安得華、菲律賓的方毛沙及我當評委(12)，並選出英國的莎夏夏迪的作品為首選。舉辦設計比賽是提高設計水平的一個最有效的辦法，但比賽要有公平的規則及大公無私的評委。不然的話，所選出來的方案會有反效果。同年學會與香港賽馬會合作舉辦了另外一個設計比賽，就是香港演藝學院的設計比賽，這次我也被邀請為評委之一。

除了舉辦展覽、研討會及設計比賽外，最能推動學會的工作，還有另一方式，就是利用媒界去表達意見，廣泛地傳播信息，通過報章爭取社會人士的支持。我在1981年及1982年間用了不少時間去接受記者的訪問及向報社投稿，兩年內在報章及雜誌上有關學會的工作及本行的報導不下105次(見附錄2一些例子)。

從1983年開始我繼續參與學會的工作並多次接待中國建築師訪港團(13)。因我在1981年始創了到京拜訪中國建築學會的先例，以後各新會長上任都安排到京，我也隨團到京多次(14,15)。1987年5月學會在北京舉辦第二次香港建築師工作展覽，我也參加了開幕儀式(16)。

7

8

9

10

11

12

13

14

我從1973年參與學會工作至1984年，積極地投入工作已有11年。1984年後因自己的設計公司的工作繁多及有其他學會及社會的工作，就決定減少參與香港建築師學會的工作。但直至今日常與學會保持一定的聯系及每次被邀參與工作的時候，也都是全力以赴。

I joined HKIA as a member in 1972 and since then started my twenty six years connection with the Institute. In 1973 I was elected as a member of the HKIA Council and served until 1984. Between 1974 and 1982 I had served in 23 . of the Institute's committees as a member and also as chairman of some of them. (see appendix (1)) To participate in professional institution's work has great effect on the development of an architect. It enables the architect to familiarize with various problems facing an architect in practice and also help to promote the aims of the Institute, raise the standard of design and the level of ethics in practice. In addition one can participate in the improvement of architectural education to benefit the society and the younger generation. The Hong Kong Institute of Architects is a professional organization and with the exception of paid administrative staff, all members participate in the Institute committee's work are on voluntary basis and without paid. Therefore all members serving the Institute does it with dedication hoping to make some contribution to the profession. I am interested in all aspects of the Institute's work and between 1974 and 1982 I participated in almost all the Institute's committees' work in one time or another. There were a few particularly interesting matters which I undertook including drafting the first draft of the Architect's Ordinance in 1976, in 1974 and 1981 I participated in the jury for the annual Silver Medal awards, in 1979 I represented the President to participate in the inaugural meeting of the Architect's Regional Council Asia in Jakarta, in 1977 and 1981 I participated in the HKIA visitation to the School of Architecture of the Hong Kong University and in 1978 I acted as chairman of the organizing committee for a large international triannual conference for the Commonwealth Association of Architects in Hong Kong.

In 1981, I was elected President of HKIA to serve for two years(1,2,3). In 1973 I founded my own design company and by 1981 the company had been operated on track for eight years. Therefore I can spare about 30% of my daily work time to devote to the Institute's work. The first major task after becoming President was to organize our silver anniversary exhibition which opened in June in the Landmark with a cocktail in the evening(4). After that I visited the Architectural Society of China in Beijing(5) and discussed with the President on the organization of the first exchange exhibition with China, thus paving the way to future exchanges between HKIA and the Architectural Society of China (ASC). In one direction the works of the Hong Kong architects were exhibited in Beijing towards the end of 1981 which included 800 panels representing the work of twenty outstanding firms. The opening ceremony of the Hong Kong architects' work exhibition was attended by the late Mr. Yang Ting Po, President of the ASC and Mr. Li Xi Ming, minister of the Urban and Rural Development and Environmental Protection Bureau(6). This exhibition was staged in various twenty other cities after Beijing including Nanjing, Shanghai, Zhengzhou, Xian, Chengdu, Kunming and Guangzhou etc. In the other direction, the ASC staged an exhibition entitled "Traditional Chinese Architecture" which opened in August, 1982 in the New World Centre, Hong Kong(7). It displayed some 400 photographs including the original rubbings of architectural design in the Warring State period and rubbings of Song Dynasty city plans which were exhibited for the first time outside of China. The exhibition also included photographs of the imperial palaces and plans of the traditional courtyard houses. The Vice President of the ASC, Mr. Huang Hua Bin, the Secretary General of the ASC, Mr. Tsang Jan and the Secretary for Housing, Hong Kong Mr. Donald Liu attended the opening ceremony(8). In 1981 the world famous architect Buckminister Fuller visited Hong Kong and I took the opportunity as President HKIA to request a meeting with him(9). The HKIA was responsible to organize the second annual meeting of Arcasia in Hong Kong in August 1982. I took the opportunity to invite architects from China and Taiwan to participate as observers to the meeting. This was the first time since 1949, the architects from both sides of the Strait met. Soon after that I attempted to organize a conference entitled "The Lanting Meeting" for architects from China, Taiwan and Hong Kong. Professor Wu Liang Yong of Tsing Hua University even wrote a calligraphy piece to celebrate the occasion (10). I also invited Mr. Tang Qi Zing, a famous seal carver in Hong Kong to carve me a commemorative seal to mark the conference(11). However due to the particular sensitive circumstances at the time, the conference never took place. In 1982 HKIA organized for the first time an international design competition. This was a housing complex and the competition was named "the Peak" as the site was situation on the peak district of Hong Kong. More than 570 submissions were received and Mr. Arata Isosaki from Japan, Mr. John Andrews from Australia, Mr. Formoso from the Philippine and myself acted as the Jury(12). We selected the design by Saha Hadid as the winner. To organize design competition is the best way to raise design standards, however, the competition must have fair rules and just jurors, otherwise it can have adverse results. In the same year HKIA organized another design competition in co-operation with the Royal Hong Kong Jockey Club for the design of the Academy of Performing Arts. This time I was also invited as a member of the Jury.Other than organizing exhibitions, conferences and design competitions, the other way to promote the work of the Institute is to use the press media to express opinions on professional matters and lobby public support on our stand. Between 1981 and 1982 I spent quite some time in talking to reporters and sending press releases to the newspapers. As a result in the two years the press has published no less than 105 times on matters related to our profession (see some samples in Appendix 2).From 1983 onwards I continued to participate in the work of the Institute and entertained many visiting architects from the mainland(13). In 1981 I set an example by visiting the Architectural Society of China on the start of my presidency and all the subsequent presidents followed suit and I accompanied them on several occasions(14,15). In May 1987 HKIA staged the second members' work exhibition in Beijing and I participated in the opening ceremony(16).

I participated actively in the Institute's work between 1973 and 1984 for some eleven years. After 1984 because of my own design work and involvement in other institutions' work, I decided to reduce my involvement with HKIA. However up to-date I have maintained my connection with the Institute and is all too willing to participate again whenever approached.

15

16

中國建築學會 (1980~　)
Architectural Society of China

1974年我參加以僑光置業有限公司董事長梁燊先生為團長的第二團香港建築師到大陸參觀(1,2,3)，從此開始了與大陸建築界的密切關係。當年我等到過廣州、上海、無錫、南京、天津、瀋陽、撫順、鞍山、大連、北京及深圳10多個大城市與當地的建築師交談。而當時正是文化大革命的後期，我等能在這個時刻到大陸各地參觀，真是千載難逢，所見所聞是我難忘的經歷。我真正與中國建築學會的接觸是在1980年，當年我的表兄郭彥弘教授、鍾華楠先生與我被中國建築學會邀請，參加了中國建築學會第六屆會員大會暨學術研討會，參與小組討論並首次聽到戴念慈先生在研討會上對現代中國建築設計的精彩理論。1980年開始每逢有內地建築師到港(4)，都盡量安排他們去參觀主要的建築師事務所及政府有關部門，因為我對與同行的交流甚感興趣，所以從此每年都接待一兩個團到港考察的大陸建築師。1980年6月中國建築工業出版社社長楊俊帶團到港訪問，我與鍾華楠合作安排他們在港的行程(5)。10月我與鍾華楠、郭彥弘和黃錫巨組團由中國建築工業出版社副總編楊永生帶領在北京短留後，到承德、成都、昆明、長江三峽及重慶參觀訪問。在昆明及成都作了學術報告及與各地的建築師交流，受到熱情的接待(6,7,8)。1981年的11月同太太麗華到南京、蘇州及北京拜訪大陸的同行。首先到南京拜訪中國建築學會理事長楊廷寶老先生(9,10)。楊老給我的印像十分深刻，他溫文典雅的風度，學者的謙虛態度，使我五體投地。我當時帶了一支普通的墨水筆送給他作為一般的見面禮，但他卻拒收，並對我說："我已經有一支了，用不著另多一支。"他的話使我慚愧萬分，對他的崇高修

1

2

3

4

5

6

養，更有望塵莫及的感覺。晚餐後他還題字送我作為紀念(見412頁)。離開南京後我與太太到過蘇州，暢遊各名園之餘，還與當地建築師們交流。給我印象最深的毛心一先生，我記得有一天到過留園之後，在晚上與毛先生交談(11)，在話別時他對我說："明天我帶您去再參觀留園"。我對他說我今天已經到過留園，但他堅持由他導遊，會別有風味，所以我只好同意。結果他一早七時左右便到我房間，當時我還在夢鄉，穿衣後便跟他到留園。在園中他邊談邊唱，有時還跳起舞來，如他所說，由他導遊的留園，真的使我另眼相看，別有風味。最後他還題字送我以為紀念(見411頁)，這次是很有意義的會面。最後一程到了北京(12,13)，主要是拜訪中國建築學會的各位領導及商討兩會合作的項目(14)。1981年我以香港建築師學會會長身份，首次正式與中國建築學會合作籌辦建築圖片交流展覽。有史以來第一個香港建築師事務所的作品展，於1982年6月在北京開幕(15,16,17,18)。在北京的展覽結束後，由中國建築學會安排在全國20多個城市巡迴展出。當年11月由中國建築學會提供的"中國古代建築圖片展覽"於香港的新世界中心揭幕。1981年8月趁亞洲建築師協會在香港舉行第二次委員會會議之機，我邀請了中國建築學會代表及台灣建築師與會，這是1949年以後，兩岸建築師首次促膝相談。到會的有中國建築學會秘書長曾堅，台灣的建築師學會代表許仲川、許坤南及蔡博安(19)。

同時提議中國及台灣加入亞洲建築師協會，這樣就開始了我就兩會入會事項8年的斡旋。1986年10月戴念慈理事長、吳良鏞副理事長及劉開濟先生就此事親自參加了亞建協第7屆在馬來西亞的吉隆坡市舉行的委員會會議，並與來自台灣的許坤南及朱祖明等商討入會事項(20)。

1981年尾中國建築學會秘書長金歐卜帶團到港訪問，我以香港建築師學會會長身份，協助安排他們在港的活動(21)。1982年4月組織有10多位香港中外建築師參加的團到大陸考察及作學術報告。從北京出發到過承德、鄭州、洛陽、西安及廣州等地(22,23,24,25)。當年5月以何廣乾為首的中國建築學會考察團到港(26)，我以香港建築師學會會長的身份，安排他們參觀有關建築及政府部門。1983年4月初我邀請了英國著名建築師丹尼士·拿仕頓及澳大利亞的著名建築師約翰·安德魯到北京做學術報告。中國建築學會負責安排並於《建築師》刊登報告全文。北京市建築設計院的劉開濟精通中、英文，所以我請他代兩位作翻譯，他本來

7

12

8

13

9

14

10

15

11

16

應做的，但因拿士頓不習慣預備講詞，而單做安德魯的報告已很吃力，結果在報告前的晚上，開濟兄告訴我他實在難以應付擔當拿仕頓的翻譯工作。所以報告當日我只好硬着頭皮，嘗試替拿仕頓做翻譯。但因為他的講詞每一句都比較深奧，及提及富有歐洲背境的引言，引到每一句都需要多句的中文來翻譯，所以最後我也不得不放棄，而由華楠兄補上，但華楠兄過了一段時間後，也要放棄，由陳占祥老先生補上。但會後得知占祥兄的鄉音，有些到會人士聽不懂，為了使拿仕頓的精彩報告得到應有的交待，我會後促請他回國後寄給我一篇講詞，結果他的報告也在《建築師》刊登了。

1983年4月下旬再帶10多名香港中外的建築師到黃山及附近考察，受到當地的建築師同行熱情的接待(27)。1984年初在港舉辦了一學術研討會，清華的吳良鏞教授也參加了，適逢台北的陳其寬先生正在港舉辦畫展，我知道他們原來是老同學之後，便安排他們兩人會面，這是他們40年來第一次再碰頭(28)，也是我對兩岸交流的興趣的開端。

1984年8月與鍾華楠等到上海拜訪陳從周教授(29)，他老人家還當場揮筆題字畫送給我們留念(見411頁)。10月到北京被邀參加了國慶觀禮團並入住釣魚台國賓館(30～33)。11月我與華楠兄及彥弘兄，以中國建築師學會名譽理事身份邀請了學會領導人戴念慈理事長、閣子祥副理事長及龔德順秘書長來港，並安排他們到各處參觀及與本地建築同行交流(34)。

1985年2月華楠兄、彥弘兄與我到京拜訪城鄉建設環境保護部副部長周干峙及中國建築學會領導，並對規劃及建築界的問題交換意見(35)。年中蘭州任震英老先生帶團到港考察(36)，當時我正在蘭州研究生土建築設計及在西安進行小型飯店的可行性研究工作，所以除陪同任老視察之餘，更向他取經。當年10月香港建築師學會組團到京拜訪政界領導人之外，也與中國建築師學會等人敘談，加深兩會之關係(37)。11月我再度到京參加學會舉辦的生土學術會議(38, 39)，首次聽取各國專家關於傳統及現今的生土研究的報告，得益不少。同年與妻麗華被邀參加了在清華大學舉行的紀念梁思成教授誕辰85周年紀念大會及創辦清華大學建築系40周年大會。

1987年5月香港建築師學會第二次在北京展出香港建築師的作品並組團到京參加了開幕式。我也隨團到京支持這次的活動。7月第一次被邀到烏魯木齊參加現代中國建築創作

17

22

18

19

23

20

24

21

25

研究小組第4次會議，並作了以"一個香港建築師眼中的現代中國建築"為題的學術報告(見306頁)(40)。同年12月應陳從周教授的邀請到上海參觀由他負責維修的豫園(41)，他不單帶我在園中各處參觀並題字送我留念，使我受寵若驚。

1988年5月到泉州市參加由中國建築學會舉辦的學術會議(42)。我的祖先原籍是泉州，所以這次到泉州是有特別的意義。可惜在從香港到泉州的旅途中，遇到不太理想的交通問題，雖然在早年到國內參加學術活動，經常碰到航班延誤事故，但這次是最複雜的一次。首先航班在港拖延起飛4小時，而從香港到廈門的飛航時間不到一小時，到廈門上空時因天氣不能降落，所以轉飛福州。到達福州機場已是深夜，機場工作人員已下班，所以被逼在機上過夜，幸好我靈機一觸，要求下機到機場大樓抽煙，因除我之外有多名搭客也有所要求，結果得到當更的機場人員同意才可在候機室內的長椅一睡。翌日早上航空公司人員告訴我們不知何時才能飛廈門，同機有一熟悉福州的香港人及一來自泉州的女士，於是我們三人在機場外乘上空調巴士，坐了6個多小時，終於到了泉州。坐上巴士後，我自問為何屢次身經百戰地到國內參加這些學術會議，不是有點兒自討苦吃嗎？有如臨盤的媽媽們，多數是覺得生完這個嬰兒後就再不生了，但過了一段時間後，生孩子的痛苦便忘記，看到可愛的孩子，又想再生了，我是借用這精神來繼續參加學術會議的。當年9月到韓國漢城參加亞洲建築師學會的週年大會，中國建築學會的戴念慈理事長、吳良鏞教授及張開濟先生等，和台灣的朱祖明及喻肇川先生也參加了會議(43)，我們便利用機會再度商談參加亞建會的事。

當年冬天應邀到杭州參加中國建築學會第7次代表大會暨學術會議(44)。大會落實了新一屆的領導班子，我幸運地再任名譽理事。1989年5月兩岸學會第一次在北京歐美同學會正式開會(45)，商討加入亞洲建築師學會事項，我以中間人的身份，也應邀參加了。經過一天的會議，雙方對台灣學會的名稱有不同的看法，但這次的會面，卻有意義極大的效果。這是台灣建築師學會與中國建築學會第一次正式交流，我能見証這一時刻，實覺得十分榮幸，中國建築學會終於1989年成功地加入了亞建會。

自從1981年起香港建築師學會新任會長，多數上任後不久會組團到京拜訪中國建築學會的領導及有關政府部門，1991年9月夏佐禮會長照慣例組團到京，我也參加了拜訪活動(46)。

26

27

28

29

30

31

32

33

1992年8月應邀到京參加中國建築師學會第8次代表大會暨學術會議(47,48)。我在會上做了發言(見316頁)，講及建築評論的重要性及提倡推動建築評論。1993年3月趁到京參加全國政協會議的機會，與學會葉如棠理事長及其他學會領導敘談(49)。1993年8月到廣州參加國際民居會議(50)，會議在華南理工大學舉行，到會有多名外國專家。他們提供了精彩的報告，使我受益不少。我對中國傳統民居的研究有濃厚的興趣，所以如有時間，一定參加有關的學術會議。11月到京參加中國建築學會成立四十週年慶祝大會(51)。1994年初我的公司被聘請為北京西單北大街西側商業區總顧問，及執行總規劃深化工作並負責設計2、4及5號地的大型商辦大樓，所以我公司聘請了北京市建築設計研究院作為本地顧問。5月趁到京工幹時與北京的張鎛老先生會面(52)，能與中國第二代的著名建築師會面我覺得十分榮幸。當年10月到杭州參加第5次兩岸建築學術交流會(53)，兩岸交流至今已有一定的基礎而且也每隔一年交流會便到台灣舉辦，使兩岸的建築師實行本來是一家的精神。11月香港建築師學會應廣州市建築學會的邀請，組團到廣州舉辦建築師註冊與開業研討會(54)。我也隨團到穗，並主持了部分的研討會，會上除香港建築師陳述香港註冊與開業的情況外，還有內地在深圳試點的私人建築師事務所的負責人提供他們的經驗。

1996年11月到京參加中國建築學會第九次全國會員代表大會暨學術年會並獲繼續委任為學會的名譽理事(55)。

1997年8月全國第8屆中國民居學術會議在香港舉行，我也參加了會議。

1998年可能因亞洲經濟不景，我在國內的學術活動也不多，只是近年尾開始與吳良鏞教授及竇以德秘書長商討，籌辦1999年世界建築師大會中亞建協的研討會及為了應邀參加原定在12月於台北舉行的兩岸建築學術交流會，我準備了一個以"古為今用"為題的報告(見338頁)。結果交流會拖延至1999年1月12日才能舉行(56)。中國建築學會團由秘書長竇以德先生帶團參加，這是我從18年前起投入兩岸事務中第一次到台北參加兩岸會，受到當建築師熱情招待，並公認為兩岸結合的媒人。

我與中國建築學會的關係從1980年至今已近20年之久。回顧這漫長的歲月，我不能不提及1980年是通過楊俊、楊永生的介紹，得到倍受人們尊敬的閻子祥副理長的重視，才得以"回歸"的。我要特別感激閻老，他為大陸、香港、台灣重新合成一家人，做了奠基性的工作。我通過學會的活動，認識了過百的建築師，走遍國內各地，吸收了不少寶

34

39

35

40

36

41

37

42

38

43

貴的經驗，並且得到不少的題字及畫(見附錄10)，實在三生有幸。

In 1974 I joined the second group to visit the mainland organized by Kiu Kwong Investment Corp. Ltd. and led by their chairman Mr. S. Liang (1,2,3). This started my close relationship with architects in the mainland. We visited over ten cities including Guangzhou, Shanghai, Wuzi, Nanjing, Tianjin, Shenyang, Fushun, An Shan, Dalin, Beijing and Shenzhen and met with local architects. At that time the cultural revolution was near an end and it was a rare chance that, I had the opportunity to visit the mainland at that time. I can never forget the trip. My first formal contact with the Architectural Society of China (ASC) was in 1980 when my cousin Prof. Reginald Kwok, Chung Wah Nan and I were invited to attend the 6th members' Congress and symposium. We also participated in the discussion groups and heard for the first time Mr. Dai Nianci's colourful opinion on modern chinese architectural design. From 1980 onwards whenever chinese architect groups visiting Hong Kong, I always tried my best to arrange visits for them to local architects' offices and relevant government departments as I am interested in interchanges between architects from different places. From then on, there were about one to two such groups visiting Hong Kong each year(4). In June, 1980 Mr. Yang Jun, Chief editor of the Architectural & Bullding Press led a groups to visit Hong Kong. Mr. Chung Wah Nan and I made arrangements for them to visit various places in Hong Kong (5). In October, I joined a group which included Prof. Reginald Kwok, Chung Wah Nan, C.K. Wong led by Mr. Yang Yong Sang, deputy editor of the Architectural & Bullding Press to visit Beijing, Cheng De, Chengdu, the three gorges of the Yangtze river and Chongqing. We delivered talks at Kunming and Chengdu and had interchanges with local architects. We also received warm welcomes everywhere we went (6,7,8).

In November, 1981 I took my wife Le Wah to visit architects friends in Nanjing, Suzhou and Beijing. We first paid a courtesy visit to Mr. Yang Ting Bo, the President of the Architectural Society in Nanjing(9,10). I was deeply impressed by Mr. Yang's gentlemanly composure and learned appearance. I brought along a very ordinary fountain pen for him as a token for the visit, but he refused to accept it and said "I already have one fountain pen and I do not need another". I felt so ashamed. After dinner he wrote a piece of calligraphy as a present to me (see page412). After Nanjing my wife and I went to Suzhou to see all the gardens and had discusson with local architects. I was particularly impressed

44

49

45

50

46

51

47

52

48

53

by a Mr. Mao Xin Yi (11). I remember having gone to the "Liu" garden, I talk to him about the visit that evening and he immediately said that he would take me over there again tomorrow. I told him that I have already visited the place, but he insisted and said that it would be an experience to visit the place again with him as the guide. He turned up at the door of my room the next morning at seven. I was then still in my sleep. After getting dressed I went with him to the "Liu" garden. In the garden he showed me around, sometimes singing and sometimes evening dancing. I must admit he had provided me another dimension of the garden which I could not have seen without his assistance. Afterwards he wrote some calligraphy for me as well (see page411). It was indeed a meaningful visit. The last stop was Beijing (12,13) and it was mainly to meet up with the ASC to discuss co-operation work between the two institutes.(14) In 1981 as President of HKIA, I co-operated with the ASC for the first time to organize an exchange exhibition. This was the first time the work of members of HKIA were exhibited in the mainland. It was opened in Beijing in June, 1982 (15,16,17,18). When the exhibition was closed in Beijing, the ASC made arrangements to exhibit it in over twenty other cities. In November, an exhibition named "Traditional Chinese Architecture" supplied by the ASC was opened in the New World Centre. In August, 1981 Arcasia was having its second Council meeting in Hong Kong and I took the opportunity to invite architects from the mainland and Taiwan Provence to attend the meeting as observers. This is the first time since 1949 that architects from opposite the Strait met. Mr. Tseng Chien, Secretary General of ASC, Mr. Shu Jong Chuan, Mr. Hsu Kun Nan and Tsai Bok Au representing the Taiwan Architects Association attended (19).

At the same time I proposed that both institutions apply to join Arcasia. This started my eight years struggle on this matter. In October, 1986, Mr. Dai Nianci, President of ASC, Mr. Wu Liang Yong, Vice President ASC and Mr. Liu Kaiji attended the 7th Council meeting of Arcasia at Kuala Lumpur, Malaysia to further discussed this issue with Mr. Hsu Kun Nan and Mr. Chu Joe Ming from Taiwan (20).

Towards the end of 1981, Mr. Jin Oubo led a delegation to visit Hong Kong and I helped to arrange their activities in Hong Kong as President HKIA (21). In April, 1982 I organized a lecture tour of about fifteen chinese and foreign architects from Hong Kong to the mainland. We started from Beijing and visited Chengde, Changzhou, Luoyang, Xian and Guangzhou (22,23,24,25). In May Mr. Ho Guang Chen led a delegation of architects from China to Hong Kong (26) and I did the usual visit arrangement for them. In June I led a delegation of representatives from HKIA to attend the opening of our exhibition "Architectural Design of Hong Kong" in Beijing. In April, 1983 I invited Mr. Denis Lasdun, the famous architect from London and Mr. John Andrews, the famous architect from Sydney to deliver lectures to the ASC. The ASC organized the event and published their talks in their magazine "the Architect" afterwards. Mr. Lui Kaiji of the Beijing Design and Research Institute is fluent in both Chinese and English and I had asked him to do the translation for both talks by Lasdun and Andrews. He agreed but as Denis Lasdun was not in the habit of preparing a written text for his talk and in addition he found doing John Andrew's talk was rather demanding, he told me the night before the talk, he could not do the translation for Denis Lasdun. Therefore I had to stand in as translator for Denis Lasdun. However as Lasdun is rather philosophical and fond of quoting passages with an european background which had to be translated with several sentences to explain fully, I gave up translating after a while and asked Chung Wah Nan to stand in. However after a while Wah Nan had to give up as well and asked Charles Chan to continue the translation. After the lecture I was told that though Charles managed somewhat, his heavy provincial accent was not appreciated by many people. In order to justify Denis's lecture, I asked him to produce a written text on his return to London and send it to me so that I can arrange publications in the magazine "the Architect".

In April 1983 I led a group of over ten Hong Kong architects on a study tour of Huang Shan and was received warmly by the local architects (27). Early 1984 there was a conference in Hong Kong and Prof. Wu Liang Yong participated. It so happened that Mr. Chan Qi Huan from Taipei was holding a painting exhibition in Hong Kong. I learnt that they were school mates and I arranged for the two to meet for the first time in forty years (28). This marked the beginning of my interest in "cross Strait" interchange. In August, 1984 I went with Mr. Chung Wah Nan to visit Prof. Chen Chun Zhou (29) and he wrote some calligraphy pieces for us (see page 411). In October I was invited to attend the National Day Celebration in Beijing and stayed in the Diaoyutai Government Guest house (30~33). In November, as honorary Council members, Prof. Reginald Kwok, Chung Wah Nan and I jointly invited Mr. Dai Nianci, President of ASC, Mr. Yan Zixiang, Vice President ASC and Mr. Gong Deshun, Secretary General of ASC to visit Hong Kong and arranged fro them to visit relevant places and meet with local architects (34). In February, 1985, Reginald, Wah Nan and I went to Beijing to pay a visit to Mr. Zhou Ganzhi, Minister of the Town and Country Development & Environmental Protection. We also met other senior member of the ASC and discussed with them matters concerning the planning and architecture design professions (35). Middle of the year Mr. Ren Zheng Ying of Lanzhou City led a delegation to visit Hong Kong (36). At that time I was involved with research work in ram earth construction methods in Lanzhou and the feasibility study of a small hotel in Xian. Therefore besides taking him round the visits, I took the opportunity to learn from him his experience in ram earth construction methods. In October I went with the HKIA delegation to Beijing to visit ASC and other relevant government officials to further cement our ties (37). In November I returned to Beijing to attend the symposium on ram earth construction (38,39). It was the first time I had the opportunity to hear papers by experts from all over the world on the subject and learnt a lot from them. Later my wife and I were invited to attend the 85th anniversary celebration for Prof. Liang Sze Cheng and the 40th anniversary of the founding of the School of Architecture in Tsinghua University in Beijing.

In May 1987 the HKIA organized another delegation to attend the opening of the second exhibition of the members' work in Beijing. I went along to lend support to the event. In July I was invited for the first time to attend the fourth symposium organized by the Committee on the Research of Modern Chinese Architecture in Urumqi and delivered a paper called "Modern Chinese Architecture in the eyes of an architect from Hong Kong" (see page 316). (40) In December, Prof. Chen Chun Zhou invited me to visit the Yu Garden in Shanghai which he had done some restoration work. I was much impressed with his hospitality as he not only showed me around the garden but also wrote some calligraphy piece for me (41).

In May, 1988 I went to Chuenzhou in Fujian Provence to attend a conference organized by the ASC.(42) My ancestors originated from Chuenzhou and therefore the trip had special meaning for me. Unfortunately on my way there, I encountered problems in the traveling arrangements. Although in early years I am used to encountering flight delays when I attend conferences in the mainland, but, this time it was the most complicated encounter. Firstly I was delayed four hours in Hong Kong waiting for the plane to arrive from the mainland and the trip to Xiamen would only have taken a little less than one hour. When we reached Xiamen we could not land because of bad weather so the plane was diverted to Fuzhou. When we reach Fuzhou it was late in the evening and the airport was closed. Therefore we were forced to stay on board for the night. Luckily I thought of a way out and asked that they allow us to have a smoke in the airport lounge. There were a few others smokers so they have to agree at the end and we spent the night stretched out on the long benches of the lounge. The next morning we were informed that they did not know when the flight will take off for Xiamen again. There was a person on the same flight who was familiar with Fuzhou and another woman who came from Chuenzhou originally. Therefore the three of us decided to take the air-condition bus outside of the airport terminal for Chuenzhou. After a six hour ride we arrived at Chuenzhou. While on the bus I asked myself why bother to participate in these symposiums in the mainland with all the associated transport problems. It was like asking for trouble. I am fond of quoting the story of the pregnant mother as a justification. A pregnant mother immediately after giving birth always swears that she is not going to have another baby, but, after a while when she sees how lovely a child can be, she wants to have another baby again. I always use this story to encourage myself to continue participation of these symposiums. In September I went to Seoul, South Korea to participate in the Arcasia annual meeting. Mr. Dai Nianci, Prof. Wu Liang Yong and Mr. Liu Kai Ji from the ASC attended the meeting and Mr. Chu Joe Ming and Mr. Yu Chao Chuen from Taiwan Provence also attended (43) and we seized the opportunity to discuss the matter concerning the entry into Arcasia for the two institutions.

In the winter I was invited to attend the 7th general meeting of the ASC and symposium in Hangzhou (44). The meeting elected the new Council and I was luckily re-elected as honorary Council member. The first formal meeting of the two institutions across the Strait took place in the Western Returned Students' Association Club in May, 1989 to discuss the entry into Arcasia matter. (45) I attended as the "middleman". After one full day of talks, the two institutions could not agree on the name to be used by the institution from the Taiwan Provence within Arcasia. Despite the disagreement, this meeting was meaningful as it was the first occasion since 1949 for the two institutions to meet. I was very fortunate to witness the occasion. ASC finally become a member institute of Arcasia in 1989. Since 1981 the new president of HKIA, upon election, will lead a delegation to visit the ASC and relevant government departments in Beijing. I went along in September 1991 when Chris Haffuer became President HKIA(46).

In August, 1992, I was invited to attend the 8th general meeting of the ASC and symposium (47,48) and delivered a paper at the meeting (see page 316) concerning architectural criticism. In March 1993 I was attending the annual CPPCC meeting in Beijing and took the opportunity to meet up with Mr. Ye Rutang, President of ASC and senior member of the ASC (49). In August 1993 I went to Guangzhou to attend the international symposium on "Chinese Traditional Houses" (50). The meeting took place in the South China University of Technology with many foreign experts participating. They delivered many papers on the subject which enlightened me. I am very interested in the study of traditional houses and that is why I always attend such conference whenever I can. In November that year I went to Beijing to participate in the celebration of the 40th anniversary for the founding of ASC (51).

Early 1994 I was appointed the lead consultant for Xidan North Street Western Commercial District and was responsible for the master plan development and architect for the commercial/office complexes on sites 2, 4 and 5. We appointed the Beijing Institute of Architectural Design & Research as our local consultant. In May I took the opportunity while I was in Beijing to meet Mr. Chiang Bor (52). It was indeed an honour to be able to meet an architect of the second generation in China. In September I went to Hanzhou to attend the 5th meeting of architects between the "Straits" (53). The meeting of architects between the Mainland and Taiwan Provence had been established as a regular event for sometime and by then they were meeting annually with alternate meeting being held in Taiwan Provence. This facilitated the family spirit amongst the architects. In November HKIA organized a symposium in Guangzhou at the request of the Guangzhou Institute of Architects.(54) The symposium discussed matters related to architects' registration and practice matters. I went along to the meeting and chaired one of the sessions. There was an interchange of talks on practice conditions in Hong Kong and Shenzhen.

In November 1996 I participated in the 9th general meeting of the ASC and symposium in Beijing and was re-elected honorary council member for another term (55). In August, 1997 I took part in the symposium on "Traditional Chinese Houses" in Hong Kong.

The economic situation in 1998 was deteriorating and my participation in academic exchanges in the mainland was somewhat reduced. Towards the end of the year I started co-operating with Prof. Wu Liang Yong and Mr. Dao Yide, Secretary General of ASC on the preparation work for the co-operating with Prof. Wu Liang Yong and Mr. Dao Yide, Secretary General of ASC on the preparation work for the Arcasia Workshop in the UIA Congress in 1999. I also prepared a paper entitled "Old lamp for new" (see page 338) in preparation for the symposium between the architects of the Mainland and Taiwan Provence originally scheduled to take place in Taipei at the end of 1998. However the symposium was evertually re-scheduled to Jan, 1999 (56). The ASC delegation was led by Mr. Dao and I received warm welcome in Taipei and was recognized as the official "go between" between them.

My association with the ASC started in 1980 some twenty four years ago. I have to mention that it was in 1980 through the introduction of Mr. Yang Jun and Mr. Yang Yong Sang that I came to know the well respected Mr. Yan Zixiang. I am much grateful for his efforts in initiating the "re-unification" of architects from the mainland, Hong Kong and Taiwan. During this long period of time I come to know over a hundred architects through the various activities in the mainland and have travelled all over China. This gave me invaluable experience which I shall treasure all my life.

54

55

56

亞洲建築師協會
Architects Regional Council Asia (ARCASIA)

1979年9月我代表香港建築師學會會長參加亞洲建築師協會在印尼雅加達的創會會議，從此開始了我與亞建協20年來的密切關系。亞建協是一國際建築學會組織(見附錄7亞建協的歷史英文版)，會員是以亞洲各國及地區的建築師學會為單位，而執行委員會的成員，是會員會的會長。亞建協是由6個英聯邦建築師協會的亞洲區建築學會會員發起，包括香港建築師學會、新加坡建築師學會、印度建築師學會、巴基斯坦建築師學會、斯里蘭卡建築師學會及馬來西亞建築師學會，並於1969年成立籌備委員會。但因籌備一個國際性的組織非常複雜，所以籌備工作到1979年才告成功，並於雅加達的例會會議上由到會的6個學會互簽了創會章，此後其他4個學會也確認了會章。1980年8月於泰國曼谷舉辦亞建協第一屆委員會會議，到會的有10個亞洲區建築師學會的會長及會長代表，並在會上推選了我為第一任亞建協主席，主席的任期為兩年。我在任內創辦了亞建協工作坊，並於1981年3月在馬尼拉舉辦了第一次會議，草擬了國際執業道德守則，第二次會議於1981年8月在香港舉行，討論事項包括有限公司執業、垮國執業、建築師的活動性、建築師的崗位及執業守則等問題。1982年10月創立了亞洲建築師論壇，並於斯里蘭卡的科倫坡市舉辦了第一次論壇，論壇主題是"建築設計創作"。從1982年起至1991年我負責主辦了6次論壇，主題包括"亞洲風格"、"亞洲建築設計的方向"、"我的建築設計"、"亞洲建築設計的新意"及"亞洲居屋設計的創新"。同時編寫了4次論壇叢書及出資印刷。亞建協每年舉行一次委員會年會及研討會，我對亞建協及其工作特別有濃厚的興趣，所以從1979年至今，20年都參加每年的年會。這樣每年都能參加年會是不容易的事，因為只有建築學會的現任會長或他的代表、每學會委任的觀察員、各學會教育委會代表、亞建協主任委員及亞建協顧問才能參會，我很幸運從1979年至今都能以不同的資格參加了20年的年會。1979年是代表香港建築師學會會長，1980至1982當任主席，1983至1996年被7屆主席委任為亞建協顧問，1997及1998是以亞建協元老會主辦人參會。而我在1996年提議創立元老會的目標，是有鑒於能參會的資格非常嚴格，所以一般會員會的代表，參加了一年或兩年的年會後，便沒有機會再參會，對亞建協的長遠發展是有一定的影響，所以我創意組織一元老會，使參加過年會的人，如有興趣還可以通過元老會參與亞建協的活動。元老會現今有50個會員從83年至90年代初期，香港建築學會的會長及委員會成員對參與國際學會工作提不起興趣，所以我當亞建協顧問的13年內，也在亞建協年會上兼任香港建築學會的唯一代表，但這兩、三年內，幸好有一班比較年青的建築師對國際學會工作有濃厚的興趣，所以這幾年內參加亞建協年會及研討會活動的香港建築師都有十個八個，我從此不再單刀赴會了。

參加亞建協會工作，收益良多，可以通過每年的年會及研討會與亞洲各地及國際的建築師交流，從中汲取他們在執業上的經驗及設計的心得，同時也在各地結交了一群同行。每個年會都在亞洲不同的地方舉行，所以在開會之餘更能深入視察各地的情況，增廣見聞。

亞建協除了執行委員會外，另有一建築教育委員會由各會員會的教育委會主席組成，每年開會討論亞洲建築教育的情況及推動有關的改革，更在每兩年舉行一次學生研討會。由每會員會邀請學生代表2人，參加一週的學習及研討會。亞建協另一活動，就是兩年一次頒發設計優異獎多個，由會員會提交參選方案。此外，亞建協也出版了些刊物，例如：亞建協新聞報、《亞洲建築》等。1992年我編制了亞建協手冊及操作規則，把亞建協會章、會規、歷史、會議程序及各主任委員的工作綜合編寫，成為亞建協的基本運作手冊。我認為，亞建協是國際建築師學會中最有意義的一個組織，所以我會繼續參與它的活動並鼓勵更多人去參與。

In September, 1979 I represented the President of the Hong Kong Institute of Architects to participate in the inaugural meeting of Arcasia in Jakarta, Indonesia and from then on commenced my nineteen years of association with Arcasia. Arcasia is an international professional organization (see appendix 7 for the history of Arcasia). Membership is restricted to architects' institutes in Asia and the council is composed of Presidents of member institutes. Originally Arcasia was founded by six asian institutes who were members of the Commonwealth Association of Architects including the Hong Kong Institute of Architects, The Singapore Institute of Architects, The Indian Institute of Architects, The Institute of Architects Pakistan, the Sir Lanka Institute of Architects and the Pertubuhan Akitek Malaysia.

In 1969 the Foundation Council was established, and as the formation work for such an international body was rather complicated, it was not until 1979 when the preparation work was completed. The constitution was signed by six institutes at Jakarta followed by the signatures of four other institutes later. In August 1980 the first Arcasia council meeting took place in Bangkok and ten presidents or their representatives attended the meeting and elected me to be the first Chairman of Arcasia. The term of the chairmanship was for two years. During my chairmanship I initiated the founding of the Arcasia Workshop and organized the first Workshop to take place in Manila in March 1981 resulting in the drafting of the International Code of Ethics. The second Workshop took place in Hong Kong in August, 1981 discussing various subjects such as limit company practice, cross boundary practice, the mobility of architects, the position of the architect in society and practice matter. In October, 1982 I initiated the Arcasia Forum and organized the first forum to take place in Colombo, Sir Lanka., The theme of the forum was "Innovation in Architecture". From 1982 to 1991 I was responsible for the organization of six forums including the themes of "Asian Identity", "Design Directions in Asian Architecture", "My Architecture", "Food for thought in Asian Architecture" and "New thoughts in Mass Housing in the Asian Context". At the same time edited and published at my own costs four books on the proceedings of forum 1 to 4. Arcasia maintains an annual meeting of its council together with either a congress or a forum. I am very interested in the work of Arcasia and since 1979 I have attended every annual council meeting for the past nineteen years. It is not easy to attend continuously every annual meeting as those who are qualified to attend are Presidents of their institute or his representative, official observers appointed by member institute, Chairman of board of architectural education of member institute, Arcasia officials and Arcasia Advisors. I was lucky to be able to act in various capacities to attend each annual meeting for the past nineteen years. 1979 I was my own institute's President's representative, 1980-1982 I was Chairman Arcasia, 1983 to 1996 I was appointed by seven chairmen as their adviser, 1997 to 1998 I was convenor of the Arcasia Fellowship. The reason I proposed to establish the Fellowship was that attendance to Arcasia meetings is rather restrictive to institute officials at the time and most people will only able to attend in one year or two years at the most. The Fellowship will enable people who had attended Arcasia meetings as official delegates to continue attending the annual meeting and participate in related activities thus benefiting Arcasia in a long run. The current membership of the Fellowship is around fifty. From 1983 to the nineties the Presidents of HKIA were not very interested in international affairs and as I have been Arcasia Adviser for thirteen years, I have to double up as the HKIA delegate as no body else would attend on behalf of HKIA. However in the last two to three years, there is a group of younger HKIA members who have taken an interest in international affairs and I am no longer the lone "wolf" at Arcasia meetings.

There is much to be gained in participating in the work of Arcasia. Through the annual meeting one can have exchanges with fellow architects in the region on design and practice matters. At the same time one comes to be friends with architects in other countries in the region. Arcasia meeting takes place in different part of Asia every year and one has the opportunity to learn a great deal from other places while attending the annual meeting.

Arcasia has a committee of Architecture Education other than the Council and it is composed of Chairmen of the respective Boards of Architectural Education of member Institutes. Every year the Committee discusses matters related to architectural education and there is a student jamboree every two years where two student representatives are invited from each member Institute to attend a symposium and workshop. Another activity of Arcasia is the bi-annual Design Award selected from works produced by member of the member Institutes. In addition Arcasia also publishes from time to time e.g., the Arcasia Newsletter, and the Asian Architecture magazine. In 1992 I edited the Arcasia Handbook and Checklist which contained the constitution, bye-laws, history and meeting procedures for Arcasia and it became the operational manual of Arcasia. I consider Arcasia the most constructive international institution and I shall continue my participation in its work as well as persuading others to join in.

1998年科倫坡大會
Colombo meeting, 1998

1986年吉隆坡大會學術研討會
ACA2 meeting at Kuala Lumpur, 1986

1987年芭里島大會晚宴
with Ken Yeang and Ejaz Ahed at the Bali meeting, 1987

1993年香港大會
Forum 7 at Hong Kong, 1993

1997年元老會成立大會，於東京
Inaugural meeting of the ARCASIA Fellowship at Tokyo, 1997

1986年學術研討會，泰國建築師作報告
Sumet Jumsai at ACA2, 1986

4位前亞建協主席在1988年漢城大會上
Four past ARCASIA Chairmen at the Seoul meeting, 1988

1994年新加坡大會晚宴
dinner at the Singapore meeting, 1994

1986年學術研討會與象設計集團成員合照
with Higuchi and Hayashi of Team Zoo at ACA2, 1986

1989年曼谷大會與泰國及菲律賓代表合照
with Apsorn Kuromarohit and Felipe Mendoza at the Bangkok meeting, 1989

1995年馬尼拉大會
ACA 6 at Manila, 1995

1987年芭里島大會期間會見當地官員
Meeting local government official at the Bali meeting, 1987

1992年巴基斯坦大會後觀光
Sightseeing after the Lahore meeting with Tay Kheng Soon, Koichi Nagashima, Felipe Mendoza and Richeto Alcordo

1996年雅加達大會晚宴
dinner at Jakarta meeting 1996 with Hisham Albakri, Edward de Silva and Adhi Moersid

英聯邦建築師學會 (1979~1983)
The Commonwealth Association of Architects (CAA)

我是從1979年初，通過香港建築師學會開始加入英聯邦建築師學會的工作。香港建築師學會是英聯邦學會的會員之一，而且成功地爭取到於1979年在香港舉辦英聯邦學會每3年一次的大會。我被學會邀請擔當大會籌備委員會主席一職，這是我第一次負責籌備國際性的大會，也是給我以後籌備學術大會打了根基。到會的有120人，當時港督麥理浩及各國使節也參加了開幕儀式(1-3)。

1981年以香港建築師學會會長的身份，參加了英聯邦會在非洲肯尼亞的奶路比市舉辦的大會(4)。會上有些學會代表，為了要影響大會的決策，不顧會議程序和大局，胡作亂為，最後我在忍無可忍之下作了一個比較轟動的發言，說服大多數的代表回到正常開會的程序。會後英國建築師學會的雜誌內刊登了一段報告，並指出我是混亂的罪魁禍首，所以我寫了一篇較長的信給雜誌的總編說明一切(見附錄4)。我意想不到因為我在大會上的表現，把我牽入學會內部的工作。我被推選為下一屆的亞洲區副主席，從此參加了多次的理事會會議。3年內為了開理事會議，還到英國、加拿大、澳大利亞及馬來西亞(5)等地。為了化解西方地區副主席的經常霸道行為，每次理事會上我都要特別提起精神，盡量游說，務求顧全大局，使大家都能在平等的情況下推行會務。

英聯邦建築師會是由25個建築學會的會員組成，全部會員除英國外，都是前英國殖民地的國家及地區。而最有價值的一項工作，就是建築教育水平的審核制度。每間建築學院，經申請後，學會便組成一個審查委員會，對學院進行審查，一經委員會的認可，便能加入英聯邦建築師學會的認可名單，這名單是全英聯邦都承認的，是一個資格的認証。以後每五年學會就對每間已在名單的學院進行復核，確保一定的教學水平。除這項工作外，學會因為資源不多，而且各會員的執業情

況有異，所以除每3年舉行大會及學術會議之外，並沒有其他活動。我做了3年充滿政治化的學會工作，在參加1983年的大會後，更淡出學會的工作。

I started to participate in the work of the Commonwealth Association of Architects (CAA) in 1979 through the Hong Kong Institute of Architects (HKIA). HKIA is a member of CAA and was successful in securing the staging in 1979 of the CAA triannual conference. I was asked by the HKIA to chair the organizing committee for the preparation of the conference. This was the first time I undertook the organization of an international conference and this provided me valuable experience in organizing conferences later. There were 120 participants to the conference including the then Hong Kong Governor Sir Murray Mclehouse and the respective consular core who attended the opening ceremony (1-3).

In 1981 I represented HKIA in the CAA Triannual Conference which took place in Nairobe, Kenya (4). During the conference some member institute representatives wanted to influence the decision of the conference and used unsavory methods to create near chaos in the proceedings. Finally I could not hold my tongue and lashed out with a rather aggressive speech. My speech managed to bring everybody back to their senses. The RIBA journal later reported on the conference and mentioned in the article that I started the "fun and game". I wrote to the Chief Editior of the Journal explaining to him what had actually happened (see appendix 4). To my surprise my performance at the conference led me to participate in the work of the Council. I was elected Vice President for the Asian region during the session. Since then I participated in several council meetings during the next three years travelling to England, Canada, Australia and Malaysia (5). In order to defuse the dominating attitude of the other vice presidents of the developed regions, I had to be alert at all council meetings and try my best in maintaining a balance for all concerned.

The CAA in 1979 composed of 25 member institutes. All the member Institute with the exception of England. came from countries which had been British colonies at one time or another. The most valuable work of the CAA was the recognition system of architectural schools. Each architectural school can apply to be included into the list of schools, CAA will send an visitation board to the school and carry out an examination of the school.. Once recognized by the visitation board as worthy of the standard of registration, the school's name can be included in the list of recognized schools. This list is recognized throughout the Commonwealth and is a recognition of high standards. Other than the list of schools, the CAA, because of shortage of funds and practice condition being different amongst its member institutes, has no other activity other than the triannual conference. I participated in the politic ridden work of the council for 3 years and after attending the triannual conference in Sydney I decided to call it a day.

4

5

國際建築師協會 (1979~1987)
Union of International Architects

我第一次參與國際建築師協會的活動，是在1979年9月在印尼的雅加達舉行的民居研究組會議，我當時是代表香港建築師學會參加的。第二次是以香港建築師學會會長身份參加協會1981年在波蘭華沙市舉行的4年一次大會及理事會議(1,2)。國際建築師協會的大會實在是一個大型的建築設計博覽會，在會期內有多項來自世界各地的建築設計展覽及數十個設計報告。而參加大會的人數由3千多至7千多人，會場十分熱鬧。對建築師來說，這是難得的機會，能夠在短短的一、兩個星期的時間，可以與很多來自各地的建築師交流，看到全世界最新設計的潮流。但國際建築師協會理事會的工作與英聯邦建築師學會一樣，充滿政治性及富有發展中國家與已發展國家的爭鬥，而強行將執業方式國際化，這實際上是行不通的。所以我總覺得國際建築師協會的工作，大多數是紙上談兵，一事無成。在1987年7月我為了要支持菲律賓建築師協會加入國際建築師協會，代表香港建築師學會，參加了在愛爾蘭德背爾舉行的大會及理事會會議。在理事會會議過程中，因為有另外一個比較有少數會員及缺乏代表性的菲律賓建築師學會與協會相爭加入國際建築師學會，而有部分理事會成員，包括主席對兩會入會的歷史不太清楚，所以大會一時又鬧起爭論。我因對這類的爭論有經驗，而且是有備而來，所以當時我向大會作了發言及提議處理辦法(見附錄5)，結果大會接納我的提議。具國際性的理事會只是隔幾年開會一次，而每次大部分的代表是第一次參加會議而且又沒有做足會前的準備，所以開會時往往對會中事項不了解，因而浪廢理事會的時間，我對這種情況已經麻木，所以這次之後，我再也沒有參加國際建築師協會的理事會工作。

My first involvement with the UIA was to represent the HKIA in the conference on Urban Village in September, 1979 at Jakarta, Indonesia. The second time was to participate in the 1981 UIA Congress and Council meeting as HKIA President. The Congress takes place every four years (1,2) and it is a large scale architectural design exhibition. During the Congress there are many exhibitions on architectural design from all over the world with many talks. The participants normally number between three to seven thousands and the grounds are full of activities. To architects, it is a rare opportunity to meet with so many architects from all over the world and see the latest design trends all within a short period of one to two weeks. However the work of the UIA Council is like that of the CAA, full of politics between the developing countries and the developed countries on issues such as to insist on architectural practice to be internationalized which is an impossible task. That is why I felt that the work of the UIA Council was a waste of time and non productive. In July, 1987 I attended the UIA Congress and Council meeting in Dublin, Ireland representing the HKIA to give support to the United Architects of the Philippines (UAP) for their entry to UIA. During the Council meeting there was another contender for the membership from the Philippines, the Philippine Institute of Architects (PIA) which had fewer members and was much smaller in size. Unfortunately some of the Council members including the President were not familiar with the background of the contest and created arguments during the meeting. I had experience of this kind of "unrest" and was well prepared before coming to the meeting. I spoke on the issue and suggested a way out of the chaos (see appendix 5). Finally the meeting accepted my motion. Council meeting of an internation body normally meets once every so many years and every time the majority of delegates do not do any homework and are unfamiliar with the items discussed at the meeting, thus wasting a lot of the meeting time. I grew tire of this kind of situation. Therefore after that meeting I did not participate in any more UIA work.

香港科技協進會 (1985~1989)

The Hong Kong Association for the Advancement of Science and Technology

我在1985年10月加入香港科技協進會，當年是陳耀華教授擔任主席。科協創會有多個目標，其中一個是在香港提倡科技作為文化的一部分。科協每個月都舉辦有關科技的研討會，並邀請有關人士作報告，同時也參加社會有關的活動。我入會後第一件事是替會設計一個會徽(見146頁)，我用了簡化的核工業的圖案及"品"字型的背境，來表達3種會員的不同行業，例如教授、技術人員及科學家。其後在1986年至1989年間被選入理事會，1987年當選為主席。隨即帶團到北京拜訪了中國科技協會主席錢學森教授(1)並與他商討合作活動，包括將於當年11月在港舉行的國際研討會"亞洲四小龍"。7月我在香港北區扶輪社午餐中作了"聯繫各界，發展科技"的報告(見310頁)。同年我到過倫敦拜訪英國皇家學會並促成與中國科技協會及我會的三邊合作計劃去推動學術及專業的交流，及與中國資訊及顧問服務中心、深圳科技局科技發展及交流中心等機構商量合作事項。8月與美國科學協進會會長商談兩會合作事務。當年科協成立了多個研究小組，專研的項目包括精密工程、基本物理程序、虛構智慧及專家系統、微電子、電腦技術、大型電腦、生物技術、未有解釋的現象、自動化工業及機械人、科技與社會等。各小組於1988年及1989年間舉行了研討會。11月科協舉辦了第一次的國際研討會(2)，主題是"亞洲四小龍的工業發展"，提交報告的有韓國、台灣、新加坡及香港的15位著名工業家及教授。我當了籌委會的主席，到會者約有120人，報章都報導科協的活動(見附錄9)。通過科協的工作，我認識了不少我本行之外的朋友，我覺得與他們交談增添了不少對事物的看法。

1

2

I joined the Hong Kong Association of the Advancement of Science and Technology (HKAAST) in October 1985 when it was founded and chaired by Prof. Chan Yau Wah. It aims amongst other things to promote and further develop science and technology as a culture in the Hong Kong community. The Association organizes monthly meetings on subjects related to science and technology with guest speakers and participates in community projects related to the aims of the Association. The first thing I was asked to do was to design a logo for the Association (see page 146). I used the simple forms of the nuclear technology and a y shape to denote the three disciplines in the compositions of our membership i.e., academics, technicians and scientists. I served on the Council between 1986 and 1989. In 1987 I was elected President and led a delegation to visit Prof. Chen Xue Xing, President of the China Association for Science and Technology in Beijing and discussed with him various joint activities for the year including the international conference on "The Four Little Dragons of Asia" scheduled to take place in November that year in Hong Kong. (1) I also delivered a talk at the Rotary Club North on 14th July entitled "Unite all front to develope Science & Technology" (see page 312). I also visited the President of the Royal Society in London and came to an arrangement together with China Association for Science and Technology to promote academic and professional exchanges amongst the three Associations, and also met with the representatives of the China Statistical Information and Consultants Service Centre and the Shenzhen Science and Technology Commission and Science & Technology Development and Exchange Centre on Joint projects. In August I met the President of the American Association for the Advancement of Science and discussed potential area for mutual co-operation. In the same year we established a number of study groups covering precision engineering, fundamental physical processes, artificial intelligence and expert system, microelectronics and IC technology, super computing, biotechnology, unexplained phenomena, industrial automation and robotics, science, technology and society. The groups held seminars on each of the subjects throughout 1988 and 1989. In November, 1987 HKAAST hosted the first international conference on "The New Industrial Development in the Four Little Dragon of Asia" in Hong Kong with 15 leading industrialist and academics from south Korea, Singapore, Taiwan and Hong Kong.(2) I chaired the organizing committee for the conference and we have over 120 participants to the conference. Through the work of the Association I came to meet a variety of people other than from my own profession. I find it stimulating from discussions with them and this helped to broaden my outlook.

建築師協會亞洲會
AA Asia

1

4

2

5

1990年初，一群英國建築師協會建築學院畢業的亞洲建築師開始聚會，並交換各人的工作經驗。3月在香港正式決定去組織"建築師協會亞洲會"，向協會建築學院畢業或執教的建築師，在亞洲提供一個組織，並提倡在亞洲地區友善、明智、藝術及學術上的交流。這個組織在新加坡註冊，每年舉行會議，由會員介紹自己最新的創作及彼此討論心得。此外，還組織考察團到亞洲各地考察，及在亞洲各地舉辦工作坊去提倡學術交流。過去的活動包括1991年在檳城的設計集會(1,2,3)；1992年在新加坡的設計工作坊；1993年日本考察；1994年孟買市工作坊、與同濟大學在上海合辦的設計工作坊及在新加坡的建築設計展覽；1996年在新加坡的現代化傳統建築設計工作坊、在吉隆坡的"空間時間"工作坊；1997年在新加坡的巴華建築師工作坊、在北京與清華大學建築系合辦的"現代化傳統建築設計"研討會(4,5,6)及1998年日本考察。並於1998年出版在清華大學舉辦的研討會文章，成為該會的第一本書。現今有50多個會員。我在過去提供了3篇報告，其中兩篇刊登在這書內，包括"回顧及前路"(見325頁)及"西西工程4及5號地的設計"(見330頁)。

3

6

In early 1990 a small group of AA graduates from the asian region decided to meet informally and to exchange information about each other's work. By March that year the group met in Hong Kong and decided to establish a society called "AA Asia" to provide a platform in Asia for graduates and teachers of the Architectural Association School of Architecture and to establish friendly, intellectual, artistic and educational ties amongst architects in the Asian region. The organization was formally registered in Singapore. The annual meeting consisted of talks given by members on their latest work which was discussed amongst the members during Design Jamboree, Penang 1991 (1-3), Design Workshop, Singapore 1992, Japan Tour 1993, AA Asia Yatra, Bombay 1994, AA Asia/Tongi University Design Workshop, Shanghai, 1994, exhibition of alternate architecture in Asia, Singapore 1994, a Workshop on Contemporary Vernacular in Singapore, 1996, a workshop on Free Time, Kuala Lumpur, 1997, a Workshop on Geoffrey Bawa, Singapore, 1997, a seminar on "Contemporary Vernacular - Modern Architecture Traditions", Tsinghua University, Beijing, 1997 (4-6), Japan Tour, 1998. The first monograph was published in 1998 for the Beijing seminar. Currently there are 52 members. I have contributed three papers so far and two of them are published in this book entitled "Reflection and the way forward" (see page 325) and "Design for Sites 4 & 5 of the 'Xi-Xi Project'" (see page 330).

中建學社 (1986~1994)
Zhong Jian Society

中建學社是於1986年6月開始籌備並於1987年10月在香港成立。我是創社社長，任職至1990年。創社的意向是希望能夠組織12位對建設有關的主要專業人士為核心社員，每個月舉行一次討論會。討論的題目包括中國建設的近況、社員在中國的工作經驗、中國建設的新法制、體制和程序、中國投資項目中的新聞和中國建設的宏觀問題等。當時因為我等以不同的角色參與在國內的建設，而覺得需要彼此交流經驗及信息，使我等的工作更加有效。組員包括建築師、會計師、銀行家、建築商、工程師、結構工程師、材料測量師、產業測量師及律師。我等也經常邀請對我等研究工作有關的專家來參加討論和作報告。學社的活動舉辦不到10年便停下來，因為此後各社員的工作繁忙，而且國內建設的信息已比較流通。

Preparation work for the formation of the Zhong Jian Society was started in June, 1986 and the Society was established in October 1987 in Hong Kong. I was the founding President. The aim for the formation of the Society was to invite about twelve professional people concerned with development work as members and to hold monthly discussion meeting. The subjects for discussion included the latest information on development work in China, the individual experience of members in their work, new regulations, systems and procedures concerning development in China, news on investment projects in China and the mega view of development in China. At the time we were individually involved in different role in development work in China and felt that there was a need to exchange experience and news in order that our work can be more effective. The members included architects, accountants, bankers, developer, engineers, structural engineers, quantity surveyors, estate surveyors and solicitors. We also invited experts related to our work to deliver talks. The activities of the society was stopped after less than ten years of the formation as the members became more busy in their work and the news concerning development in China became more easily available.

中國建築論壇(1995~　)
China Architectural Forum

7

舉辦中國建築論壇的設想是在一個很湊巧的場合裡產生的。1995年3月22日，我正在北京參加鍾華楠先生為了新任香港建築師學會會長一職，到京與中國建築學會及有關單位的領導人進行學術交流活動，晚上在下塌的西苑飯店咖啡廳與好友《建築師》雜誌編委會主任楊永生先生把酒言歡，談及通過學會辦學術活動，在籌辦工作上需要很多人力、物力，如果能夠舉辦一些小規模而又有深度的研討會，通過私人贊助，便能對推動中國建築設計起一些實際的作用，如古人的蘭亭會。當時楊先生提議先在深圳舉辦一小型研討會，邀請10至12位建築師、學者出席並提交論文，用兩三天的時間去詳細討論提交的論文，會後可把論文出版公諸於世，使更多的人受益。結果，《中國建築論壇》就這樣產生了。

第一次論壇於1995年10月在深圳舉行(1)，天津科學技術出版社同意研討會及把提交論壇的文章出書。論壇的主題是"建築與評論"。到會的除擔當主持的永生兄及我之外，還有10名國內著名的建築師及教授，他們提供的報告有：蔡德道"建築評論隨感"、鄒德儂"從先鋒建築手法的標新立異看建築創作的進步和倒退"、左肖思"身在此山活此山"、王天錫"繁榮建築評論的前提"、曾昭奮"從大屋頂的束縛中解放出來"、邢同和"從'標誌性建築'和'里程碑'談起"、劉力"建築師的參與"、張孚佩"建築的多元、傳統與現代評論"、孟建民"建設熱潮中的建築創作"及陳世民"在新的歷史條件下看建築評論及其標準"。

1996年8月得到天津科學技術出版社同意再度主辦，第二次研討會論壇就在天津舉行(2,3)。論壇主題是"比較與差距"，會

1　　3

2　　4

上提交的論文有沈克寧"中西建築之'异'偶見"、顧孟潮"論紀念性建築的環境藝術"、吳耀東"中日現代建築發展歷程的比較研究"、董衛"北歐城市——過程與經驗"、鄒德儂"在國際性和國家性建築框架里的中國現代建築"、呂富珣"前蘇聯建築及其對中國建築的影響"、王貴祥"略論科學革命與西方建築空間發展"、王小東"中亞建築折射中的思考"、聶蘭生"現代建築創作與傳統的回潮"、韓林飛"莫斯科建築學院的建築學教育及對我國的啟示"、馬國馨"建築評論的評論"、賴德霖"楊廷寶與路易‧康"、齊康"比較與差異"，彭一剛"傳統建築文化與當代建築創新"、仲德崑"中華民族的視覺特徵與中國建築空間的組織"、、鍾華楠"放下袍袱，腳踏實地，克勤克儉，建設美好的未來"及楊永生"中國需要比較建築學"。

1997年8月第三次論壇研討會得到東南大學建築研究所和建築系的安排在南京舉辦(4,5)，主題是"現狀與出路"。提交報告的有陳志華"趕緊培育健全的建築學術界"、彭一剛"悅目與賞心"、鍾訓正"給城市多一點綠地和公共活動空間"、張欽楠"永恒的十字路口"、戴志中"建築教育要求觀念性變革"仲德崑"我看中國建築教育"、劉管平"正視人居環境，增強全民意識"、鄒德儂"中國現代建築理論的解困"、潘祖堯"下真功夫，走向世界"、左肖思"對建築設計創作環境、現狀的淺見和展望"、齊康"城鎮建築現象的窺視"、程泰寧"地域性與建築文化"，戴复東"山窮水盡愁無路，柳暗花明覓新村"及羅小未"精品!精品!精品"。天津科學技術出版社不辭勞苦繼續支持和出書。當年"建築報"更多次刊登了有關論壇的報導(見附錄6)。

1998年10月得到重慶建築大學建築城規學院及天津科學技術出版社的支持，在重慶大酒店舉辦第四次研討會(6,7)，主題是"建築與社會"。提交報告的有鍾訓正"近現代經驗教訓的管見"、彭一剛"從建築與社會看模仿與創新"、程泰寧"折射與尋回"、姚兵"呼喚全社會重視建築美學知識教育"、崔愷"建築創作與社會環境"、吳耀東"扎根的建築、健康的建築及其他"、張為耕"高揚中國民族建築文化的旗幟"、齊康"建築文化現象的態勢"、戴志中"地方特色是現代建築創作的主題"等。

從1995年至今，能每年都舉辦一次論壇，全靠有關單位及天津科學技術出版社的大力支時，而又能得到永生兄的熱心投入，他不單是負責每年安排主辦單位，多方徵求意見，大家商定主題，更負責邀請參會者的工作，論壇的成功全靠他助予一臂之力。論壇至今可算已奠下基礎，而且有多位常客給予支持，希望以後能在建築學術上成一支柱，透過出版叢書，使更多的年青一代的建築師受益。

5

6

The thought to establish the China Architectural Forum came by chance at a meeting on the 22nd of March, 1995 in Beijing. It so happened that I was taking part in an HKIA visitation to the Architectural Society of China (ASC) upon Mr. Chung Wah Nan becoming the President of HKIA. That evening I was having coffee with my good friend Mr. Yang Yong Sang, the Chairman of the Edition Board of the Architectural & Building Press at the Xie Yuan Hotel and we were talking about the problem of organizing academic exchanges activities through the ASC. We felt that if we could organize some smaller scale symposium using private funding, it would probably provide an easier way to advance the standard of architectural design in China, such as the historical "Lanting meeting". Mr. Yang suggested at the time to organize a small symposium in Shenzhen to start with and invite about ten to twelve architects and academics to attend with a paper each. The symposium should last about two to three days discussing about the papers delivered and the papers will be published after the symposium to benefit a wider public. Thus the China Architectural Forum was born.

The first forum took place in Shenzhen in October, 1995(1) and the Tianjin Scientific and Technology Press agreed to publish the proceedings. The theme for the forum was "Architectural Criticism" and ten famous architects and academics participated with the following paper, Cai Dedao on "Thoughts on Architectural Criticism"; Zou Denong on "A Review of the Rise and Fall of Architectural Innovation based on the Latest Architectural Design Methodology"; Zuo Xiaosi on " Living in the Mountain thrive on the Mountain"; Wang Tianxi on "The Prospect of Popularizing Architectural Criticism"; Zeng Zhaofen on "Liberate from the Constrain of the Tradition Roof"; Xing Tonghe on "Monumental and Leading-edge Design"; Lui Li on "Architects' Involvement in Architectural Criticism"; Zhang Fupei on "Criticism on the Multi-facet, Traditional and Modern Architecture"; Meng Jianmin on "Architectural Innovation in Boom Time"; Chen Shimin on "Architectural Criticism and its Standards under New Historical Conditions"; Ronald Poon on "Liberate Architectural Criticism".The Tianjin Scientific and Technology Press agreed to organize the second Forum in Tinajin in August, 1996. (2) The theme was "Comparison and Differences" and the papers delivered included Shen Kening on "Snapshot on the Differences of East/West Architecture"; Gu Mengchao on "Commemorative Architecture"; Wu Yaodong on "Comparison between Chinese and Japanese Architectural Development"; Dong Wei on "The Development and Experience on Northern European Cities"; Zhou Delung on "Modern Chinese Architecture within the International and National Context"; Lu Fuxun on "The Influence of Soviet Architecture on Chinese Architecture"; Wang Guixiang on "The Periodic Specialties of the Science Revolution and the Development of the Western Architectural Space"; Wang Xiaodung on "Thoughts on the Development of Middle Asia Architecture"; Nie Lansheng on "Modern Architectural Creation and Traditional Rebirth"; Han Linfei on "The Architectural Education in the Moscow Architectural School and its Effects on China"; Ma Guoxin on "A Critique on Architectural Criticism"; Lai Delin on "Yang Tingbao and Louis Kahn"; Qi Kang on "Comparison and Differences"; Pang Yikang on "Traditional Architectural Culture and Modern Architectural Creation"; Chung Dequan on "The Visual Capability of the Chinese People and the Organization of Space"; Chung Wah Nan on "Put Down the load on the back, foot on the ground, with hardwork create a wonderful future"; Yang YongSang on "China Needs Architectural Criticism". The Tianjin Scientific and Technology Press agreed again to publish the proceedings.

The third Forum was held in the Architectural Research Institute of the South East University at Nanjing. (3,4,5) The theme was "Now and the Future". The papers delivered were Chen Zhi Hua on "Urgent Development of the Architectural Learned Circle"; Pang Yi Kang on "Visual Delight and Appreciation"; Chung Zun Zheng on "Give the City a Little Green Area and Community Space"; Zhang Qing Nan on "Crossroads Forever"; Chung Dequan on "My Views on Chinese Architectural Education"; Liu Guon Ping on "Take Note of the Living Environment and Increase People's Awareness"; Zhou Delung on "The Liberation of Modern Chinese Architectural Theory"; Ronald Poon on "Show the Real Capability and Advance Towards the World"; Jaw Shao Shih on "The View and Aspiration on the current Architectural Design Creation"; Qi Kang on "Views on Current Urban Design"; Chin Tai Ning on "Regionalism and Architectural Culture"; Dai Fu Dong on "There is a Way Forward"; Luo Xiao Wei on "The Best, the Best and the Best". The Tianjin Scientific and Technology Press once again agreed to publish the proceedings and provide us the necessary support. In addition there were some reporting in the newspaper about the Forum (see appendix 6).

The fourth Forum took place in Chongqing in October, 1998 with the support of the school q Architecture, Chongqing Jianzhu University and the Tianjin Scientific and Technology Press. (6,7) The theme was "Architecture and Society". The papers delivered included Chung Zun Zheng on "Lessons to Learn"; Pang Yi Kang on "From Architecture and Society to Imitation and Creation"; Chin Tai Ning on "On the Attack and Path Finding"; Yao Bing on "Call on the whole Society to take the Education of Architectural Esthetics seriously"; Cui Kai on "Architectural Creation and the Environment of the Society"; Wu Yaodong on "Architecture with Roots, Architecture with Health and others"; Zhang Wei Gang on "Raise the Flag for Chinese Architectural Culture"; Qi Kang on "The Position of Current Architectural Culture" and Dai Zhi Chung on "The Main Theme for Architectural Creation is Local Identity".

We were able to maintain one forum per year from 1995 to now is mainly because of the support from the local institutions and the Tianjin Scientific and Technology Press. In particular I am indebted to Yong Sang's devotion and his effort every year to identity the local supporting institution and to invite all speakers. He had contributed to the success of the Forum. The Forum has by how established itself with some regular speakers. I hope it can provide support to the pursuit of better architectural design and through publication benefit the younger generation of architects.

現代中國建築創作研究小組
(1987~   )
The Research Group on Modern Chinese Architectural Creation

我是於1987年8月應吳國力等的邀請,參加了在烏魯木齊舉行的小組第4屆年會(1,2)。小組是由一班我國中年建築師在中國建築學會的直接關懷下於1984年4月在昆明組織而成。第二屆年會於1985年香山舉辦,第三屆年會于1986年在武漢舉辦。到1987年已有組員大約70人。我是當時唯一的海外成員,並負責推薦聯繫港、澳、台及日本方面的有關專家參加第三屆於1986年在武漢舉辦的學術討論會。第4屆年會討論會的主題是"傳統建築文化與現代中國建築創作",我在會上作了以"香港建築師眼中的中國現代建築"為題的報告(見306頁)。到會的正式代表有64名,做報告的有24位。這次學術討論會,不論在深度、寬度上還是理論與創作上,都是在較高的層次上進行的。通過大會發言,尤其以熱烈的對話和辯論,更深化了討論的實質,也使得會議開得生動、活潑,是一種值得提倡的、有學術爭鳴和民主氣氛的會議方式。1988年10月小組在杭州召開第5屆年會及討論會(3,4),主題為"文化環境與現代中國建築創作"。我參加了年會並參與討論建立國際華人建築師協會或現代中國建築創作研究會,會後"建築雜誌"第5期也有報導(見附錄8)。小組第6屆年會於1990年11月在深圳大學舉行(5),主題是"回顧與展望"。我邀請了香港的鍾華楠、嚴迅奇、歐中樑及何弢到會並做了報告,而我負責主持大會中的一個會議。到會的組員有60人之多。之後因種種我們不知的原因,活動不那麼多了。直至1995年在小組成立10周年之際,在蘇州召開了第7屆年會,主題為"現代中國建築設計的思路",但因會期改動,因公幹沒有到會。

小組的年會活動,集中了組員的近年設計成果及理論,彼此交換意見,而且人人平等,沒有一般研討會的嚴肅及硬性的儀式,所以是充滿學術氣氛,而且非常有創意性。我對小組的活動十分支持,而且有空一定爭取參加。我國的現代建築設計,正在經過一段探索的時期,小組的存在是必要的。

1

2

3

4

5

It was Mr. Wu Guo Li who invited me to participate in the fourth meeting of the Research Group in August, 1987 (1,2). The Research Group was formed in April 1984 in Kunming by a group of middle aged architects in the mainland with the blessing of the Architectural Society of China. The second meeting took place in Xang Shan in 1985. In 1987 there were about seventy members. I was the only oversea member of the Group and was responsible to liase with fellow architects in Hong Kong, Macao, Taiwan and Japan to participate in our fourth meeting at Wuhan in 1986. The theme for the fourth meeting was "Traditional Architectural Culture and Modern Chinese Architectural Design". I delivered a paper on "Modern Chinese Architecture in the eyes of an Hong Kong Architect" (see page 306). There were sixty four participants to the meeting and twenty four of them delivered papers. The breath and depth of the discussion and the quality of the papers bored higher standards than the normal conference. There were heated debates which explored in depth the matters related to the papers creating a lively and energetic atmosphere for the meeting. This kind of democratic and learned gathering should be encouraged. In October 1987, the fifth meeting took place in Hangzhou (3,4). The them was "The Cultural Environment and Modern Chinese Architectural Design". I participated in the meeting and also took part in the discussion regarding the formation of an international chinese architects association or a society for the research in modern Chinese architectural design. The meeting activities were reported in the fifth edition of the magazine "Architecture" (see appendix 8). The sixth meeting of the Group took place in November, 1990 at the University of Shenzhen (5). The theme was "The Past and Future Aspirations". I invited Chung Wah Nan, Rocco Yim, Joanlin Leung and Tao Ho to participate and deliver papers. I was session chairman for one of the sessions. There were about sixty participants to the meeting. From then on Mr. Wu Gou Li, one of the prime movers was posted abroad and for one reason or another the Group suspended its activities for a number of years until 1995 the seventh meeting took place in Suzhou on the tenth anniversary of the founding of the Group. The theme was "The Direction of Modern Chinese Architectural Design". However as the date for the meeting was changed at the last minute I was not able to participate in the meeting. The annual meeting of the Group concentrates on presentation of the recent work of its members and discussion on the works. Every one present is on equal basis and the meeting is without official ceremonies and stiff procedures unlike other conferences. As a result it is a highly academic and creative gathering. I am very supportive of the Group's activities and I always make an effort to attend. The formulation of modern architectural design in China is currently going through development and the existence of the Group is essential.

社會工作

Community work

中國人民政治協商會議全國委員會 (1993~　　)
The National Committee of the Chinese People Political Consultative Conference (CPPCC)

1

5

早於政協第6屆(1983~1988)會議之前，新華社即派人與我商談參予全國政協工作一事，我考慮多時，因為我太太是台灣人，而在當時的政治氣候條件下，當了全國政協委員，肯定不能踏足台灣。當時我雖然已投入與國內建築界的交流，但對政協的工作不甚了解，所以最後婉拒了新華社的好意。1992年新華社舊事重提，我喜出望外，便馬上答應加入第8屆(1993~1998)的委員會，從此開始了我與全國政協的不解之緣。

2

6

全國政協是中國人民愛國統一戰線的組織，在1949年9月成立。政協的委員會是由各黨派、海外同胞及特邀人士，大約2千多人組成。歷屆主席包括毛澤東、周恩來、鄧小平、鄧穎超、李先念及李瑞環。每年舉行全体會議，近年的會期大約9天至兩星期，政協主要的職責是政治協商、民主監督及參政議政。

1993年3月我第一次到京參加全國政協第8屆第一次會議(1, 2, 3)。承蒙霍英東副主席的安排，全體香港委員入住北京飯店貴賓樓，得到6星級的招待，我實在有受之有愧之感。會

3

7

議期間，除了聽取國家領導人及主要政府部門的工作報告，旁聽人大開幕，還有小組會議討論各報告及進行參政議政的討論。會期中送到我們委員手上的資料之多，如要完全細閱，兩星期不睡也看不完。而總理工作報告等文件甚多，所以開會10多天，沒有一天能有足夠的睡眠。在參加1993年第8屆第一次大會時，所有事物對我來說都有點新鮮，例如到人民大會堂聽總理、政協主席等讀出他們長

4

8

達1個多小時的工作報告，聽取其他委員讀出他們已發給我們的發言稿及小組會議(4)。小組會議主要是讓委員們對各工作報告提出意見，由秘書組將每個委員發言中他們認為最重要的紀錄下來，並且編印在簡報上，發給所有2千多位委員閱讀。

政協的活動除了每年一次的會議外，有多項其他的活動，例如到國內各地考察。政協每年都安排香港委員一至兩次到國內考察，一般我都爭取積極參加。這幾年到過貴州、廣西、浙江等地(5,6,7,8)。這些考察給我一個難得的機會去深入了解各地的情況，增加我對國情的了解。另外一種活動是在香港參與政協委員專題小組討論會。專題小組的討論範圍，包括香港與國內有關的事務，國內發展的種種問題等。小組討論的紀要是提交政協有關部門作參考。第8屆內的專題小組是由6位委員召集籌辦，而第9屆因香港委員人數大增，所以召集人也增加到10位，我是兩屆的召集人之一，"香港植樹日"就是由第8屆的6個召集人發起的，間接為香港做了點好事(9,10)。此外，政協委員也參與第一屆香港政府推選委員會及特區第9屆人大代表選舉會議等工作(見附錄(12))。

通過參與政協的工作，我認識了一批熱愛祖國的香港成功人士(11,12,13)，使我得益良多。而且更能夠與國內的建築師委員，每年最少碰頭一次(14)。全國政協就好像是一個大家庭一樣，委員們都是志同道合，希望能為祖國做一點事。第9屆慶幸獲得連任，以後會更加努力，為人民服務。

9

12

10

13

11

14

It was approached by the Xinhua News Agency shortly before the 6th Meeting of the CPPCC in 1982 to see if I was interested in participating in the work of CPPCC. I considered the invitation at length and because my wife came from Taiwan and the political climate at the time would certainly prevent me from visiting Taiwan if I became a member of the CPPCC. Though I was already involved in academic exchanges with architects in the mainland, I was not familiar with the work of the CPPCC. At the end I did not take up the invitation. In 1992 the Xinhua News Agency raised the matter again much to my surprise and I accepted immediately to become a member of the 8th National Committee of the CPPCC (1993-1998) as the political climate had improved by then. There began my association with the CPPCC.

CPPCC is an organization involved in peoples' patriotic unification in China. Established in September, 1949. The membership consists of people from all political parties, oversea Chinese and specially invited persons. There are over two thousand members. The past chairmen include Mao Ze Dong, Zhou Enlai, Deng Xiaoping, Dang Ying Chao, Li Xian Nian, and Li Rui Huan.

The National Committee meets once a year lasting nine to fourteen days. The main task of the CPPCC is to act as consultative body to the National Congress, exercise a supervisory role and participate and discuss about national politics.

In March, 1993 I took part for the first time in the first conference of the eighth session of the CPPCC (1,2,3). Through the kind hospitality of Mr. Fok Ying Tung, the Vice President of the Congress, we stayed at the six star Grand Hotel. I felt a little bit uneasy about such lavish treatment. During the Conference other than attending the deliberation of reports by the senior members of the Government and attending the opening session of the National Congress, we had discussion groups on the various reports. The papers delivered to us during the Conference were in such a quantity that even without sleep one cannot finish reading them within two weeks. As a result I did not have one restful night. When I attended the first Conference in 1993, I had difficulty getting use to the events during the Conference, for example, attending the meeting at the Hall of the People listening to the Premier, the Chairman of the CPPCC reading out their reports which took over an hour or so and listening to other members reading out their reports which were already distributed to us earlier. The discussion groups (4) allowed members to express their opinions on the various reports and their comments were selectively recorded and published in a summary information sheet which were then distributed to all two thousand odd members.

The CPPCC has other activities other than the annual conference, such as, study visits to the various regions in China. There are normally one or two such visits arranged for the members from Hong Kong and I make a point of participating in such visit whenever possible. In the last few years I have been to Guizhou Provence, Guangxi Provence and Chekjiang Provence (5,6,7,8). These visits gave me a chance to familiarize myself with the conditions of the various places and increase my knowledge on the mainland. Another activity takes the form of a briefing session by members of the Executive Committee on deliberations of the Executive Committee in Beijing. In addition special meetings are arranged to discuss matters concerning Hong Kong and the mainland.

The summary reports of these special meetings are passed onto relevant departments of the CPPCC for reference and further actions. There were six convenors for such special meetings during the eighth session and ten for the ninth session as the number of members were increased. I was one of the convenors for the two sessions and the Hong Kong "Tree Planting Day" activity was founded by the six convenors of the eighth session which indirectly created activity for the community (9,10). In addition the members in Hong Kong also participated in the China-Hong Kong Special Administrative Region First Government Selection Committee and the China- Hong Kong SAR Electoral College for 9th Peoples' Congress Deputies (see appendix (2)).

I made friends with a group of patriotic and successful persons in Hong Kong through the work of CPPCC (11,12,13) and through attending the annual conference I have the chance to meet up with the three architect members from the mainland every year (14). CPPCC is like a big family, the members share the same goal wishing to make some contribution to the motherland, I was fortunate to be re-appointed a member in the ninth session. I shall from now on be more diligent in serving the people.

香港房屋協會 (1981~　)
The Hong Kong Housing Society

香港房屋協會是一獨立自主非年利的民間組織，致力於以一般香港市民能負擔的價格，為他們提供住屋。1947年倫敦市長以市長的"空襲救災基金"向香港社會福利議會捐贈了一萬四仟英鎊，該會同意以該筆款項成立一個委員會，以及拓展房屋計劃，為工人和其家屬提供住屋。其後該會成立的有關小組委員會，成為香港房屋協會的前身。1948年由一班熱心公益的社會人士建議成立房屋協會，創會初期，適逢第二次世大戰過後，大量居民重返香港，房協的主要工作，因而成為提供極為需要的獨立居住單位，以低於市值的租金給收入低微的家庭。房協由1951年成為一法定團體，其所有的運作受房協自有的法規，由一執行委員會監管。執行委員會成員是由不同行業的社會精英組成，他們以義務性質參與。執行委員會屬下有多個小組委員會，例如建築及招標小組委員會、物業發展小組委員會、物業管理小組委員會、財政小組委員會、行政小組委員會及審查委員會。小組委員會的成員也是由社會精英組成，以義務形式參與工作。執行委員會的決策是由一班受薪的職員負責，由一執行總幹事統領。1988年有員工大約500多人。房協自1952年第一個出租屋村上李屋在深水落成後，開始自負盈虧以市值三分之一的價格向政府買地，興建其他發展項目，而建成的房屋是以市值的三成至六成出租給需要的市民。1974年房協得到政府低息貸款一億港元，推動了市區改善計劃，重建了多項市區舊樓。1977年在當時落成的祖堯村創辦了年長者宿舍，提供了130個單元。1988年推出"住宅發售計劃"，以低於市值的價格，為中下收入的家庭提供優質住屋。

1981年經鍾逸傑爵士的推薦，我加入了房協，並參加了執行委員會的工作，至今已17年。1991年當選為主席，連任至1994年。在任期間，推動改革，創辦了"智囊集會"與各委員會成員及高級行政人員進行一日的討論會，會上把房協的工作作一回顧及定下展望將來的方向。1993獲香港政府委託進行"夾心階層住屋貸款及住屋興建計劃"(1)，協助中等收入家庭置業。同年舉辦了房協首次的大型酒會，慶祝房協成立40週年(2,3,4,5,6)。1994年策劃首次向外融資發行20億港元的浮息票據(7)，使房協的建屋項目規模大增。此外，更推動房協利用儲備購買位於世界貿易中心28至30樓的三層辦公室為總部，把散佈各處辦公室的管理層，集中一處。我參與房協的工作已有10多年，在任主席期間，參與各委員會的工作及其它與房協有關的活動(8~18)(見附錄13)。最近幾年只參與建築及招標小組、提名小組、審查委員會及執行委員會的工作。總括來說，我對房協的工作，很有滿足感，因為通過我及其他委員及行政人員的努力，部分的市民能得到有瓦棲身。但因最近幾年政客當道，對房協的工作產生了不必要的障礙，使房協的工作倍加負擔，而我的心情更有"好人難做"之感。

1

3

2

4

The Hong Kong Housing Society is an independent non-profit making private organization, aiming to provide affordable house to the public. In 1947 the London Mayor's "air raid relief fund" donated to the Hong Kong Welfare Association £14,000. The Association established a committee using the fund to expand housing development for the workers and their families. That committee was the fore runner of the Hong Kong Housing Society (HKHS). In 1948 a group of warm hearted citizens founded the HKHS. It was right after World War Two and great number of residents were returning to Hong Kong. The main task for HKHS was to provide housing to low income families at lower than market rent. HKHS was established by ordinance in 1951 with its own constitution and supervised by an Executive Committee. The members of the Executive Committee compose of the elites from various professions and businesses from the community. They participate in the Society's work on voluntary basis and without pay. Under the Executive Committee there are several sub-committees such as the Building and Tenders Sub-Committee, The Development Sub-Committee, the Estate Development Sub-Committee, the Finance Sub-Committee, the Administration Sub-Committee and the Audit Committee. Members of all the sub-committees are also selected from the elites of the community and serve on voluntary basis. The policies of the Executive Committee are executed by the salaried administrative staff led by an Executive Director. In 1988 there were about five hundred staff. On the completion of the first rental estate at Sheung Li Uk, the HKHS became self supporting financially and start to buy land from the Government at one third the market value to develophe further estates renting them to needy families at thirty to sixty percent market rent. In 1974 HKHS received a one hundred million dollar loan from Government to carry out urban renewal projects. In 1977 HKHS pioneered the home for the aged in the then newly completed Cho Yiu Estate providing some one hundred and thirty units. In 1988 HKHS introduced the "home for sale" scheme offering lower than market price flats to middle and lower income families. I joined the HKHS in 1981 through the introduction of Sir David Akers-jones and immediately became a member of the Executive Committee and had been involved in the HKHS for the last seventeen years. In 1991 I was elected Chairman served for several terms until 1994. During my term of office I pushed for reforms and founded the "brain storming" session to allow members of all committees and the senior staff to participate in a day's seminar to review past performances and decide on future directions. In 1993

5

10

6

11

7

12

8

13

9

14

the HKHS was appointed by the Government to carry out the Sandwich Class Housing Schemes (1) to provide sale flats for the middle income families. In the same year we held a large cocktail reception for the first time to celebrate our fortieth anniversary (2,3,4,5,6). In 1994 I initiated our first ever outside loan in the form of a floating rate note issue for $2 billion (7) to facilitate increase production of housing projects. In addition we ultlised our reserve fund and purchased three storeys of office space in the World Trade Centre i.e., the 28th to the 30th floor partly as an investment and partly to use it as our headquarters housing all our central administrative staff who were otherwise scattered all over different locations. I have been involved with the work of the HKHS for the past seventeen odd years and have been involved in the work of all committees and sub-committees when I was Chairman (8~18) (see appendix 13). The last few years I was only involved in the work of the Building and Tenders Sub-Committee, the Nomination Committee, the Audit Committee and the Executive Committee. All in all I find satisfaction in the work in HKHS because through the work of all the members and the staff, some people get a roof over their head. However in recent years because of the activities of some politicians, the work of the HKHS became complicated and more troublesome and I cannot help feeling disheartened.

15

17

16

18

香港市政局 (1991~1995)
The Hong Kong Urban Council

市政局的職責是為市區居民保持環境衛生及提供文娛康樂設施和活動。市政局的決策由其行政機構市政總署執行。市政局的決策範圍甚廣，差不多涉及所有有關市區生活的事項。概括而言，市政局的職責分為下列三大方面。第一方面是公眾衛生服務，例如清掃街道、收集垃圾、檢驗食物及巡視食物店、防治蟲鼠、管理街市、屠場、墳場及火葬場，以及管制街上小販。第二方面是康樂服務，包括興建及管理公園、花園、公眾休憩用地、泳池場館、港島的公眾泳灘、各種戶內戶外的體育設施，以及推廣體育活動。第三方面是文化服務，管理在市區的主要演藝場所、公共圖書館及博物館，以及推廣和主辦文化與公眾娛樂節目。

市政局是經由12個專責委員會作決策，以及運用行政上的管制力。這些委員會分別是：政務委員會、建設工程委員會、文化委員會、娛樂委員會、財務委員會、圖書館委員會、酒牌局、街市及販商事務委員會、博物館委員會、公眾衛生委員會、康樂委員會、大球場董事局、市政局及覆檢委員會。所有專責委員會的會議均開放供市民旁聽。各專責委員會均每月開會一次，其轄下小組或工作小組通常亦如此。小組或工作小組是特別設立的，以深入研究所屬委員會複雜的或其爭論性的問題。各項提議往往要先由兩個或以上責任互有關聯的委員會通過，然後提交每月開會兩次的全局常務委員會核准。市政局的公開月會每月在市政局大樓舉行，而所有的決策由市政總署負責執行。40位市政局議員中，15位是委任議員，15位是民選的，以及10位來自區議會的議員代表。市政局議員都有本身的工作或事業，擔任市政局議員，他們並不支取薪金，但可每月領取津貼。我是在1991年被政府委任為市政局議員，直至1995年因政府取消委任制度，而我對競選的活動提不起興趣，所以就結束了我做議員的生涯。我總覺得在大庭廣眾前講你自己的優點，甚至講你競選對手的缺點要有相當的勇氣，而我是自己知自己事，是永遠提不起這樣的勇氣來，但總算"曾經擁有"，為市民做了點事。在任期間參與了多個委員會及工作小組的工作(見266頁)。我也對多項市政的建設提了修改意見，包括大會堂重建、茶具博物館新翼、軍事博物館及中央圖書館等。在中央圖書館項目中，我提議公開設計比賽，而且受主席所托，進行了比賽的籌備工作，得市政署及建築設計署同意認為設計比賽是可行的。但因當時圖書館委員會主席堅持為了盡快開工，所以設計比賽計劃告吹。但中央圖書館因種種原因直至1998年才開工，而且所設計的後現代派外貌引起高度爭議。從這一個工作效率的例子，可以看出我對市政局的工作興趣不大的原因之一，同時也表現出我在市政局能發揮我的專業經驗而使市民受益的作用不大。主席當時為了安撫我對失去舉辦中央圖書館設計比賽的不安，同意了我另一提議，就是舉辦兩個公廁的設計比賽。這個比賽順利舉行並建成了兩座設計優美的公廁，一座位於維多利亞公園內，另一在山頂(1)。我通過市政局的工作，認識了一群熱心公益，各行各業的朋友(2)。而且通過會議工作，使我在這方面的經驗倍增，確是一個難得的機會。單在寫作，每年一次在年會上的陳詞對我以後寫這類的文章有很大的幫助，我寫了兩篇比較滿意的陳詞(見322頁)，當時報章上也有表揚(見附錄11)。市政局的活動包羅萬有，包括到日本考察(3)、到北京考察(4)、順德考察(5)及以主禮嘉賓的身份代表市政局參與多項市政局主辦的活動(6,7,8,9,10,11,12,13,14,15,16,17)。1995年功成身退後，與幾位市政局認識的議員好友組織了一個叫"無得撈噃"會，敘餐談歡，是我在市政局工作中的最大收穫。

The functions of the Urban Council are to maintain a healthy environment and provide sports and cultural activities within the urban area of Hong Kong. The policies of the Council are wide ranging and are executed by the Urban Services Departments. The policies involve matters related to the living requirements of the city. In general the duties of the Urban Council can be divided into three major categories. The first category concerns public health such as, street cleaning, collection of refuse, food inspection, food outlet inspection, pets control, management in food markets, abattoirs, cemeteries incinerators and hawkers control. The second category concerns sports and recreations such as, the building and management of parks, gardens, public open spaces, swimming pools, beaches, indoor and outdoor sports facilities and promotion of sports activities. The third category concerns cultural activities such as, management of all major urban performing centres, public library, museums and promotion of cultural and entertainment programmes.

The Urban Council has fourteen select committees which decide on the operational policies and maintain supervisory functions. They include the Administration Select Committee, Capital Works Select Committee, Cultural Select Committee, Entertainment Select Committee, Finance Select Committee, Library Select Committee, Liquor Licensing Board, Markets & Street Traders Select Committee, Museum Select Committee, Public Health Select Committee, Recreation Select Committee, Board of Governors of the Hong Kong Stadium, Public Accounts Committee and Review Select Committee. All the meetings of the Select Committees are open to the public and take place once a month. The Sub-Committees set up under the Select Committees are also operate in the same manner. The Sub-Committees and work groups are set up to work on special and complicated issues. Any proposal will normally be studied and passed by two or more related sub-committees before the presentation at the twice a month Standing Committee of the Council for adoption. The monthly Council meeting is open to the public. There were forty councillors, fifteen were appointed, fifteen were elected and ten came from the District Boards. Most of the Councillors have their own work and profession and they participated in the work of the Urban Council without salary but with a monthly allowance.

I was appointed an urban councilor by the Government in 1991 until 1995 when Government abolished the appointed system. As I was not interested in the election process, there ended my urban councillor's work. I also feel that you need great courage to boast about your ability in public and even

1

6

2

7

3

8

4

9

5

10

to point out the shortcomings of your opponent. I know my own weakness and I will never have enough such courage. In any event I already had the taste of an urban councillor and perhaps had the opportunity to do a little good deeds for the community. That is enough for me. During my term as councillor I participated in the work of a number of committees (see page 266). I had also make suggestions to revise the designs of the City Hall, the Tea Museum, the Museum of Defense and Central Library. In the case of the Central Library I even proposed a public competition and was instructed by the Chairman to prepare for such a competition. I even managed to convince the Urban Services and the Architect's Services Department that it was a feasible task. Unfortunately the Chairman of the Library Select Committee at the time insisted on early commencement of the construction work and my proposal for the competition was dropped. However because of many other reasons, construction work on the Central Library was delayed and was only commenced in 1998. The final design which is of post modern nature led to many criticism. From the handling of this matter, one can easily see why I became disinterested in the work of the Urban Council and at the same time one can see that there is limited opportunity where I can contribute my professional expertise. The Chairman was most concern over my disappointment at the time and to give me a slight comfort, he offered to me to organize a design competition on two public toilets instead. This design competition went off smoothly and produced two beautifully designed toilets at the end, one was situated in the Victoria Park and one at the Peak (1). Through the work of the Urban Council, I came to know a bunch of friends from various walks of life who were devoted to community work (2) and through the committee meetings, my experience in polities was greatly enlarged. It was indeed a rare opportunity in life. In writing ability alone I had benefited greatly as one has to produce an annual speech to be delivered at the annual opening meeting of the Council. I was quite satisfied with two of such speeches (see page 322) and one of which was reported in the press at the time (see appendix 11). There were all kinds of activities with the Urban Council including study visits to Tokyo (3), Beijing (4), Shunde (5), and representing the Urban Council at the openings of various functions (6-17). I left the Urban Council in 1995 and formed a group called the "Motorola Fellowship" with a number of ex-councillor friends to meet now and again over a glass of wine. This was my best "gain" from the work of the Council.

11

15

12

16

13

17

14

敏求精舍 (1980～　　)
Min Chiu Society

1

6

2

7

敏求精舍乃由本港中國文物收藏家組成，於1960年創立，旨於交流文物鑑賞的經驗及心得，並志於弘揚中華文化，提高社會人士對文物鑑賞的興趣。

"敏求"一詞源於《論語.述而篇》："我非生而知之者，好古敏而求之者也。"本會的學術活動繁多，除每月舉行定期聚會兩次，讓舍員及同好展示新購藏品或暢談研究心得以外，並經常邀請著名學者作專題演講，又安排海外團體、訪客到會舍相聚。近年來更為舍員舉辦了多次學習團往英國、美國、法國、日本、新加坡、台灣及中國大陸等地參觀藝術館，使舍員對世界各國重要中國文物收藏有更深入認識，並藉以加強本舍與各地藝術館之聯繫(1,2,3)。敏求精舍過去曾多次與香港藝術館及香港大學合作舉辦舍員藏品展覽，希望藉此向普羅大眾推廣文物鑑賞的風氣，發揚我國的藝術及文化。

3

8

我是於1980年加入敏求。1990年是敏求成立30週年，我也參加了當日的紀念晚會(4)。1990年及1991年當選為副主席，1992年及1993年任主席。在任期間推廣文物收藏之交流及統籌與香港市政局合辦的兩個大型文物展覽，1990年展覽主題是"歷代文物萃珍"，1992年是"清朝瑰寶"。1990年因市政局經費有限，編寫與展覽有關的大型圖書，就交由我會負責，而我就擔當了主編。1993年統籌與美國Arthur M Sackler Gallery合辦的敏求會員藏品展，此展覽為敏求精舍第一次在海外舉辦展覽，把中國文物交流推廣到海外。

我和太太在收藏現代畫的過程中，認識了一群國內的藝術家，而且建立了永生難忘的友情，給收藏的意義增加了色彩。有幸認識的有朱紀瞻(5,6,7)、黃永玉(8,9)、黃冑(10,11)、唐雲(12,13)、程十髮(14,15)、黃苗子、王世襄(16,17)、劉海粟(18)、關山月(19)及馬伯樂(20)。

4

9

5

10

The Min Chiu Society was founded in 1960 by a group of chinese antique collectors in Hong Kong. The founders were not only interested in the exchange of information and connoisseurship of chinese art but also in the promotion and appreciation of chinese antiquities in society. The words Min Chiu were taken out of the Analects of Confucius in which the Master said, "I am not the one who was born in the possession of knowledge, I am one who is fond of antiquity and earnestly seek knowledge there in". The Society has two regular meetings a month in its premises in which members and friends gathered to show each other their latest acquisitions or simply to exchange views on the various aspects of their collection. On many past occasions eminent speakers were invited to give talks on specific subjects and there is a regular stream of visitors to the Society's premises from various museums and academic institutions around the world. In recent years, the Society has organized study tours abroad to visit museums such as those in the United States, United Kingdom, China, France, Japan, Singapore and Taiwan. These tours enabled members to view the major collections of chinese antiques in the world as well as to establish relationships between the Society and the Museums. (1,2,3) The Min Chiu Society has in the past co-operated with the Hong Kong Museum of Art and the Hong Kong University gallery in mounting numerous exhibitions of its members' collection to promote connisseurship as well as general understanding of chinese art and culture amongst the general public.

I joined the Min Chiu Society in 1980. In 1990 Min Chiu Society celebrated its 30th anniversary and I participated in the commemorative dinner (4). I was elected Vice-Chairman in 1990 and 1991 and Chairman in 1992 and 1993. During my term of office, I continue to promote exchanges in antique collection and organized two major exhibitions with the Urban Council. The theme for the 1990 was "Selected Treasures of Chinese Art" and for 1992 was "Splendour of the Qing Dynasty". In 1990 the Urban Council was short of funds and I was asked to act as editor for the exhibition catalogue. In 1993 I organized the joint exhibition of our members' collection in the Arthur M. Sacker Gallery. This exhibition was the first oversea exhibition of our members' collection extending the cultural exchange to outside of Hong Kong.

In the course of collecting modern Chinese paintings, my wife and I came to know a number of artists in the Mainland. We have established long lasting friendships with them giving the art of collecting another dimension. We are fortunate to have met Zhu Qizhan (5,6,7), Huang Yongyu (8,9), Huang Zhou (10,11), Tang Yun (12,13), Cheng Shifa (14,15), Huang Miu Tze and Wang Shi Xiang (16,17), Lui Haisu (18), Guan Shanyue (19) and Ma Bola (20).

11 16

17

12

18

13

19

14

20

15

香港政府屬下委員會(1977~　)
Hong Kong Government committees

我從1977年開始被邀參與香港政府屬下的委員會工作，多數的委員會是與建築有關。參加委員會的工作，一方面對社會做些事，另一方面可以通過各委員會的工作，充實我對本行運作有更深入的理解。所以逢政府有關部門邀請我參與任何委員會，我都樂於接受。這20多年間，我參與了不下18個政府委員會的工作，直至現今只有一、兩個還擔當委員的工作。因為這些委員會的工作有部分是要保密，有部分十分沉悶，所以我也不想多提，只把下列主要委會記錄下來。

1) 香港律師紀律委員會委員 (1993~　)
2) 香港政府旅館業及社會上訴委員會委員 (1993~　)
3) 中國香港特別行政區第一屆政府推選委員會委員 (1996~1997)
4) 中國香港特別行政區第9屆人大代表選舉會議成員 (1997)
5) 香港市政局議員 (1991~1995)
6) 香港市政局博物館委員會副主席 (1992~1994)
7) 香港市政局全局常務委員會會員 (1991~1995)
8) 香港市政局文化委員會委員 (1991~1995)
9) 香港市政局土地用途小組委員
10) 香港市政局重建大會堂工作小組主席
11) 香港市政局綠化香港工作小組委員
12) 香港市政局架空鐵路計劃工作小組委員
13) 香港政府土地供應特別委員會委員 (1982~1985)
14) 香港政府建築設計上訴法庭委員會委員 (1982~1997)
15) 香港政府法庭審裁顧問 (1980~1995)
16) 香港政府註冊建築師及工程師執業委會委員 (1978~1981)
17) 香港政府修改消防條例委員會委員 (1977~1980)
18) 香港政府消防員及註冊建築師執業委員會主席 (1977~1980)

I started to participate in government committees' work in 1977. Most of the committees are connected with architecture. Participating in government committee's work can serve the community as well as through the work can have better understanding of the workings of one's profession. That is why I always gladly accept any approach from the Government for such work. In these twenty odd years I had taken part in not less than eighteen of such committees. Even to-date I am still involved in one to two committee's work. Part of the work of these committees are confidential and part very boring. I shall not go into the details here except to list below their names for record.

1) Member, Solicitors Disciplinary Tribunal Panel (1993~　)
2) Member, Hotel & Guest house Accommodation Appeal Board and Club (Safety of Premises) Appeal Board (1993~　)
3) Member, China-Hong Kong Special Administrative Region First Government Selection Committee (1996~1997)
4) Member, China Hong Kong SAR Electoral College for 9th Peoples' Congress Deputies (1997)
5) Urban Councilor, Hong Kong (1991~1995)
6) Vice Chairman, Museum Select Committee, Urban Council, Hong Kong (1992~1994)
7) Member, Standing Committee of the Council, Urban Council, Hong Kong (1991~1995)
8) Member, Cultural Select Committee, Urban Council, Hong Kong (1991~1995)
9) Member, Land use Sub-Committee, Urban Council, Hong Kong
10) Chairman, Redevelopment of the City Hall Sub-Committee, Urban Council, Hong Kong
11) Member, Greening of Hong Kong Sub-Committee, Urban Council, Hong Kong
12) Member, Overhead Rail Development Sub-Committee, Urban Council, Hong Kong
13) Member, Government Special Committee on Land Supply (1982~1985)
14) Member, Government Building Appeal Tribunal (1982~1997)
15) Lay Assessor, Magistrate Court (1980~1995)
16) Member, Authorized Persons and Registered Engineer Registration Committee (1978~1981)
17) Member, Government's Revision of the Fire Code Committee (1977~1980)
18) Chairman, Fire Services Department/HKIA Liason Group (1977~1980)

建築設計評委工作 (1974~　)

Work of design competition juror

我對建築設計比賽，十分感到興趣。因為通過比賽，在同一條件及要求下，可以看到不同的建築師所作出的設計，從而挑選出最優秀的建築師來執行項目的設計。而且年青的建築師也可以利用設計比賽的機會，來揚名及獲得工作，多位世界著名的建築師都是從設計比賽得獎後起家的。除了有興趣參加設計比賽之外，我對當評委及籌備設計比賽的工作，也有相當的興趣。自從1974年參加了評選香港建築師學會銀牌獎至今已參加了大約17個設計比賽的評委工作，及中獎項目實施的專家顧問和負責籌辦兩個設計比賽，但我總覺得，國內的設計比賽雖然在比賽條件上已可以追得上國際水平，但在比賽的運作上，業主有時對評委的決定不太尊重，且作出與原來條件有異的安排。加上因有些組織比賽的有關單位，對國外的建築師單位不太認識，多數依靠個人關係介紹，所以往往參賽單位的水平低劣，有國內建築師在海外註冊的假外國單位，今後應該由一專家小組研究，提供一個認可的外國建築師單位名單提供給各省各市作為參考。另外有關單位應發指引材料，促使業主單位依照比賽條件辦事，以免外國設計單位對我國的設計比賽失去信心。

以下是我從1974年起參與的項目：

1) 香港建築師學會銀牌獎評委 (1974及1981)
2) 香港國際"山頂"建築設計比賽，評委及籌辦人 (1982)(1)
3) 香港演藝學院建築設計比賽，評委及發起人
4) 中國福州元洪城設計比賽，評委 (1992)(2)
5) 《建築師》雜誌在哈爾濱主辦的"建築師杯"全國中小型建築優秀設計評選，評委 (1992)(3,4)(見附錄(14))
6) 香港市政局公廁設計比賽發起及籌辦人 (1992)
7) 中國深圳市航天廣場設計比賽，評委 (1993)
8) 《建築師》雜誌第二屆"建築師杯"優秀建築設計評選，評委 (1994)(5)
9) 中國深圳市中心區購物公園設計比賽，評委主席 (1996)
10) 中國深圳市中心區中軸綫公共空間黑川紀章的設計，專家顧問 (1997)(6,7)
11) 中國深圳市中心區李明義設計的水晶島及廣場規劃，專家顧問 (1997)
12) 中國深圳市中心住宅小區設計比賽，評委 (1997)
13) 中國深圳市廣播中心國際設計比賽，評委 (1997)
14) 中國深圳市中心區市民廣場設計，專家顧問 (1997)
15) 中國深圳市中心文化中心國際設計比賽，評委 (1998)
16) 中國深圳市沙頭角海濱區規劃國際設計比賽，評委 (1998)(8,9)
17) 中國北京國家大劇院國際設計比賽，評委 (1998)(10, 11,12)

1

2

I am very interested in architectural competition because through competition under one set of conditions and requirement, one can see a variety of solutions and can pick the best design for execution. Young architects can make their names by winning competitions and many world famous architect started their career by winning competitions. Other than being interested in participating in design competition, I am also interested in acting as competition jurors. I started my juror "career" in 1974 participating in the HKIA Silver Medal Award. Since than I have acted as juror for about seventeen competitions and twice as adviser to winning entries and was responsible as organizer for two competitions. However I felt that the competitions in the mainland although the conditions are near to international standards, the client body sometimes does not respect fully the deliberations of the jury and make alternate arrangement. In addition the competition organizers sometimes are unfamiliar with architects' practices overseas and rely mainly on personal recommendation for selection of practices for invited international competitions. As a result many participating oversea practices are of poor standards. There are those practices with architects from the mainland registered overseas and pretend to be foreign practice. I suggest that one should ask an expert committee to form an approved list of oversea practices so that it can be a general reference for competition organizes in the mainland and to ask relevant government body to issue a directive to the client bodies to honour the Jury's deliberation to avoid overseas practices getting disinterested in the participation of competitions in the mainland.

The following is a list of design competition I have participated since 1974.

1) Juror, HKIA Silver Medal Award (1974 & 1981)
2) Organizer and Juror, the "Peak" international design competition (1982)(1)
3) Proposer and Juror, The Academy of Performing Arts design competition, Hong Kong
4) Juror, the "Yuan Hung City" design competition, Fuzhou, China. (1992)(2)
5) Juror and Sponsor, the National competition for medium and small projects designs organized by China Architectural & Building Press, Harbin, China (1992)(3,4) (see appendix (14))
6) Organizer, Urban Councils Public Toilet competition, Hong Kong (1992)

3

8

4

9

5

10

6

11

7

12

7) Juror, the "Hang Tien" Plaza design competition, Shenzhen, China. (1993)
8) Juror, the second "Architect's Cup" best design award organized by the China Architectural & Building Press, Shenzhen, China. (1994)(5)
9) Chairman of the Jury, the Shopping Park design competition, Shenzhen Central Area, China. (1996)
10) Adviser, Central Axis design for the Central area by Kisho Kurokawa, Shenzhen, China. (1997)(6,7)
11) Adviser, Crystal Island and Plaza design by John Lee, Central Area, Shenzhen, China. (1997)
12) Juror, Residential District design competition, Shenzhen Central Area, China. (1997)
13) Juror, Broadcast Centre international design competition, Shenzhen Central Area, China. (1997)
14) Adviser, Citizen Centre by John Lee, Shenzhen Central Area, China. (1997)
15) Juror, Cultural Centre international design competition, Shenzhen Central Area, China. (1998)
16) Juror, Master plan of Shatoujiao Seashore Area of Yantien District international design competition, Shenzhen, China. (1998)(8,9)
17) Juror, National Grand Theatre International design competition, Beijing, China. (1998)(10,11,12)

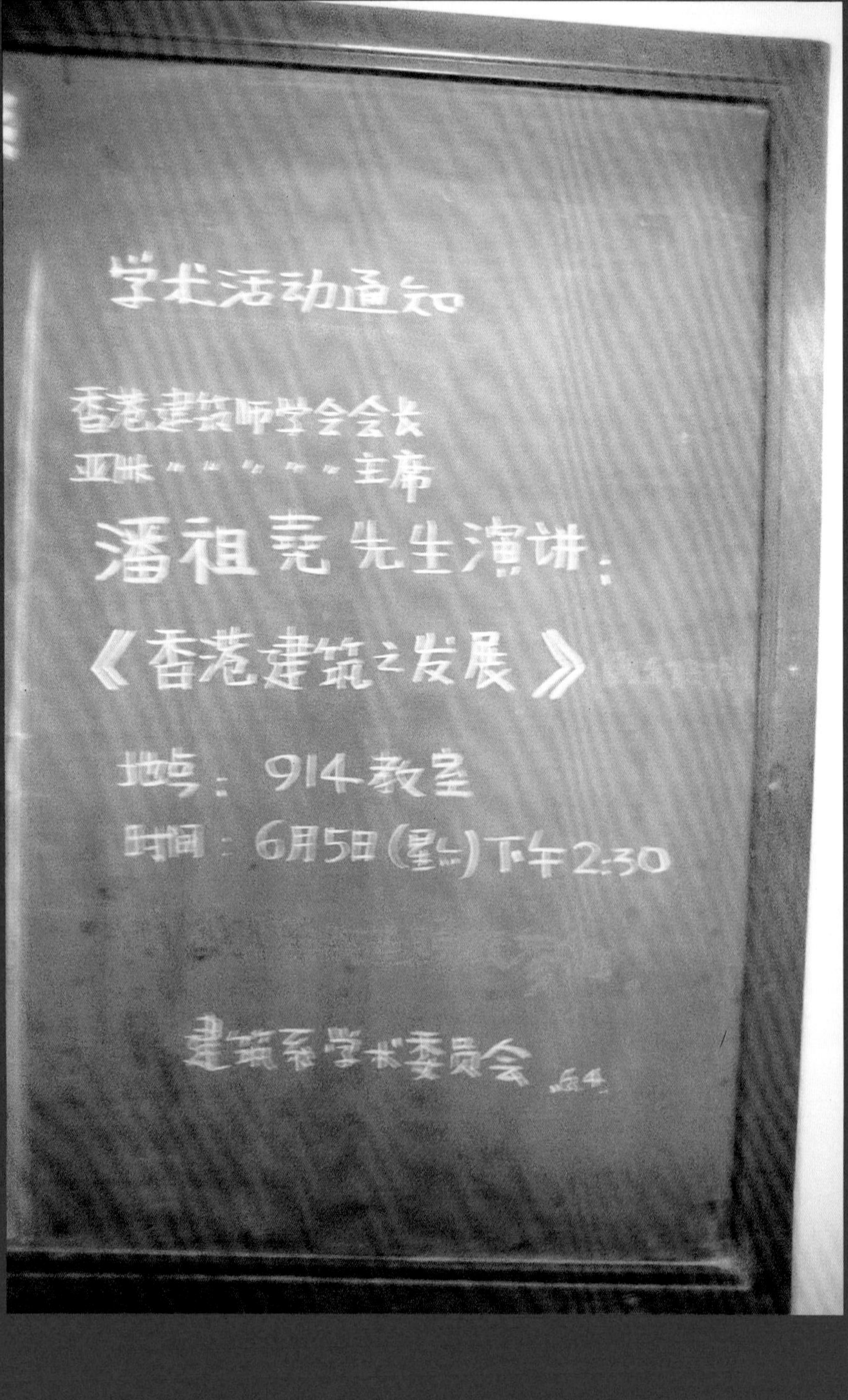

学术活动通知
香港建筑师学会会长
亚洲 〃 〃 〃 〃 〃 主席
潘祖尧先生演讲：
《香港建筑之发展》
地点：914教室
时间：6月5日(星六)下午2:30
建筑系学术委员会 6.4

建築教育

Architectural education

我對有關建築教育的活動十分感到興趣。從60年代后期，我從英國回港不久，就參加了由一班熱心的朋友統籌的夜間再教育的業餘課程，我負責執教建築歷史及設計課程。我覺得作為一個建築師，可以從教學的活動中，不只能把自己的經驗傳送給青年一代，更可以從准備教材及與學生交流中檢討自己的理論與經驗。1978年及1979年我在香港大學建築系任兼職講師，1977年及1981年兩次參與了英聯邦建築師協會、英國註冊建築師委員會及香港建築師學會合辦的審核香港大學建築系課程委員會的工作，這個工作是每5年進行一次，目的是評審香港大學建築系的課程是否達到一定的水平，使建築系能保持在英聯邦建築師協會的核準建築學院名單內。此外，我也曾在1977年及1978年參加了香港建築學會的建築教育委員會及執業考試委員會的工作，在80年代參加了英聯邦建築師協會在斯里蘭卡的建築教育會議及亞洲建築師協會在香港的建築教育會議。從1980年開始到大陸多個城市的大學建築系作學術報告，包括：清華大學(1,2)、同濟大學、華南理工大學、西安建築大學、深圳大學等。90年代因工作繁忙，只能抽空作學術報告，包括1997年9月在清華大學講述"西西工程"4及5號樓的設計及1977年8月在香港大學建築系以"回顧及展望"為題的報告。1993年獲聘為華南理工大學顧問教授(3,4)。

I am very interested in architectural education. In the late 60 shortly after my return from England, I joined a group of friends who organized some evening classes on continued education and I was responsible to conduct classes on history of architecture and design. As an architect one can benefit from involvement in architectural education not only in being able to pass on your knowledge and experience to the younger generation but also through the preparation of class material and exchanges with students, one can re-examine one's theory and review one's experience. In 1978 and 1979 I took up a part time lectureship in the School of Architecture at the Hong Kong University. In 1977 and 1981 on two occasions I was a member of the Hong Kong Institute of Architects (HKIA)/Commonwealth Associations of Architects (CAA) and Architect Registration Council U.K. Visitation Board to the school of Architecture, University of Hong Kong. This Visitation Board review the academic standards of the School every five years and see if it has maintained a standard suitable to continue being listed in the CAA list of Schools. In 1977 and 1978 I participated in the HKIA Board of Education and the HKIA Professional Examination Committee. In addition in the early eighties, I participated in the CAA Board of Architecturalo Education conference in Colombo, Sir Lanka and the Architects Regional Council Asia's Board of Architectural Education conference in Hong Kong. My lectures tours in China started in 1980 and since then I had given lectures in architectural schools in several major universities including Tsing Hua University, (1) Tong Chi University, South China University of Technology, Xian Metallurgy University, Shenzhen University etc. In the nineties I could only manage to deliver a few lectures due to heavy workload including a talk on "Design for Sites 4 and 5 of the Xi-Xi Projects" at the Tsing Hua University in September 1997 and "Reflections and the way forward" in the Hong Kong University in August 1997. In 1993 I was appointed Advisory Professor by the South China University of Technology.

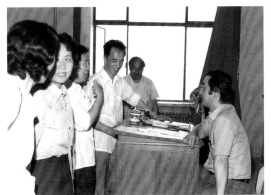

2

3

4

已發表文章及學術報告選輯
Published articles & academic papers

1. Architects and environment of Hong Kong, 1982 — 276
2. 香港建築設計現況及未來發展，1982 — 278
3. 淺談香港現代建築的發展，1983 — 279
4. 淺談世界第三代建築師及後現代派與中國建築設計發展問題，1984 — 282
5. 香港建築特刊序言，《世界建築導報》1985 — 295
6. 淺談香港及內地的現代建築風格，1985 — 297
7. 淺談香港私人建設前景，1985 — 304
8. 一個香港建築師眼中的現代中國建築，1987 — 306
9. 聯系各界，發展科技，1988 — 310
   Unite all front to develope Science & Technology", 1988 — 312
10. 《香港建築》序言，1989 — 314
11. Obituary of "Mr. Dai Nianci 1920-1991", 1991 — 315
12. 在中國建築學會第8次全國會員代表大會暨學術年會上的發言，1992 — 316
13. HKIA - ASC Linkage, 1992 — 317
14. 擺脫束縛、開展評論，1995 — 320
15. 建築風格與古都風貌，1995 — 321
16. Speeches delivered at the Urban Council annual debate, 6/1/92 及 12/01/1995 — 322
17. 在建築論壇第二次研討會上的開幕辭，1996 — 324
18. Reflections and the way forward, 1997 — 325
19. 譜寫香港的新歷史，1997 — 329
20. Design for Sites 4 and 5 of the "Xi-Xi Project", 1997 — 330
21. 建築設計——現狀與改革，1998 — 334
22. 下真功夫，走向世界，1998 — 336
23. 《香港著名建築師作品選》前言，1998 — 337
24. 古為今用——現代化傳統建築設計探索，1998 — 338

Architects and Environment of Hong Kong
Originally published in Asian Architect & Builder, 01/1982

The environment in Hong Kong is hardly in keeping with its being one of the most successful financial centres of the world. There is a lack of proper planning, be it in housing provision, education or transport. We are now faced with constant housing shortage, traffic jams everywhere, acceleration of crime rate and above all, a growing discontent amongst the middle and lower incoming groups who make up 80-90% of our population.

Though most of the architects here in Hong Kong are willing and able, we are prevented from making a significant contribution to the quality of our environment by two major stumbling blocks:

The Client
The Government

THE CLIENT
There are many types of clients and of course many types of architects to serve them. This perhaps applies to the medical and legal profession as well, but very few patients will tell their doctors how best to operate on them and very few clients will interfere with a lawyer's work normally. Most of our clients, however, come with their ill informed preferences and preconceived and outdated ideas and insist on cut fees, or partial payment of fees. More often than not we are treated like draftsmen with magazine clippings and other architects' design thrown at us for copying work. Most clients in Hong Kong are still very much insular, and few are informed of the latest developments in the more advanced countries such as USA, England, Canada etc. That of course applies to some of our architectural design in Hong Kong is inevitably about 10-15 years behind time. Most developers do not have a social conscience. Their only concern is quick profit. Little do they realise that with such acute shortage of land supply, a chance lost in capitalising on the maximum potential of a site is a loss forever. Not many projects can follow the recent Wanchai example of a residential block being demolished right after the issue of occupation permit for an office block replacement. Little do some of the developer realise that by cutting the proper fees for our architects, he is set to lose more than the amount of fees saved. More people in Hong Kong now recognise the pestige of luxury goods brands such as Dunhill, Dupont, Cartier and seldom bargain on such purchase but most developers still have not waken to the fact that less fees inevitably means less quality, less service and above all less profit. Those architects or signature architects as they are sometimes referred to as such, more often than not work only as draughtsmen of general plans and their type of offices are rarely in a position to produce proper working drawings let alone act properly as the co-ordinator of the whole development work. In a booming property market like Hong Kong most of the developers are not concerned with the quality of their buildings and all the defects of a cut fee project are often passed onto the purchasing public. In Hong Kong through tradition and ignorance very few property purchasers are claim conscious and therefore most developer get away with it. When one erects buildings such as the Causeway Bay Centre and some of the buildings down in Western and North Point, he is adding further instant slum in the already deteriorating urban fabrics.

It is laughable when you do a simple calculation and find out that nowadays the average architect's fee on a $100-200 million project is but only 1.3% of the total development cost excluding land cost etc as architect's fees are based on total construction cost which works out nowadays to about 1/3 of total development cost.

For a project with a $100 million construction cost and $200 million land and other costs taking roughly about three years' development time, architect's fees amount to 0.43% per annum on total development cost while interest alone would cost the develop well over 20% per annum or $170,000 per day. For the cut-fee-architect, it is certain that he can not afford to devote sufficient time to what a project demands. (A typical set-up would be an office of 10 with one principal running 60-80 jobs at the same time, 20-30 of which may be under construction ie, he could barely spare one hour's attention to each job even if he runs a 60 hour week, 40 hours per week) Under such circumstances, delay and omission are inevitable. With $170,000 per day interest, delays are detrimental to the profit margin. In other words, the proportion of architect's fees to the cost of the overall development is like the proportion between the cost for a set of standard tyres to that of the whole Rolls Royce. Some mean owners might want to make a saving in getting cheaper replacement tyres which might cause a collision. The saving in the tyres cannot possibly pay for the repairs.

Only by using the service of a qualified architect and paying him proper fees can a developer be assured he best profit for his development. Naturally one stands a better chance of securing the service of a proper architect if the developer selects him out of our membership as all our members are governed by a very strict code of professional conduct and proper service agreement.

THE GOVERNMENT

With Government, it is not so much the problem of outdated legislation though such has definitely affected the quality of our environment but the interpretation and procedure as exercised by Government servants that is detrimental to quality development in Hong Kong. Unfortunately the Building Legislation was originally written to be executed with a good degree of discretion as no one building site is exactly the same, and as the art of building design is not an exact science.

The present attitude of Government officials has caused chaos in the building industry. During the last two years the attitude of Government officials has changed from bad to worse: whether it is towards us the professional or towards the interpretation of regulations. The sense of co-operation has gone. The reasons they try to explain away such attitude include shortage of staff, pressure of work and the hint of the ICAC in the background. The shortage of staff is acute but even with that problem dealt with, their attitude is unlikely to change. Shortage of staff has one humiliating effect on experienced professionals who have to spend more time now arguing with less experienced officials and getting nowhere.

It can be safely said that not a single active architect's firm in Hong Kong can claim that they have not experienced some `unreasonableness' in their dealings with Government Departments in the past two years. We professionals are not born overnight while some officials can materialise themselves within 24 hours from abroad. Our experience is always there, the pattern of development may have changed a little over the years but the Government official must bear the full responsibility for the wind of change. It's often hinted that the success of ICAC has something to do with it. If anti-corruption encourages work to rule, disrespectful to professional people and careless attitude, then there is definitely something very wrong somewhere.Upon realising the gravity of the situation, the Institute of Architects appointed a working party to study the problems involved with Government Departments at the beginning of this year and received submissions from member firms of well over a hundred cases within a short period of one month. A report has been submitted to the Secretary of the New Territories as well as the Director of Public Works. So far with he New Territories Administration we have been offered some very constructive remedial measures. As regards the Public Works Department, they have given us the impression that in the majority of the cases the complaints against the Building Ordinance Office were not justified and only 10 cases `appear to have some justification'. Bearing in mind that this is the first time ever a professional Institute has produced an `en-mass' complaint to Government and that many of us cannot afford, understandably, to have the courage to complain and that there are enough valid cases to warrant a constructive review, the PWD gave us a typical bureaucratic reply. We are naturally disheartened.But we are not alone in this. This edition of the South China Morning Post remarked in his Editorial on the 20th May 1981 "In the Public Works Department, for example, there is said to be an increasing trend to `work to rule' ostensibly because even the slightest departure might incur questions or investigation. Major projects in which many millions of dollars are involved are being held up for weeks and in cases months by an insistence on a punctilious observation of all the various requirements where in the past a degree of flexibility was permitted, without contravening standards. This is costing developers considerable sums of money." And lately theoutspoken Legislative Councillor Miss Lydia Dunn made a plea in Leggco "for improved and simplified procedures in the administration of the Building Ordinance to improve productivity in the building industry." She also directed our attention to "the many complaints voiced about the time it takes for building plans to be approved and occupation permits issued. " What she implied there is obvious but what she may have realised but not in a position to point out is that even with improved systems the problem of `human frailty and error' to quote from a recent Commission of Inquiry report will invalidate all constructive renovation. Upon registration as an Authorised Person, the architect would have at least seven years of professional education and training behind him, more than any other professionals concerned with the building industry and yet our work and our opinion have still to be examined by a fellow professional in the government service who on occasions might be less qualified than us. While Government Architects working for Government building projects are exempted from such procedures, private practitioners working on private or semi public projects have to subject ourselves to such scrutiny. Surely academically and practical experience-wise, Government architects are the same as private practitioners. Though many would say that commercial pressure will discourage proper conduct, every architect has a prison sentence possibility hanging over him should he misbehave. I would further urge Government to establish an Architects Ordinance to regulate proper behaviour of architects, so that Government need not apply double standards for their own architects and private practitioners.

With the constant shortage of staff and a deteriorating quality of performance of the Government Departments, it is about time Government take an urgent and serious look at the issue and save the tax payers' money by avoiding a possible `Suicide and a Commission of Inquiry'.

香港建築設計現況及未來發展

原載《經濟導報》總1751‑2期,1982年1月1日出版

近年香港的建築水準不斷提高,但設計方面只不過是逐漸跟上歐美、日本等先進國家的技術水平,香港建築界仍未有什麼創舉。這種情況除了因時間及則師費的限制外,業主對建築設計方面的認識及接受能力也是原因之一。很多時,建築師往往只是充當業主的繪圖員,被指派抄襲某雜誌或其他則師的設計。因為部分香港業主及建築師眼光短淺,所以香港一般的建築設計都比先進國家落後了一至十五年。更有不少業主只求急功近利,沒有想到他們的草率建設,對香港的環境及居民生活都有所影響。在寸金尺土的香港,實在浪費了很多有用的土地資源。

舉例來說,現時在香港十分流行的玻璃幕牆,早於十多年前已經在外國使用,而且發現很多弊處,例如抗熱和抗冷效果不好,以致耗費能源。處於熱帶地區的香港,寫字樓並不適宜採用玻璃幕牆作為建築材料。其實,玻璃幕牆的採用,需要考慮實際環境。背向陽光而又希望可以看到海景的話,玻璃幕牆是適宜採用的,但在對向太陽的位置,還是使用石屎水泥為佳。香港業主為求省時間、省金錢及省功夫,未有顧及實際環境而濫用玻璃幕牆,致使住戶及小業主蒙受損害。

好的建築設計並非奢侈,而是必需的。但業主們卻要討價還價,不接受"一分錢一分貨"的道理。減收費用的建築師或簽字則師只能應付工務局批則的要求,而對施工圖及管理工程的責任不能兼顧。當然,由於業主的寒酸苛求態度所引致建築物設計和施工的缺點,全都需由買樓者負起。香港一般買樓人士,暫時仍未察覺到這方面的損失,所以業主仍然可以偷工減料,不顧環境和設計水平問題,以致在灣仔、西環、北角某些樓宇有徙置區形式的設計。

香港仍保留有中國民族的特色。同時,香港居民具有特別的生活方式,結合熱帶地區的氣候環境,應該可以在建築設計中,創造出香港獨有的特點。太平山頂的"老襯亭",可算是富有香港地區風味的代表建築物,香港在這方面的建築設計,着實需要大力發展。

另外,建築師費用在整個物業發展費用當中是佔很小的比例。例如,以一個一億港元的建築工程來說,建築師的費用只是佔總投資額百分之一點三。因為建築師的費用是根據建築工程的費用而定,並不包括地價,現時的建築費用約佔總投資額三分之一。這類工程多數需時三年,而每年建築師費用只等於全部投資額的百分之零點四三,但銀行貸款息口方面已超過全部投資額百分之二十以上,或者是每日十七萬元的利息負擔。建築師費用若果低於建築學會訂下的最低收費率,建築師往往不能細心處理每一工程。同一則師樓可有六十至八十個工程,亦可有二十至三十個是在施工中,就算建築師能夠在一個星期內工作六十小時,也不可能於一個星期內花費一小時在每一工程上。因此,錯誤百出便很自然,以每天息口十七萬元來算,業主實在負擔不起這樣的花費。以簽字則師比喻某些富翁購買勞斯來斯房車,車主希望在車胎費用方面省錢,而改用雜牌車胎。這樣很容易因為車胎壓力不平均及旋轉率不標準,以致發生交通意外。車主雖然省下一筆費用,但這還不足交付十分之一的修理費。

香港的建築設計一直都是向高空發展,對市區居民來說,這是無法避免的。但在地價比較便宜的郊區,港府應該多發展低層的住宅樓宇。現時,港府對新界的開拓和發展,仍未見一套正式和完整的計劃,只有幾個區域性的研究。從長遠發展的觀點來看,新界郊區還有很多可供發展的土地。而且,吸取外國的經驗,我們可以發現一般人都有居住在低層樓宇的傾向。香港在建築發展上,亦應該顧及這一點,在新開闢的郊區衛星城市中,着重興建低層的住宅樓宇。這種城市建築設計的發展,正好像一個埃及金字塔,中間繁榮的中心區佈滿高樓大廈,但建築物的高度正向外逐漸減低。

香港居民的收入日漸提高,對居住環境的要求也隨而增高。中產家庭希望入住環境較好的低層住宅樓宇,而木屋區居民渴望轉往設備較完善的高層大廈。為了配合社會大眾對不同居屋的要求,港府在發展高層住宅樓宇的時候,低層的也不能忽視。

在建築方法上,有預製形式的鋼架建築法。這種方法可以節省時間,但結構鋼的費用比傳統的混凝土及鋼筋高。而

淺談香港現代建築的發展

1982年6月5日在清華大學的學術報告
原載《建築師》14期. 1983年3月刊

香港被公認為全世界高樓大廈最多的城市,高層住宅和辦公大樓櫛比林立。市區內人行道上和空曠的地方,人之密度可以同北京的王府井大街相比。改建之繁密使得港變成了一處大工地,24小時不停地嘈雜。

許多地方的現代建築是從本地的古建築發展而成的。但香港卻不同。傳統性的香港建築有兩種:

一是廣東一帶的古建築,如廟宇、圍城、民居等;
二是19世紀英國殖民地式建築。

在香港的現代建築中尋找這些傳統的建築因素,實屬渺茫。但如果看看外國實例,應不難找出答案。多數香港的著名建築師是受外國教育的,如英國、美國、澳大利亞等,而且多數又對本地的風土人情缺乏研究。50年代,香港的人口膨脹起來,工業得到發展,產生了現代建築。當時,在上海開業的一些建築事務所跟隨紡織業遷移香港。當時的建築設計反映了20世紀初期歐美的建築形式。全部的建築好像是由一班陌生的外國建築師所作,對當地的風土人情全不理會。香港大學的建築系是由白高頓(Gordon Brown)教授領導下,從1951年開始招生的。其畢業生逐步充實了香港建築師隊伍。

1953年石硤尾窮人木屋區發生大火,促使當局積極興建了第一批平民住宅區。這就是本港著名6~7層高工字形樓。這些樓房的設備簡陋,但有公廁所及沐浴設備。到1957年這一類房屋容納了12萬人。

香港房屋署是1954年成立的。他們專為月薪100至900元港幣的貧苦家庭提供住宅。

由於房屋缺乏、工業發展,建築業也開始蓬勃發展。因為建且,香港消防局認為鋼材燒至一定的程度便會彎曲,所以不接受鋼架建築法。對於預製形式的建築法,在香港私人建築界,發展還受到很大限制。因為這種建築法需要連續不斷和大量的房屋需求支持,而且建成的樓宇都是千篇一律,沒有多大變化。外國大型的醫院及標準工廠建設計劃,可以提供預製建築法必須的條件。香港房屋署的廉租屋村計劃,也可以採用這種方法。1978年時,香港政府便有類似八層高的公共屋村計劃,但不太成功,現在又重發展,據聞有一屋村已在策劃。

近年,歐洲各地及澳洲的建築師都來港開業,而香港建築師亦有向東南亞發展的趨勢,這種情況除了有技術交流的作用之外,香港的建築設計界也有得益。因為外國建築師的設計水準比香港的高,通過同行內的競爭,可以提高香港的建築設計水平。

最後,香港政府應該向新加坡政府學習。新加坡政府每年撥出十多塊土地,給建築師作為設計比賽之用。透過評判團的挑選,具有高水準的設計而又投價合理的話,政府便交給得勝者落實發展,這對新加坡的建築設計貢獻很大。香港建築設計若要有更大的發展,香港政府實在需要出力從旁支持。

築業的發展及缺乏建築師，當局在1955年修改了1935年制定的建築條例。產生了"注冊建築師"為一稱號，結構工程師、建築測量師也可以兼負建築師的責任。從而，修建了很多非建築師設計的樓房。盡管解決了建築暫時短缺的難題，但產生了很多不太理想的建築物。這種情況，時到今日未改變。

從1962年當局又開始修改建築條例，以應付建築業的過分發展。但新的條例直到1966年才開始實行。然而，在短短的4年間，香港便從5層樓的面貌改變為20層樓的聳高姿態及高密度建築。

此時產生了些常見的設計特徵，如"飛機型"的"平面圖。這種平面是針對香港有個人社會及有污穢廚房的情形而設計的。這種設計有很經濟的公共通道，自然通風，臥室與浴室之間的聯系以及廚房和傭人住房的安排都很理想。後來，因為社會環境的變化，缺乏傭人，又產生了背對飛機型及無尾飛機型的平面圖。

另外一個常見的設計特徵就是樓梯形式的屋頂。這種形式是為了應付街道透光角度而產生的。此外，准許15米以下的商業建築佔有地段的全部面積而產生了上窄下寬的樓房外形。這樣一來，戰前騎樓設計的特點化為烏有，而顯現出參差不齊的中心區醜態。

60年代由於建築業及旅遊業的蓬勃發展，建立了許多新的建築設計事務所。這一批建築師多數是受外國教育的，但他們的設計已逐漸脫離外國建築設計的影響。此時開始修建香港第一個大規模住宅區——美孚新村，它對後來的大型住宅區設計有很大的影響。在公共住宅區設計上，彩虹新村又創造了新的住宅區形式。

60年代中期的一段時間，建築業的發展中斷了。但到60年代後期，70年代初期，因為証券市場活躍及經濟增長，建築業又回復原狀。由於開挖海底隧道，修建地下鐵路，修建大型的住宅區、飯店、文教醫療單位，以及實施"居者有其屋計劃"，形成了新的建設高潮。與此同時，當局也開始籌備大規模的地區性發展計劃。

近20年來，由於香港特殊的經濟社會條件，建築形式也千姿百態，有模仿，有創造，有成功，也有失敗。下面暫且分做四大類，加以介紹。念及每一大類都包含若干種不同的形式，不同的背景，統冠一詞，恐難貼切，姑且稱之為第一類、第二類…。

第一類：這一類建築在香港最為流行，立面多系盲目抄襲外國設計，就像人們所議論的那樣，穿起龍袍也不像皇帝。設計水平參差不齊，有些工程師和建築測量師設計的建築也屬此類。其多數是為了應付建築條例，缺乏想像力。大部分是由低級助手和繪圖員做的。對建築朝向、窗戶防水及適應氣候條件的細部處理多屬不當。住宅平面多採用"飛機型"。結構多數笨重，而且是設計好平面才加上去的。

這一類所包括的大型建築群，多由以上一類形式的樓房組成。在大型住宅區設計過程中，不少功夫是用在研究如何充分利用大自然的條件做完滿的樓房分佈規劃。在這方面，香港房屋署設計的建築群是一成功的實例，如華富村。他們為了完成其十年計劃，節省設計時間，不得不選擇幾種標准的住宅設計，在建築工地的平面圖上來擺佈。近年來，房屋署在執行"居者有其屋"計劃上，已委託多間私營建築事務所來進行設計。因為建築投資比"廉租屋"的略高一些，所以在設計上可能會多下些功夫。這類房屋往往用高層方式去處理，這是由於地價高漲，人口稠密。

第二類：這一類建築，總括來說，比前一類設計增加了一點活力。例如，木球會的設計，雖然在屋頂形狀上下了一些功夫(取材自柯布西耶)，但這種做法反而破壞了屋頂與屋身的一體表現。又如，香港仔游艇會的外形是想模仿船的外形做抽象的表現，但由於處理不當，反而弄假成真，變成了一只水泥船。再如，愉景灣規劃，雖然房屋是根據地形佈置的，然而產生了硬性的擺佈形式，缺乏鄰里的生活氣氛。沿道路兩旁佈置建築，頗有一水隔天涯之感。在低層房屋設計上，建築的設計上對傳統卻全無反映，黯然失色，以致使整個度假村設計不倫不類，完全缺乏地方色彩。

另一個典型的例子，就是聖士提凡書院的設計，建築師用了天然混凝土及基本色彩模仿了路易斯·康的金寶美術博物館(Kimbell Art Museum)的圓塔式的屋頂。這形狀與工地上臨樓的尖形屋頂很不相稱破壞了新舊和諧，甚至鐘樓也是取材於柯布西耶20年前在法國設計的一座寺院。

下面再介紹幾個比較優秀的例子。美國銀行大廈，其圓形設計對三角洲形式的工地及城市環境有適當的交待。

藝術中心獲得香港建築師學會1978年的銀牌獎，它吸取了路易斯·康20年前的多種設計技巧。如結構外露反映了路易斯·康之耶魯大學書廊的天花設計。暴露的空調系統是取材於Olivetti廠。建築師吸收了世界著名大師及其他亞洲同行(如泰國的Sumet Jumsai、日本的F.Maki等)的心得施用於自己的設計。

由於香港經濟繁榮，路易十四式家俱及西班牙式房屋應運而生。顯然，這件舶來品——西牙房屋是適應地中海氣候，適合戶外生活。後現代派的影響也波及香港，盡管這一充滿消極態度的流派在全世界已產生激烈的爭論。這一派的產生起源於有些建築師對高度工業化(High Tech)的不滿而跟隨了時裝設計趨勢，向古建築取材。這一現象在建築史上已屢有發生，如18世紀的洛可可派。當時這一派拋棄了後期巴洛克的作風，而表現出一種消極的、復古的作風，全神貫注於"五官"的感受，趨向比較回復自然的環境。但不同的地方是這一派表達的設計意念是暫時的、不恆久的，後現代派是同一作風。所以，這一派是建築發展期的尾聲，而不是開端。這一派的作品在香港及世界上暫時大多數還是紙上談兵，只有一小部分才真正地變成建築物。在香港比較年輕的建築師經不起這種誘惑，已經開始仿效外國的例子，可惜的是在香港開始仿效的還有相當有聲望的設計事務所，可見它具有多麼大的魔力！

第三類：歸入這一類的是帶有創舉性的建築設計。沙田穗禾村設計得到香港建築師學會1981年銀牌獎。這個住宅區設計對建築與地形的關係、居民購物要求、綠化、公共通道與居住單位的關係，都處理得有獨到之處，是香港近年來最高水平的設計。灣仔學校對不規則的工地作出了合理的安排，從樓房組合形式來看，建築師採用了中國傳統的組合前後院的方式來處理課堂的環境要求，並加上窗前的擋陽光的設備，反映了關懷本地風土氣候的特點。曉廬高層大樓(High cliff)設計改進了飛機型的房屋佈局。事實上，他解剖了傳統的飛機型，將左右相稱的平面作一個不平衡的推動，形成一個很突出的設計。

第四類：這一類是具備了香港地方特點的設計。對於高層建築，我們還要下一番功夫才能創出香港獨特的風格。但在低層建築方面，我們可算有些成就。香港山頂有一組依山而建的建築群，反映了本地居住者的生活方式，用混凝土牆很簡單地表達了結構形式並反映了依山而建的美姿。香爐山山頂纜車總站很強烈的表達出中國傳統的對衡式哲學，外型則反映出中國香爐的風姿。總括來說，這是混用中國傳統形式來設計的現代的塔。另一個例子是殘疾兒童院，這設計是用前後院形式來組合樓房，表達了中國傳統庭園設計的優點。斜形的屋頂及樓房之間空與實的對比，是用傳統風格來創造新的建築形式。

以上簡單地談了香港現代建築的發展道路和概況，很多地方都沒有作詳細及深入的討論。例如這兩三年流行的玻璃幕牆的設計對香港氣候的適應問題，高層建築對人生活的影響問題等等，都沒有涉及。因為這篇文章的出發點是向內地同行簡單地介紹一些現代香港建築的總概念，便於內地同行從中分析利弊，權衡得失，汲取經驗與教訓。

淺談世界第三代建築師及後現代派與中國建築設計發展問題

1984年6月在西安冶金建築學院的學術報告

(一)大綱

這報告主要是介紹幾位世界著名的第三代建築師，從他們的工作經驗中，希望能尋找出如何建立中國之第三代。報告內所講述之建築師主要是從英、美及日本三國選出來。首段集中講述幾位第二、三代之間有影響的著名建築師，例如：

路易斯·康Louis Khan, 詹姆士·斯特林James Sterling, 丹下健三Kenzo Tange。從他們的工作中，反映到他們對第三代的貢獻。其他第三代的著名建築師有菲利浦·約翰遜Philip Johnson, 巴克文斯特·富勒Buckminster Fuller, 阿基格拉姆小組Archigram Group, 保羅·魯道夫Paul Rudolph, 李察·邁耶Richard Meier, 鄧尼斯·拉斯頓Denis Lasdun, 伊力雲斯Ezra Ehrenkrantz, 派司Cedric Price, 史密遜Peter Smithson, 科士打和羅渣Foster and Roger, 丹下健三Kenzo Tange, 黑川紀章Kisho Kurokawa。

報告之尾段會討論後現代派對建築設計影響及對現代中國建築設計的關係。

以上所提及之第三代建築師因受社會制度、地方風俗及工農業等的影響，在他們設計中反映出第三代的特色。希望可以從他們的設計過程中，我們可找出如何去創中國第三代及第四代的答案。

(二)開場白

各位老前輩、同代朋友及年青的一代的同行，我很榮幸在此三代同堂談一談第三代的事。我雖然是一個建築師，但對建築歷史只有膚淺的認識，對國內的情況更一知半解，而且我做學術報告的經驗有限，當貴院的劉寶仲教授來信請我到這裏來作學術報告，我接到信後實是受寵若驚，正是卻之不恭，受之有愧。但沒有報告又不知如何找藉口再來西安，而且劉教授更提議帶我參觀未開放的韓城，我不得不硬着頭皮在各位面前獻醜，如果有甚麼錯漏的地方，各位要多多原諒及指正。

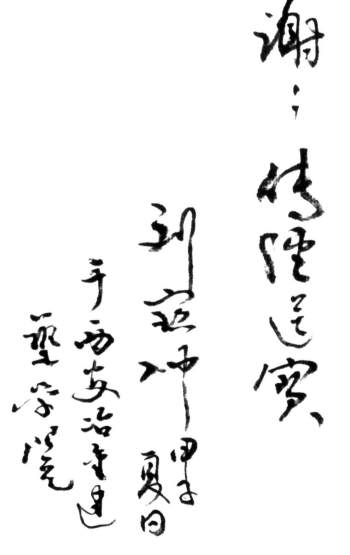

(三) 報告的宗旨及範圍

如果說19世紀末期20世紀初期勃起的佛蘭克‧羅‧萊特F. L. Wright, 密斯‧凡‧德‧羅Mies van der Rohe, 柯布西埃Le Corbusier, 跟格羅庇烏斯Walter Gropius是第一代，那麼沙利寧Sarinen, 馬素‧布亞Marcel Bruer, 前川國男Maekawa, 是第二代，第二次大戰後在不同的國家裡出了一群第三代的建築師，在第二與第三代之間，亦有一群傑出的，如拉斯頓、丹下健三、約翰遜等，因為他們是承先啟後的緣故，亦作為第三代介紹，但第三代建築師是一個人材輩出的時期，我的報告會集中於幾位我敬仰的建築師，然後從他們的作品中引論到現在面對我們國家的一、兩個難題：(一) 這就是怎麼樣可以將民族傳統表達到現代中國建築的風格；(二) 怎樣復興我國的第三代，栽培第四代。我這個報告是一個理論式的報告，並不是一個建築歷史的報告，所以建築師的名字是不重要。

(四) 美國

從第三代建築師中我選擇了在美國、英國及日本作業的幾位來作一個簡介。他們之間當然有密切的影響與交流。

在美國來說，我認為路易斯‧康是一個突出的人材(1)。雖然有人認為他不屬於第三代，但是他開始在建築事業上有成就的時候正是第三代的開端。而且我認為明白了前因就會領略到後果，明白了路易斯‧康的建築理論就能了解年青一派的建築師的作品。他的一生可說充滿災難，例如他在童年時受過嚴重的燒傷，又受過被迫移民之痛苦並經歷過建築業的低潮，加上他要到達60高齡之時才在事業上有成就。就他的例子來看，在長江後浪推前浪的條件下，有些年過60的仍是可以有成就的，不過他的成就實屬舉世驚人。他推翻了當時國際通行的傳統方式，而創造了新的思想。在50年代當西方正受工業化影響而產生了"形式服從功能"的傾向時候，他就主張建築物要符合個別的需要及能發揮個人藝術的觀點。他這個新的思想，影響到很多當時青年的建築師，例如羅伯特‧文丘里Robert Venturi, 及查爾斯‧摩亞Charles Moore等等。雖然在他創作時期的中段，他的學生曾經有青出於藍之舉，但以他豐厚的天賦才能，就在70高齡之際捲土重來，創出另一個事業上的高峰。

耶魯大學的藝術館(1951~1953)(2)是他成就的開始，這時他已是50歲。這一座建築物，至今仍是耶魯大學中最突出的建築物之一，但從這建築物中，我們可以看出他的設計方針仍然是猶豫不決。他早年的創作是受到柯布西埃的影響，繼後又受巴克文斯特·富勒及他的空間構架(3)及密斯·凡·德·羅的方形設計影響(4)，合組成藝術館中鋼筋水泥多角結構(5)及方形樓宇的矛盾情形，在各種影響下，從而產生了一種複雜的美觀。建築物的天花結構是有壓力性的，使得在室內之陳列品與天花有互相輝映的效果，這種效果就產生了後來的新野性主義。在同一時期，菲利浦·約翰遜所設計的玻璃屋(6)也是受到密斯·凡·德·羅的影響，但是所產生出的設計是同路易斯·康的藝術館相同。密斯·凡·德·羅往往喜歡表露結構的形式，而路易斯·康及菲利蒲·約翰遜卻將結構在外形上隱藏，而着重表達牆、地面及天花。密斯·凡·德·羅往往用中綫形式來管制他的設計，而路易斯·康跟菲利浦·約翰遜則用隱藏結構方式來遮掩他們設計中的雙影形式。路易斯·康多數用磚來表達，但利浦·約翰遜則用玻璃反影的效果。在兩人的設計中，有很多相同之處，最顯著的就是利用圓形和方形來分析空間。例如在藝術館的圓形樓梯，在這個玻璃屋的火爐及浴室。雖然這樣，只有路易斯·康在後期的作品重用這個原理來表達他"服務者與被服務者"的設計方針。在李察研究實驗室大樓的設計表達樓梯、水電系統為原則，產生了一種在形狀上近似歐洲中古時期的城樓之感(7、8)。室內空間的層次表現了它的"服務者與被服務者"的原理。

事因當時建築設備的重要性逐漸增加，使設備顧問及專家們受到器重，在李察研究實驗大樓的建築費中，一半是用於建築設備方面。當時路易斯·康曾對此事作評論，他說，我並不喜歡管道及管子，甚至很憎恨它們，所以我要控制它們的安置辦法，如果我只是憎恨而不好好地安置它們的話，我想它們會霸佔了整座大廈破壞整個大廈的設計。

美國著名的建築師維森特·斯卡利 Vicent Scully 曾經同路易斯.康在一個美麗的仲夏夜在克里姆林宮附近散步，欣賞由浪漫的意大利所設計的城塔。維森特·斯卡利就稱讚它的尖銳的雄姿，但路易斯·康卻相反地敬慕由這尖的形狀

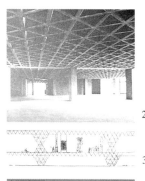
2

3

4

5

6

7

8

而表達出設計者對力學有恰當的表現。維森特·斯卡利當時只是欣賞到塔的嬌姿，但路易斯·康卻是敬仰塔的結構及形像的穩實，這就是路易斯·康對建築設計獨到之處。靜的、堅實的、不動的，這就是他設計的基本精點。這個就是路易斯.康在1952年所設計的費城大會堂(12)，這是與安·太Ann Tyng 所共同設計的。其中受巴克文斯特·富勒的影響十分顯著。

加上在1956年他所設計的費城城市中心建築大樓(13)。

這兩座設計照我看來是對後代有很大的啟示。例如:英國的阿基格拉姆小組(14)及日本的新陳代謝集團Metabolist(15)最受他的影響。

從李察研究實驗室大樓後，他的作品更加明顯地表達出他的理論。例如印第安納的藝術劇院(16-19)。請大家注意這個圓拱的結構及引透光度進屋的設計。這兩點在他後期作品上有顯著的重用及變化。孟加拉國家大會堂1962(20-23)、德加城國立醫院1963(24-25)、菲臘亦士打學院之圖書館及飯堂1967-1972(26-30,32,34)、印度管理行政學會1963(36-41)、加州之生物研究學會 1959-1968(42-45)。

在這個設計上路易斯·康用木製百葉來躲藏玻璃的反射(46)，這一特點表明了他對實體及空間的相反表現有肯定的處理方式(47)。

這個方式在他的印度及孟加拉所設計的建築物明顯地表露出來。玻璃在他來說，簡直是可惡的必需品，難登大雅之堂。

保羅·魯道夫的設計有很多方面，是和路易斯.康的設計同出一轍，"服務者與被服務者"在他的建築設計表現得很透徹，例如他耶魯大學的藝術及建築系大樓1964(48)。他的建築物外牆用人工打出來的粗綫條之混凝土的效果(49)，反映出美國社會當時的奢侈作風，世界上很少地方能夠容許如此浪費人力資源。

他在建築物設計中用的各種方式與路易斯·康在李察研究實驗室大樓所用的方式大同小異。他也很注重在建築物外形上表達結構形式和利用大自然的各種資源，例如引進陽光及遮擋烈日(50)。如紐約高路潔大學之藝術中心1964(51)、東南麻省理工學院(52)文科及社會科大樓(53,54)、圖書館(55)、黎加拉瀑布的圖書館(56-59)及水牛城住宅立視圖(60,61)。他在住宅設計中對"服務者與被服務者"的哲學也有適當的交代。

李察·邁耶畢業於哥尼大學。他於1963年設立自己事務

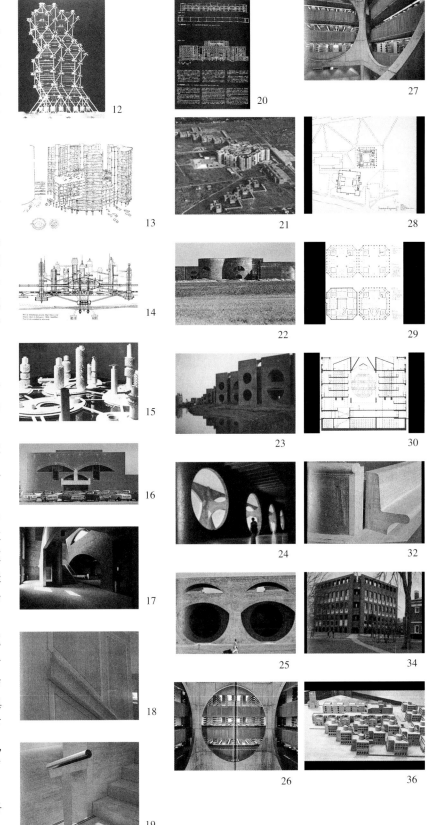

所,他是橫跨第三代第四代兩代,而且是在這青黃不接的時期中,世界上最成功的建築師之一。他的設計推動了現代建築思想到新的一個嚴肅及實用的境界。當大多數建築師正在對現代建築設計抱有懷疑及不能接受的情況下,李察·邁耶對現代思想十分忠心,但他運用的手法是史無前例的。他對柯布西埃和阿爾托Aalto十分敬仰,但他克服他們的缺點及利用現代技巧來發揚他們崇高的設計構思,所謂青出於藍而勝於藍。布朗斯中心(62-65)是他第一個比較著名的公共建築物,是在70年代初期建成的。在美國印第安納州的新和諧派的神舍是一雪白色的建築物1975-1979(66)。這個突出的設計以美國農村為背景,與柯布西埃的西和別墅(67)用法國農村作背景有異曲同工之效。這個中心設計裏面的幾何理論、形狀及空間加上他從柯布西埃吸收得來的精華,產生了在這過渡期中的經典之作。他將柯布西埃的光線與空間理論、阿爾托的自由及有規律的簡單組合的體形,及高度機械派的粗線條作風等歷史性的構思混合成一個新的美國式的風格(69-71)。他在多數的建築物外牆上蓋上一層雪白的外皮(72-74),更反映出其美國化的特色。他的設計不單只是滿足用途上的要求,更同時將傳統派的作風推動到新的境界。他不單只接受傳統現代建築設計的思想,而且更集中去發展歷代相傳的幾個建築設計的理論。例如光線與空間的互相比對,及實用主義與感性之間而產生的效果。這種設計的突破是正宗過渡時期的標誌。

李察·邁耶的建築物多數能單從外表可略知內部一二,例如在美國克佛神舍(76)的外牆設計,在正門的右方有一箱形部分,裏面原來是一個小禮拜堂。外牆的抽像形狀恰當地反映了小教堂在整座建築物中的重要性,更形成了整座建築設計的曲尺形狀。玻璃磚牆反映出裏面的樓梯而薄薄的屏風牆卻表露出進口的走廊。但這各種用途及特點,是從細心研究實與空的關係遂漸演變而成。從入口方向走,經過這薄薄的屏風牆,之後便是一個小庭園。這屏風牆將這一段空間變成與世無爭的過渡地帶,其它的各種屏風牆形成不同角度的平面。有時從屏風牆角伸展出外,來改良室外空間的佈局。

李察·邁耶最新的作品就是在美國的新高藝術博物院(77-78, 81-83)。對自己的設計有以下的說話:"這個設計是同我們事務所以往所設計的不同,其它的公共建築物是供應給很多人在不同的空間活動,但這個博物院將會被很多人在同一相連的空間同時使用,這個建築物將會是供應給觀眾及藝術用途。結構的格式對室內空間的安排是很重要的,這個

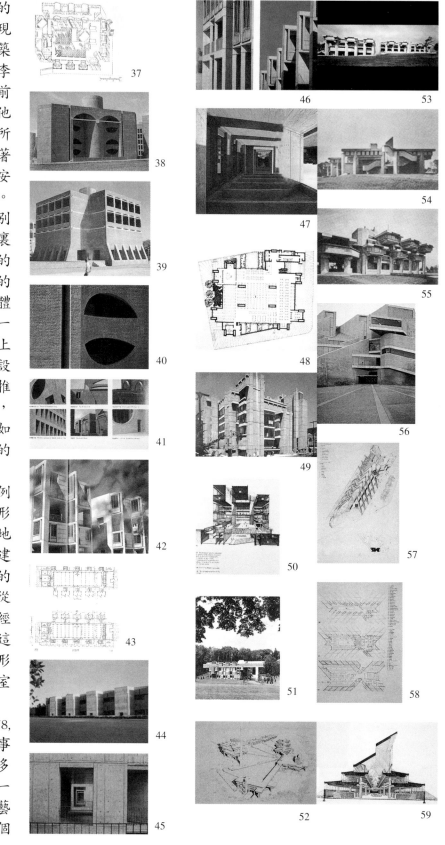

設計是我所有設計中的一個突破。整個設計以中間一個多層空間為主，光線因地形由建築物之屋頂中部引進，所有陳列室的空間是與這個中間多層空間相連，而可以從一個陳列室望過中間的空間而看到另一個陳列室。雖然它們是相連一起，但因各有不同特色，所以變成幾個氣氛不同的個別空間。"

(五)英國

新的獸性主義是英國在第三代建築界常用的名詞。在50年的中段，詹姆士·斯特林(84)就開始加入這個獸性主義，雖然日後他否認是屬於這一派。1955年，詹姆士·斯特林首次去參觀由柯布西埃所設計的Maison Jaul(85)，使他十分驚奇，覺得這個設計與自己的設計竟不謀而合。這個反應可以由他在當年所設計的咸鎮平民住宅群(86)反映出來。雖然在這設計中，對磚牆的運用，兩人都有不同之處。英國獸性主義的和合點即是將傳統性與普通性的反義，混合在十九世紀中工業結構所產生的磚和玻璃的特徵，就是斯特林和他的拍檔占士.高雲James Gowan (這是金柏威的學校校堂1958)(87-88)。

在1955年所設計的劍橋蘇雲學院(1959)(89)的宿舍表現出來。在這個蘇雲學院設計中不單是初次表現到他們早期的玲瓏與生動式的作風，更顯現出他們"前後相稱"的設計主題。從他們設計的理斯達大學工程系大樓(90-93)中我們更可以看出他們怎樣受柯布西埃的巴黎瑞士大樓的影響很大。在這個大樓設計中，表現了不少斯特林的利物浦家鄉背境，例如有些受工商影響下的建築材料、欄干、玻璃架。這個建築物可公認為集中國際獸性主義的精華，例如結構上運用多角形樓面的設計，是反映路易斯·康在1958年設計的李察研究實驗室大樓。

斯特林的設計的精華處，在於他的強硬性及建築方式，使用他的建築物的人往往為其空洞之感所吸引，而不知所措。這反映出建築設計是一個自治的機器，劍橋大學歷史系大樓(1964)(94~97)及西門子的電器工廠就是一些例子。慕尼黑的西門子廠房(1969)(98)、聖安魯大學之宿舍擴建部分(99~103)、牛津的皇后學院(1966)(104~106)、新市鎮住宅群1967(107~108)、好利獲得訓練學校(英國，1969)(109~113)。斯特林往往不近人情地放棄當代傳統的規範，用他自創獨特的風格而把觀察者吸入他的旋渦，而這旋渦本身亦不能自主，他可說是一代宗師。

鄧尼斯·拉斯頓在第三代來說，他可能是超齡，生於1914

60

67

76

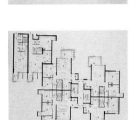
61

68

77

62

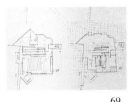
69

78

63

71

81

64

72

65

73

82

66

74

83

年，但因他的建築事業從1935年起至今已超過40年，而仍然名聞於世。在建築學報1980年第一期已有報導，我在這裏只想點綴下他的精華。

他對英國建築業有獨到的貢獻。他所設計的樓宇，表達出空間及形狀的意義，屬於優秀的設計。他的設計哲學是受30年代各種影響，例如他的教育背境，畢業在倫敦建築協會學校，專修柯布西埃的作品及他同露比根及德頓集團Lubetkin and the Tecton Group 的關係。鄧尼斯·拉斯頓的建築語言是產生於30年代，發揚於50年代，成熟於60年代，到70年代已得超然之感(114-117)。

建築物對他來說，是多層的平台供給人類活動的，這些多層平台是土地的擴建，也是用來連接室內與室外。這種理論在國家歌劇院(1967年)有明顯的表示(118-122)。這建築物位於泰晤士河畔，室內人多聚集的地方，與室外的行人通道及湖畔散步的地方，混為一片。他從柯布西埃學到混凝土的軟膠化及莊嚴的美質，他常用這建築材料來表達他的理論。

他的事務所雖小，但仍能夠對每一建築物的角落都花心思，他的樓宇不是充滿機械化的設計，而是多數用簡單而普及的材料，創造出極富想像力的設計。這一種優點是現代化機械化者所缺乏。

但這不是說他是一個保守的傳統主義者，例如他在西南大學(123-124,126-127)盡用預製的結構及外牆，這個設計是由建築物的需要而產生的，而不是用建築物來配合這個預製計劃的。

去年三月我安排了鄧尼斯·拉斯頓爵士來北京作學術報告，在建築學報1980第七期裏有他的報告"建築藝術的延續性和變化"的譯文。

倫敦大學改建方案(1965)(128-129)、教育學會(130)、教育及法律學會(131)、東方及非洲研究院(132-133)、預製混凝土細節(134)、耶穌學院學生宿舍陽台進口(135)、立視及模型(136)、樓的東面(137)、混凝土的細節(138)。

在50年度之初，英國有一個組合分區學校的建設系統叫加柏CLASP。從這個系統，產生了一個很經濟而簡單的學校建設方法。主要的建築器材就是鐵架、門窗、外牆及室內設備都是來自通用的產品。因為這個系統的結構是很生動的，所以適合各種不同的學校設計(139-143)。

在1962年初在美國加利福尼亞洲，根據英國的計劃伊力雲斯創出另外一種系統叫SCSD學校建設發展方案(他曾在英國的加柏系統機構做事)，這個設計系統，包括有隱藏在天

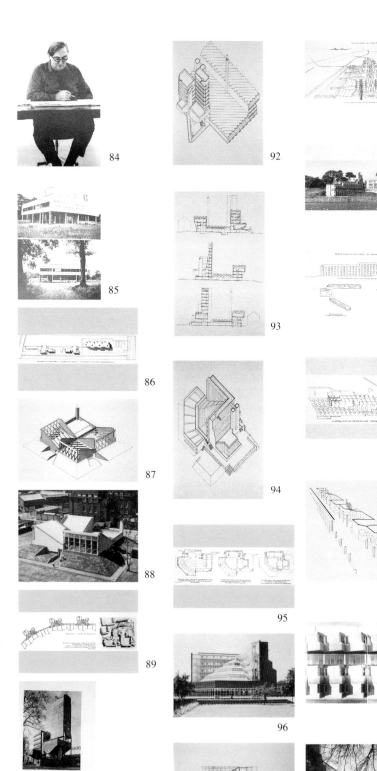

花及天面結構裏面的活動性空氣調節及水電系統，連結構的鐵架也是由一活動原理來組合。這樣的系統對建築設計完全沒有影響，因鐵架能跨過很廣闊的空間，使得房間的間格有大的改變功能。

美國的學校建設發展方案(144)，英國的派司建築師，加上美國的巴克文斯特·富勒、查理斯·嚴氏Charles Eames (145)、米斯同英國的史密遜就產生了英國的科士打和羅渣集團來(146-151)。

威簾費伯和杜馬有限公司辦公大樓在英國伊普斯、維荷1970-1975(152-155)外貌、聖斯比利博物館(156-163)、蓬皮杜文化藝術中心(164,167,169-172)。

(六)日本

前川國男可算是近代日本建築設計的源流。他吸收了西方第一代建築師的精華(173-175)，萊特及柯布西埃，所以他的作品是很有西方國際色彩的。丹下健三離開大學後，便在前川國男事務所工作了四年。丹下健三首次帶進日本"廣場式的空間"這個名詞。因為在日本傳統建築設計中，沒有這一回事，而他是從研究古典希臘建築構思所得的成果。在傳統日本的城市規劃，建築物是沿馬路而建，居民只可以在自己室內及庭園式的空間聚集，所以這個可以供居民聚集的大型空間的設計方式，在當時日本可說是反傳統。丹下健三在他的城市設計方案中，運用了這個構思。在50年代丹下健三設計了省立辦公室大樓(177)，這個設計有古典平衡之感。空間的安排有受日本古代建築的影響，加上他吸收了國際形式設計的點綴，這傑作形成了他在戰後日本的崇高地位。他了解當時工業發展的衝力及傳統，對這社會解放的助力，他完成了這個省立辦公室大樓後曾說："直至最近日本是受獨權主義的政府所壓，平民的文化發展力量受壓迫，這力量是不產生新的形式的。現在我所講的力量已開始發揮，雖然這力量暫時是無固定方向，需要更大的努力才能納入軌道，但我很有信心，肯定力量終會將日本的傳統改成的新建設。"這一番話，可能對我國現代建築設計有頗大鼓勵。

1960年丹下健三及他的事務所設計了反傳統性的東京規劃大綱(178,183)。用龐大形的建築物橫跨東京灣，這樣對在擴建中失去控制的大城市，創造出另一解決方法，這方案是由於避免讓建築物及工業作無規律的擴展到郊外。在傳統日本城市中引進了有規律的規劃方式，及市區性的聚集空間。

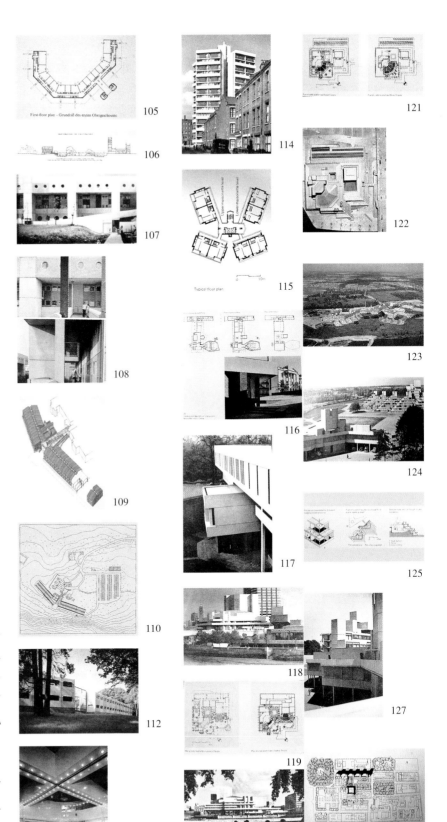

很多人都認為丹下健三是日本的新陳代謝集團的創始人，從他設計東京灣大計(1960年)時開始，他就因東方人口澎漲而提倡插入型的龐大建築物(179-180,184)。這些建築物掛滿預製的住宿箱，這個設計方式很簡單來說，就形成新陳代謝的主要理論。

當丹下健三正在設計他的東京灣方案的時候，4位年青的建築師及一位建築評論家經互相研究及討論後，共同發表一篇"新陳代謝(1960)"被命為新陳代謝集團(186)。這集團的成員有黑川紀章(187)、菊竹清訓、槙文彥，大高正人。黑川紀章當時是26歲，剛從京都大學畢業只3年，同時在丹下建三的事務所工作。8年前當黑川紀章40歲時，他已經完成三35座主要的建築物及寫了17本理論書籍，同時為一社會工程研究機構和智囊團的主席。他每天工作18小時，當時已在全日本最受歡迎的著名人中排第三位。

當時黑川紀章認為創造出一個新的國際性的設計形式是時候了，這設計形式應反映出地區性的文化，同時保存了他們的獨特風格，使各種文化形式能夠互相影響，而創造出一個新的風格。換句話說，從地區性的獨特風格，可產生成為國際性溝通的媒介。

他提倡研究日本14至18世紀的文化，及高度讚揚在19世紀中京都的街道及民眾集合的地方(188)。他想研究這些聚集點的理由，是想找尋在國際性的現代建築中缺乏之民族風格，同時為了去發掘在傳統及現代日本建築設計中的過渡空間，含蓄、不明確和變化多端的設計技巧(189-191)。

他在1970年大阪博覽會設計了東芝和德加拿座椅公司的展覽館(193,195-197)。設計中充份發揮了代謝論的理想，他用重復組合數百個立體十字形的構架，表達了代謝理論的變化和伸展特性。

在1972年他設計了中銀艙式住宅大樓(198-204)，這可算是世界上最早一個插座式的原理建成的建築物。另外一個例子，就是他的新力大樓(205-208)，他將電梯及電動樓梯等的不能代謝的設備組成一角，而將能代謝的廁所等用預製貨櫃箱形式放在另一角，可算是表達代謝哲學最好一個樣本。在70年代的中期，他在中東、東歐及非洲設計了幾項大型工程，例如阿在達哈比會議城(209,210)、滕路國際總署(211,212)。在這個時期，他開始將他對佛教及傳統日本建築的原理溶匯在他的設計中，例如福岡銀行1975(214-216)及田立人類博物館(217,218)。這些作品是表現出他已經開始脫離代謝論，或從代謝論變新，而這一轉變在設計上有了濃厚的東方傳統風格，可算不辜負他對日本傳統建築及佛教的

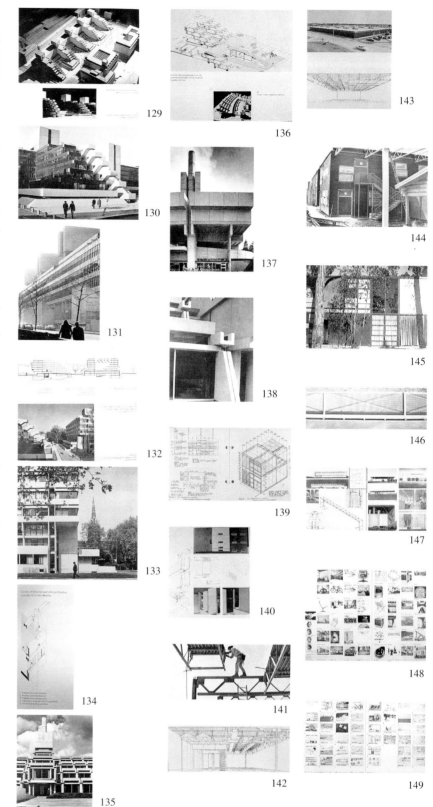

研究心得。我對他的理論只有膚淺的認識，他對外國哲學思想及東方佛教的理論，實有深厚的研究，在此實不敢提供太多的評論，只有再介紹一些他的近期作品：
日本之 Sartama Prefectural Museum of Art(1981)(219)、日本之 Ando Residence(1980)(221)、柏林之 International Bauaurstellung 1986(222)利比利亞之高級商業管理及銀行服務學院(1981)(225)。

(七)後現代派

在這大多年內，外國建築設計已產生了新的一面，這就是後現代派的作風(226,227)。我想趁此機會略略作一淺談，因這一派是仍然在發展中，而且它的影響力已吸引了無數的建築師，包括世界著名的，例如美國的菲利浦·約翰遜、英國的詹姆士·斯特林等，也有不少建築設計評論及歷史研究者沉迷於評論這一派，例如在英國的美藉評論家詹克斯 Jencks、荷蘭的范艾克 Aldo Van Eyck 及羅伯特·文丘里等。這一派的產生，起源於有些建築師對高度機械化的作風不滿，而跟隨了時裝設計的趨勢向古建築取材。這一個現像在建築歷史中，已有多次發生，例如在18世紀的洛可可派(229,230)，當時這一克派拋棄了後期巴洛克派的作風，而作出一種消極及復古的作風，全神貫注於"五官"的享受，例如視覺、聽覺、味覺、觸覺等，趨向一個比較回復自然的環境，但不同的地方就是，這一派表達的設計意念是暫時及不恒久的。後現代派是同一作風，所以這一派是建築發展期間的尾聲，而不是一個開始(231~236)。這一派的作品暫時來說，是大多數紙上談兵，只有一小部分可真的變成建築物，其中以查爾斯·邁耶的意大利廣場是值得一提(238)。這個廣場是建造在美國密西西比州的新奧爾良(是在1978年至1979年建造)，這個廣場是由美國居住的意大利人集資建造，而邁耶的設計正符合供給他們一個懷念故鄉的聚集地方。在這情況下，邁耶的設計是相當適當的，但是講及菲利浦·約翰遜的紐約市辦公大樓(239)及米高.極士Michael Graves 的砵蘭市的公共服務大樓就難以接受(240)，因為如果是可以接受的話，我們最近在香港拆除了1930年代建成的舊汇豐銀行大廈(241)，此舉豈不是向潮流倒退?因為新汇豐銀行大廈是一個高度機械化的設計(243)。

(八)結論

我國已經有第一及第二代的建築師，總括來說，近代的建築設計是屬於第一第二代的作品，與日本的前川國男是同

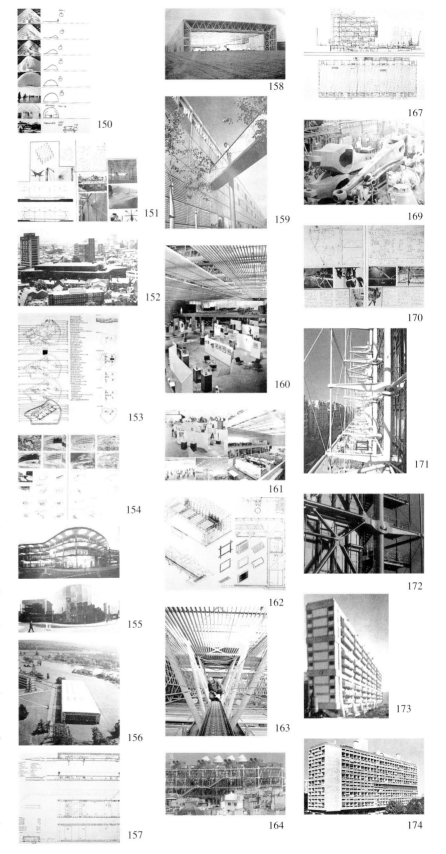

一時代，這時代受歐洲的國際派影響，從而對本風土人情、地方風格不去理會。在歐、美、日本的第三代是從反映當時社會及經濟轉變而產生，歐洲的工業革命及日本人口澎漲，助長了第三代的產生。

路易斯‧康吸收了柯布西埃的光綫與空間原理，巴克文斯特‧富勒的空架結構、密斯‧凡‧德‧羅的國際玻璃及鋼鐵派而產生出自己獨特風格，他對建築物內各種用途的處理是很嚴格的。而造型往往是壯觀的，能反映出建築物的用途，他的設計反映了室內空間複雜性的增長率，及建築設備的要求。但到邁耶的時代，高度機械化將建築設計拉過去太空發展科技的一面，同時後現代派卻將建築設計倒流。邁耶卻啟示我們將建築設計推回到較正宗，富有人情味及傳統性的道路，同時更創造了新的一派，在這一派中，用途與形式互相配合，因生產生出一個富有溫暖及有推動力的建築流派。

詹姆士‧斯特林在他的設計中反映了當時社會及經濟的轉變的氣氛。充分利用工業革命所產生的產品，來用於他的建築物上，同時保留了傳統英國的風格。

科士打及羅渣是處於一個沉醉於太空科學及電腦的世界，所以他們的設計反映到高度機械化及節約的風氣，甚至運用建造船隻的技巧，來實踐他們不尋常的建築要求。(243)

丹下健三吸收了柯布西埃，格羅庇烏斯、前川國男，國際現代建築設計大會CIAM和傳統日本建築創造出第一種可算是真正的日本現代建築的風格。這風格完全擺脫了國際流派的影響，吸收了日本傳統的風格(251)，同時反映了一個日本的新朝氣。

黑川紀章運用他對傳統日本文化根源尋求的心得，和把握了現代最新科技的知識，將丹下健三的設計風格更去蕪存精，創造了一個有深度的現代日本風格。

我覺得從研究世界第三代的發展中，可以找尋出如何去栽培中國的第三代的答案(252)。

我國已經在50年代開始，建造了不少由蘇聯及中國古建築影響下設計的樓宇，尤其是公用樓宇方面，例如候機樓(253)、大會堂(254)、劇場、博物館等等，所以身受其害情形下，中國是能夠對現在世界上流行的後現代派的浪潮加以抵擋。中國很幸運地能夠避免後現代派的破壞，可以迅速地開始建立第三代，如成功的話，可以跳班，趕上世界的第四代，因為他們正在一個青黃不接的時期，飽受後現代派及高度機械派的干擾。

我提議以下幾點折衷辦法：(一)不要盲目地仿效海外作品。

175

177

178

179

180

183

184

186

187

195

188　196

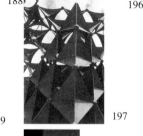
189　197

190　198

191

199

193

200

201

建築設計的詞彙中，應刪去玻璃幕墻一字。(二)研究世界各第三代、第四代著名建築師的作品。(三)深入研究中國傳統建築，尤其是清朝及民國的一段時期，不應理會建築細節技巧，而集中於研究其設計的哲學及風格。這樣一來，兩個朝代的研究已經足夠，應讓其它的朝代交由建築歷史家處理。(四)研究現代及過去的地方性社會制度、文化及經濟架構，建築業的情況，建築材料的種類及供求等等。

在創造出一個中國現代建築風格的過程中，應該徹底了解各種社會文化及傳統的背境、地理環境的限制、建築技巧等等，才能執筆設計。當我們能夠徹底執行了以上的各項提議，這枝筆會自動帶引你走上成功的道路。

以下有幾個中國傳統建築例子，例如：天壇(256)、陶屋(257)、西藏的布達拉宮(258)、民居(259)、四合院(260)、承德普寧寺(261)、深圳的正大街(262)、川西樂山附近的金鎮、羅城(265)。

以下有幾個對創新現代中國建築的嘗試品。
(一)中山溫泉的映山湖畔建築群(266)(二)台北市仁愛路葉財記皇家大廈(267)(三)桂林的蘆笛岩接待室(268)(四)北京圖書館(271)(五)新加坡的皇朝酒店(272)(六)台北的圓山大飯店(273)。
你可能已經體會到他們全是抄襲傳統中國形式的，並不是我們心目中的答案。在過去兩年我在國內作報告往往喜歡引用一句慣語"問題不大，只是技術上有點困難"，我們在創造出一個新中國建築過程中，技術是絕對無困難的，只要我們有恒心、耐性及心得，便可成功。

我的報告就說到這裏。

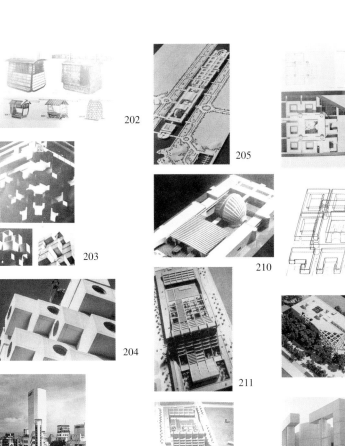
202

203

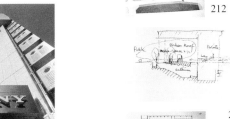
204

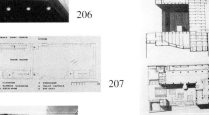
205

205

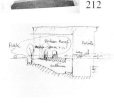
206

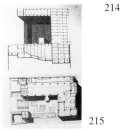
207

208

210

211

212

214

215

216

217

218

219

221

222

215

226

227

229

230

231

232

233

234

235

236

238

239

240

241

243

251

252

253

254

256

257

258

259

260

261

262

265

266

267

268

271

272

273

294

香港建築特刊序言

原載《世界建築導報》，香港建築特刊 03/04/1985 期

這一期裡，我們不但討論了香港建築之過去，也討論了香港建築的現在，同時，也包括了香港建築師在公屋設計、新城市設計和建築教育三個方面的文章，更請幾位香港著名建築師，對將來的設計發展，做了些預言和推斷。

在世界的建築流派中，暫時還沒有一類可以稱為香港建築形式的，因為這裡還是聚集和呈現着各國的建築樣板，沒有獨特的香港形式。但，在最近幾年裡，香港建築上開始有點本地風格了，所以我覺得這是一個好機會，利用觀察的結果，作一個介紹。我希望利用這個觀點，來鼓勵我們的建築師，以適當的建築設計，創造一個既符合香港文化歷史背景和生活方式，又有強烈地方彩色和風格的地方建築。

關於"本地的建築風格"這個題目，不單是香港，也是亞洲建築界的一個迫切問題，去年在菲律賓馬尼拉舉行的第一屆"亞洲建築師大會"上，曾以這個題目為主題之一，進行了學術研究和探討。

在亞洲有很多以本地風味來設計的嘗試，有些卻很幼稚，只是表面的，如新加坡一個高層綠瓦頂飯店，在台北的圓山飯店，馬來西亞檳城用傳統樣式做成屋頂的建築等等……，也有許多設計只取引人的外衣，開初雖然使普通人覺得新鮮，好像適合口味，但它們被歡迎和接受的程度是暫短的，猶如時過境遷一樣，在視覺上、心理上慢慢就產生了一個相反的作用，甚至那些俗不可視的地方風味樣板，後來就成了人們的眼中釘。

放眼亞洲，我們就可以看到亞洲快速發展的城市裡的一些情況：所有對建築風味有影響的，都有着濃重西方風格，例如在建築材料(既便是本地出產，也仿西方同類產品)、施工方法、生活情況等等都可以看得出來。而且也會發現，雖然本地氣候與亞洲其他地區是不同的，但在高層設計方面，最有影響的是建築條例、結構要求、施工程序、業主要求和實際用途等因素。難怪亞洲的高層建築物設計是千篇一律的。

我們建築師不能單獨扭轉和更改以上因素，儘管具體地區的建築條例有別，而地方風味也有所不同，但對整個建築設計來講，只有一小部分影響的。總括來講，在亞洲建築設計方面，能夠表達本地風味的因素是很小的，所以我們不應當盲目地只向本地風味去做徒勞無益的苦工，應該從地區性風格下手，唯一可以充份表達地區風味的地方，就是從建築物以內和以外的空間設計構思中表達出來，中國在建築上有"方中有圓，圓中有方"這句話，也可以使我們舉一反三。

要在建築設計上達到一個優異的空間構思、首要的就是徹底了解當地人的傳統觀念、人生哲學、文化背景、軼聞趣事，以及他們的生活方式、最新的建築科技發展情況，還有氣候條件等，這些都是建築設計上的悠關問題，我認為，有了對這方面的了解，就必然胸有成竹，建築上的地區風格，會從設計圖中自然表現出來，也就是說，當你熟習所有客觀條件，頭腦裡形成地方風味的概念，動手設計時，隨之而來的，就會水到渠成，躍然紙上，而單單靠憑空冥思苦想的方法，是不能達到預期目的的。

中國近代的一些建築設計，是被外國設計有不良影響的最好例子。清末以後，中國建築尤其是公共建築，很多受外國設計的影響，本世紀中期，突出的就是蘇聯的影響。在最近幾年又有旋轉餐廳、玻璃幕墻等等一類建築物出現，這些都是從外國設計，或從歐美大城市的建築設計例子上翻抄過來的。在國內很多大城市裡，有外國建築的"複製品"，可能有人認為這是大城市裡必不可少之物，並以此標誌出"現代化"。這些不中用的建築物，慢慢變成西方人所說的那種"白象"，雖然悅目，但又與本土格格不入，不但失去了真正的實用價值，而且在經濟上造成了不必要的負擔。

有很多國內的建築設計師，看來他們往往以為香港的建築設計是好看不好用，建築物本身似乎只說明香港表面上的繁榮和緊張擁擠的現實，可能認為對中國是不合適，因而不去認真研究了解香港建築設計方面的一些因素。雖然香港的建築設計，不能說是世界的群首之一，但，它們是建築條例、經濟要求和國際上的文化交流影響下產生的成果，從事物本身的發展來看，這種情況也往往是產生建築

上各種流派的必經之路。

香港跟大陸有密切的關係，是因為有着許多共同點，諸如倫理道德觀念、相通的語言、一致的文化背景等。香港距離中國最近，也是資本主義的一個最好櫥窗，這裡又有很多國內的外駐機構，因此，對中國考察人員來說，香港變成了去其他西方國家考察的一個最好跳板。所以從中國去外國考察的建築師，最好開始先研究香港的建築設計，不要單獨觀其外表，不然的話，直接去美歐考察西方極端進步的科學設計，是不能完全理解和接受的。這種辦法，也可以使中國在現代建築發展中，避免許多浪費和不必要的錯誤。同時，我也希望中國有關部門的領導人，能了解關鍵之所在，放棄盲目地複製和仿效外國的行動，最好能讓建築師們去充分發揮他們的智慧，創造一個在生活要求上，在文化要求上，對中國更加適合的地方風味的新建築。

至於香港的建築師，在近幾年內，已經有點地方風味了，但，要蔚然成風，要達到成功的目標，為時尚遠。今後我們要對我們的文化和其他形成地方風味的因素，要多方面進行研究，並增加這方面的興趣。要放棄仿效別人的做法，跟隨其他亞洲建築設計師一道去尋找具有真正的、永久性的地區建築風味的規律來。否則，亞洲的全部大城市建築，就會面目雷同，張李不分，變為西方建築設計的翻版，如果不幸而言中，其情況將會是令人觸目驚心的！

淺談香港及內地的現代建築風格

1985年12月 廣州學術報告

各位前輩、各位同行，我在國內從1979年起，多以演講為名，來爭取參觀不開放的建築古蹟的機會，雖然我自為廣東番禺縣人，到現在為止，用廣東話作報告這實在是第一次，所以，如果我的廣東話講話有可笑之處，請不要笑出來；同時我報告的內容是淺顯的話，希望也得到你們嘲笑一番。我對地方風格問題是特別感興趣，在去年舉行的馬尼拉亞洲建築設計論壇，我就以地方風格作主題，主持了半天的討論。從那時起我就開始向這一方面做了些初步研究，今日趁此機會講出我的一些看法，來跟大家一起探討。

我將會用香港及一些國內的現代建築設計例子，來分析兩地的地方風格。我首先從歷史背景及風土人情入手，之後從一些現代建築設計的例子中，找出何為地方風格。

因為我對香港的情況是比較熟習的，我會在這方面敘述出深一點的理解。我也感到國內的建築師們，對地方風格已開始有濃厚的興趣，並已做了一些可喜的嘗試，這是一個很好的開端，尤其是在台灣方面，所以我不得已也大膽地在國內方面作一初探。

香港並不是好像很多人意念中滿佈高樓大廈的一個地方，我們也有很多的大自然風景。當然，從另一個角度看香港，是那些好像混凝土森林的中心地帶(附1~4)。

香港的傳統建築有兩大類，第一類是中國的傳統古建築。當時的建築設計，無論是廟宇、圍屋或民居(5)，都是充滿了中國南方的地方風格(6)；綠瓦是適應地方氣候的要求；圍屋的構思是來應付保安及防衛的需要。

第二類是英國殖民地式的傳統建築，雖然這一類的建築風格沒有很濃的地方性，但在功能方面，香港會所(7)的騎樓及窗外的木百葉是完全符合本地氣候的。此外有當時的中國

1

7

2

3

4

5

6

銀行(8)、海事處(9)等建築物，都可算是當時的代表作。同時在結構設計上常用的拱型(10)，這也是受材料及氣候等要求而產生出來的。

我自從對香港的建築設計的發展做了一個總結的初探後，我發覺到現在為止，還沒有一個建築物是能夠很全面地表達出香港獨有的地方風格。但是從戰前到現在已經有些建築物，在設計構思中表達了對地方風格有交代的地方，我在這報告裡會一一提出來供給各位參考。

戰前的商業區建築設計，也是對地方氣候及商人的起居有所交代的。例如舖前的大騎樓(11)，能夠使購物人士不會被風雨所吹打，同時在烈日當頭時，也有遮蔭之地。當時的商人多數是住在舖位的樓上，因而產生民居與商店混合的一個設計方式。

50~60年代的建築設計，多是被一些外國到港執業的建築師及國內移民來的事務所所影響，對地方風格是完全脫節(12)。衹有兩方面是較有一點創建：第一就是金字塔形的外表(13,14)，這是反映建築條例街道採光的要求；另外一點就是平台式的大廈外形，可惜的是，這兩方面都是由不太理想的建築條例而形成的不良後果。而香港的大會堂，更反映出負責設計的英國建築師的英國背景，因為這座建築物的外形，正與當時英國所謂"縣政府型"的設計一模一樣，當然他這樣的設計，可能有一個作用，這就是當你在遠方看這沿海港的景色時(15)，會覺得它是突出的，有點"萬綠叢中一點紅"的感覺，換句話說，對週圍的環境是息息相關的。

反過來，在50年代因為大火災而產生的臨時工字形徙置大樓的設計(16,17)，我覺得是對當時的要求有極完整的交代。背對背的單元，騎樓通道，集中的公廁，短速的施工時間，雖然這一方式是不能作永久設計的典型，但是這一設計對當時的各種要求的處理方式，是值得我們去參考。

在60年代，產生了一些我們香港常見的"飛機形"平面設計(18)。我覺得這類的平面設計是反映地方生活方式的一個成功的例子。建築師把污穢的廚房及傭人住房放在單元的後面，而將起居室放近大門處，主人及他家人的睡房，就放在起居室的一旁，這樣的安排是同國內傳統的四合院設計有相反的做法。在四合院的平面，傭人房及廚房多數是在前院，但飛機形的平面對當時香港人的生活方式是有適當的交代的。照我看來，這飛機形式，是一個我們可叫"絕"的設計典型(19)。香港的建築師，其創舉有時是相當可嘉的，建築師為了要將飛機形的設計加以改良，來適應高級

8

9

10

11

12

13

14

15

16

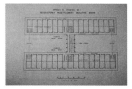
17

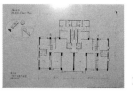
18

19

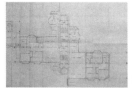
20

住宅大廈的要求,就將飛機的一翼推前,作出搓摩形式的平面(20)。因為社會環境的變化,缺乏佣人,這就產生了背對背的飛機形的平面(21),取銷了佣人房,甚至發展到我叫"直升機"形的平面(22)。另一香港建築設計的特徵,就是因為依出建築而產生的結構地基支架(23)。

香港是一個人煙稠密的地方(24),我認為香港的公共房屋設計,反映很多地方的風土人情及社會特色。彩虹村的設計(25),可算是公屋總體規劃的首次創作,開創了高層公屋發展的典型;美孚新村(26,27)是私人大型高密度房屋群的首作,雖然這是一高密度、高層次的建設,但其中公共場所的設計是有沙漠中的綠洲之感。從此之後,更有花樣百出、不同圖案的大型房屋群的設計,例如:香港仔中心(28,29)、綠楊新村(30)、置富花園(31)等,真是百花齊放。其中我認為祖堯村(32,33)的設計是有代表性的,外形設計充分表達出公屋基本要求,毫無多寸浪費,而單元的設計,對地方氣候、通風條件都有好的交代。

沙田穗禾村(34)把祖堯村的設計推向另一高峰,三幢高層的大廈組成一群(35),地下各有綠化的中心庭院,而每一大廈的中央走廊是露天的,令人產生出一種天空的庭院入口感覺(36)。屋村商業中心的設計,是別出心裁的,充滿了現代熱鬧街市的氣氛(37),把傳統騎樓的商業樓宇中特點,用現代化的手法很成功地表達出來(38)。

其他新型的屋村商場,有些是多層及充滿了濃厚色彩的設計(39),有些更反映了歐美及日本的作風(40)。可惜以上兩種與香港的地方風格及風土人情,沒有直接關係。

另外一組的設計,我們俗稱為"行貨"(41,42,43),搬字過紙,是沒有心思的設計。這種設計大多數是事務所的繪圖員,而且純粹以他們對建築條例一知半解的看法,去作基本的設計。加上在外牆上畫上一些裝飾性的花邊,來點綴一下外表,有很多這一類的設計是工程師和測量師所為。1982年,當我在北京清華大學做報告時,我把這一類的設計命為"青春豆",或俗稱暗瘡,因為當時在北方的同行不能領略這名詞的奧妙,就沒有說出來,但今次難得到廣州來獻醜,所以趁此機會說出來給各位一笑。何為"青春豆"?乃是消化不良的後果,多產生在經驗及技術不夠的年輕人身上,可惜的是青春豆在年輕人身上是過渡性的,而在建築設計上就造成了無窮的後患。

玻璃幕牆(44)的設計是屬於較成熟的"青春豆",有些設計是頗為美觀,但因為這一類設計,對氣候及室內空調都不能符合要求,而且對地方風格更是完全不理會,可以說是"好

看不好吃"的奢侈裝飾品。

香港最早的玻璃幕墙設計，可算是阮達祖老前輩的恒生銀行(45)。從此之後，各色各樣玻璃幕墙雨後春筍般地建起來了(46,47,48)。玻璃幕墙對建築師來說，會節省施工圖，因為供應商可以承包這一方面的工作，而建築費用會比有些外墙材料貴得多，因而會有多些設計費用，對建築商來說，玻璃幕墙的安裝時間比較短，而且工地較為清潔。對甲方來說，除了美觀之外，對於實用及外墙保養方面，實有很多不理想的地方，可惜的是這些常病是建成後才發現的，而甲方因為"家醜不可外揚"，所以一般領導人也不知道這一回事，而香港的投資者奮然繼續大行其道，變本加厲，使玻璃幕墙的大廈規模越來越大，例如新的港澳碼頭(49)及未來的香港展覽中心等(50)。

對國內現代建築設計有影響力的，除玻璃幕墙外就是旋轉餐廳了。香港富麗華酒店(51)，可算是香港最早及最聞名一個例子，這真是一個害人不淺之作。

西班牙的設計，也是對地方風味及風土人情沒有交代(52,53)，因為西班牙式是適應地中海氣候，適合戶外生活而設計的，對香港來說，堅持要西班牙式的甲方，實際有點"牛嚼牡丹"不知其味也!

在低層的設計中，能反映到地方風格是不難的，例如這一群度假村房子(54)。但，對高層來說，我們還有一段日子(55)，將它兩放在一個工地上，有同床異夢的感覺(56)。從這個建築物的設計來說(57)，我們可以看出建築師飽受西方"獸性派"的影響，盡量將建築物的內臟表露出來(58)，雖然可以說建築師表露的結構，是中國傳統的構思，從這設計的半遮掩式的表露來看，說是受中國傳統的影響，未免實在是有點小題大做。話分兩頭來說，這建築設計表現出有份量的仿效，而不是一般無名之輩能做得來的。

美國的路易斯·康，在20年前，在Olivetti廠的空調系統設計中，首創了暴露性的作風(59)，這是很適合美國的社會及廠房的氣氛。但是在藝術中心的大堂(60)，來複製這個手法，實屬有點兒太戲劇化了吧?!

建築師往往在發展他的建築設計的過程中，受到外來的影響，例如這個學校的入口拱門及屋頂的小亭(61,62)，甚至利用刺眼的顏色來作主題(63)。有些建築師就用中國傳統的圖案及題材，例如牌坊等(64,65)，作一地方性的交代，就算用傳統的拱門形來仿門，我總是覺得祗是雕蟲小技，不能登大雅之堂。而這著名的老襯亭餐廳大樓(66,67)，在反映中國傳統及西方結構上面，做出了一點值得我們鼓舞的開端。

43

44

45

46

47

48

49

50

51

52

53

54

55

56

57

58

59

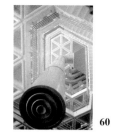
60

61

63

62

在海洋公園裏，有一群小館子(68)，從遠處看來，好像是完全仿古的做法(69)，但細心看，建築師並不是用陳從周老先生所喜歡的"仿古仿到底"的手法(70,71)，而只是表達了傳統的構思精神，用現代化的對比顏色及簡化的結構，來交代遊樂場所的戲劇化要求，把中國傳統設計帶到另一環境(72)。當然這一手法只是適應這一個特殊情況，並不能廣泛地在其他情況下發揮。雖然如此，這個手法的精神，是值得我們深入研究和發展的。

此外，香港有一種常見的具有風土人情的建築附件，就是我們俗稱的"百鴿籠"(73)。這附件反映了香港人的好霸佔性和貪便宜的壞習慣。這一現實造成了在中下層地區百鴿籠群的那種醜態，同時，這也是構成了地方社會的一個特徵。雖然政府已用違反建築條例為理由，有許多被清拆，但是在建築設計上我們也可以用一種外牆處理手法去解決(74)。這樣，一方面用混亂的顏色去掩蓋這個醜物；另一方面也發揚了一點地方風格出來。

這個辦公大樓是70年代的作品(75)，當時的建築師還關注到遮太陽的需要，可惜從此之後，在辦公大樓的設計中，就很難找到這樣適應香港氣候的必需品了。在香港所有的玻璃幕牆辦公大樓中，沒有一間是有此設備的，只有新的匯豐銀行(76)是例外。這位外國建築師是初次在東南亞大施手腳，而且花了不少心思及用大比例的模型來做實驗，去研究地方氣候的需要，可惜我們本地的同行，百思不得其解，要靠外人去指點迷津。

地方風格的另一個因素，就是建築要能夠反映生活方式和實際用途，以下的幾個建築物(77)，都能符合這一點。這三層高的排屋，充分表現了高尚住宅的形式，中間的起居室有半露天的小騎樓來供給一些戶外的活動，上下睡房表達出一家之主與孩子們分割的風俗，斜形的結構牆，很抽象地反映了依山而建的構思。

這間經營飛機餐的大型廚房(78)，充分表現整個生產及經營的過程，用很簡單的手法，去反映及應付好複雜的工業生產過程，實有真材實料，物盡其用之感。

這一殘疾兒童院方案(79)，為了甲方要保留工地當中的一幢19世紀的西式別墅，建築師便利用舊樓的形狀，做出了三個不同氣氛的小庭園，借用了中國傳統庭園的格式，來符合殘疾兒童院的不同的室外空間的要求(80,81)。斜形的屋頂是適應引進陽光，增加自然空調，及安裝太陽能設備而形成的。整個方案是充滿了濃厚的大家庭氣氛及中西合壁的地方風味。

64

65

66

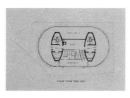
67

68

69

70

71

72

79

73

80

74

81

75

76

77

78

雖然在低層建築中，我們是比較有些表現，但在高層設計裡，我們更要走一段很長的路，誠然，香港是有些值得我們共勉的例子。

老實說，我對高層的地方風格的定義，正處在朦朧中，早上提到的穗禾村、半山的高尚住宅公寓，從這兩者之中，我們可以很清楚地分別出以下各點：這些建築物是住宅單元，穗禾村比較繁密(82)，單元也比較小，密度比較高；另一間的單元是複式的，反映了比較高尚的生活方式(83)。兩者都是用最經濟的結構和材料，反映了香港人對事論事、實事求是的作風。

這幢市區內的住宅商業混合大樓(84)及小學(85)的設計，在外形及單位佈局方面，反映出繁密和緊湊的特點，無一浪費的地方。在這幾方面，都是香港風土人情及地方風格的因素。當然，我早上已說過，要從高層設計中表達出地方風格，還要走一段道路的。

最後我借用以上的分析，來談談國內的情況。中國傳統建築包羅萬有，例如布達拉宮(86)、四川的羅城(87)、四合院(88)、民居(89)及木結構(90)...。

50年代飽受蘇聯建築影響的文化宮(91)；60至70年代有歐美的污染(92,93,94,95,96)；80年代又有不倫不類的香港風氣也吹進來了。如果用傳統的屋頂就可以交代了地方風格的話，那麼這新加坡的高層飯店(97)，也可算是中西結合的一個傑作了。

台北的圓山飯店(98)，看來真的是"仿古仿到底"，但它的誇大比例，會氣壞陳松同老先生的。另一方面，台北的美術館(99,100)，就走了仿日的極端。南京的金陵飯店(101)，不單單是對市區的交通增加了麻煩；而且對市容和地方風格也是格格不入的，形成一個強烈的對比。

台北的中正紀念堂(102)，就像香港海洋公園中的小館子群一樣，因為從特別用途的條件出發而且這用途是包括了一些戲劇化的因素，所以能夠採用半仿傳統式的手法來充分反映傳統的風格，這樣一來，增加了一些地方風格的氣氛。台北一幢地方法院的局部設計(103,104)，對地方風格有一點嘗試。而最近幾年來，台灣產生了一個中國式以後的現代派(105,106,107,108)，建築師將18、19世紀殖民地式的風格，搬到高層住宅大樓上，雖然這些設計像香港的藝術中心及老襯亭，是相當有水準的作品，但對反映地方風格方面實是有一段的距離。

地方風格是從用途、風土人情、民族傳統中產生出來的，不能夠只用傳統圖案，及其表面形式來表達。我們要認真

82

89

96

83

90

97

84

91

92

98

85

93

86

94

99

87

100

88

95

101

地、徹底地深入了解現代風土人情，和吸收傳統建築的精神，才能有成功的希望。

香港的鍾華楠大師有幾句話"方中有圓，圓中有方，光後有暗，暗後有光"(109)，風格的精神可能就是這四句的真言中找到的。

雖然我的報告沒有發表實質的答案，我只希望通過我的初探，來刺激國內同行的興趣，彼此攜手合作，向創造現代化的建築風格進軍!

淺談香港私人建設前景

1985年9月4日在希爾頓酒店
—香港地產行政人員協會午餐會上的致詞

會長、各位嘉賓：

我首先是有兩件很抱歉的事要向各位交待：第一件，為了振興中華，促進學術交流，在中英聯合聲明未發表之前，我請了一位從北京來的男秘書，以保我在文字上用普通話發表之文章的不足之處，可惜聯合聲明之後產生了港人治港的口號及設想，現在我發覺我常常要花更多的時間及心思去更改男秘書從我口中記錄下來的文稿，因為他用普通話的寫法去寫我準備用廣東話講的演詞，但今次因為你們行政人員的高度管理效率，要預先交稿，而且昨天我又要趕去參加你們劉副會長有份發起的另一專業組織的研討會，所以只能把稿匆忙的審閱一次，雖然發覺了很多用普通話的詞句，也沒時間去更改，所以你們手上的文稿有好多錯處，現在只有用即時現變的方式來補救，希望各位原諒。

第二件事是紹均兄在事前只吩咐我準備20分鐘的演講，我接到你們書面的通告時，上面是寫30分鐘。可惜我雖然向太太請了一個週末的晚上加上一個週日的大半天時間的假，我只可以寫成15分鐘的演詞，所以我會慢慢地讀出這一篇淺見，請你們不要以為我的演講的技術是有缺憾。

原本紹均兄是要求我講土地發展公司及市區重建的題目，因為種種關系，而且土地發展公司正是路經一個敏感的階段，所以我就搬龍換虎改換了題目，希望你們能夠在我發言之中領略到一些我對這敏感問題的意見。

我想首先多謝第一副會長劉紹均先生的邀請，使我今日有機會來參加你們協會的午餐會以及談談我一些淺見，我感到非常榮幸。同時我也覺得受寵若驚，因為通常我們建築師們只有聽從地產商的指示和提議，很少是有地產商要聽建築師講的，無論怎樣，過去我們建築師的意見很少被接納，特別是建設提議。當然，多數的地產商，都是熟習市場要求的，而有的建築師只是埋頭設計工作，與工務局爭論意見，但是今時不同往日，好多大型的建設都由專業人士去做鋪路的工作。我們建築師是最適合做這類工作，因為建築師對建設有影響的每一個賺錢或慳錢問題都很熟悉，而其他一般的專業人士只對建設的某一部分或某一環節有專業深入知識，例如地質、土地問題、結構、建設費用估價、市場需求、城市設計和綠化等等方面。我們建築師好像全科醫生一樣，對整體的建築計劃是最適合擔當統籌的角色。

私人建設的前景，主要是有兩個範圍：一是市區重建，二是新界發展。

現在在市區重建中，缺乏大規模的建設，到今日為止，市區重建多數都是小型投資，產生了很多圓筆形的建築物，這類建築物雖能夠供給更多的住宿單位，但缺乏其他有關的公共設施，同時，多數設計水準很低，這類建築物很快會變成城市的眼中釘。

地下鐵路上蓋的建築，可算是近年市區重建的驕驕者。這類建設多數是在地鐵公司和私人發展商合作建成的，香港政府已解散的"土地供應特別委員會"提議成立土地發展公司，去和私人地產商合作，實現大規模的城市重建計劃。這個建議，現在已交由新組成的土地及建設諮詢委員會去從長計議，準備立法程序。

我知道政府準備劃出一個地區，來做試驗性的處女作。私人地產商最好能夠開始研究和選擇適當的地區，利用現在市面的時機去開始做收購的工作。當土地發展公司成立之後，就會和私人地產商合作，收購適當的地區所剩餘下來的樓宇，幫助解決未能收購的房屋問題。當然，收購的條件要相等於市面的一般情況。前土地供應特別委員會，成立的一個研究小組，曾經提出了各種不同的收購方式，我希望現在由新的土地及建設諮詢委員會所成立的小組，能夠徹底研究這些提議。

我很有信心，在選擇一個地區作試驗性建設時，政府或者將會成立的土地發展公司會諮詢地產商們在一起做一個決定，因為沒有地產商們的同意和合作，這個試驗是不會成功的。正所謂"牛唔飲水唔襟得牛頭低"，在這方面，我希望地產商能夠經過他們的協會成立一個工作小組，去分析、選擇市區的範圍，提供給土地發展公司參考。這個工作小組的成員，可包括對地產發展有經驗的專業人士，這個小組也都可以向這一方面有經驗，以不牟利的發展機構去諮詢，例如香港房屋協會。

今時今日小型中型的地產商已覺得生意難做，這正是去考慮合組經營的最好時機。這樣一來，幾個中小型地產商可以合作，經營較大的市區重建計劃。

市區重建其中一個代價，就是將原來居民遠遷到其他地區，市區重建的地盤，多數居民屬貧戶階級，沒有能力去重新返回新建的單位。我希望地產商同發展公司能夠在將來的發展建設中，供應一些廉價的單位，同時在每一地盤能夠分期建設，使有些原本居民能夠重歸"家園"，這樣就能保存一些市區居民的傳統的民族色彩。市區重建計劃只是能做有限度發展，不能應付人口膨脹跟生活環境要求的各方面，只有從新界發展才是能解決這個方面的問題。有人提議在香港兩岸增加填海來解決，這對我們的不足夠的交通綱，更增加壓力。

就算你對中英協議—1997之後沒有信心，我都認為新界是發展房屋的最好的地區。這幾年來新市鎮的居住環境和房屋的賣價，與舊市區比較，公認新市鎮是理想得多，新市鎮只是缺乏周全的交通系統和工業的就業機會。

到今為止，我們多數居民是居住在高層高密度的居住單位，根據先進的地區的趨向，大部分人的願望都是想改住在低層的環境裡，所以我認為在不久的將來，要發展低中層和中等價錢的居屋，在新界是有利可圖的。但是，這些居屋建設應該包含較為完整的公共設施，諸如商場、文娛活動場所、小學和醫療中心等等。

在新界的西北部地區，比較適合發展低中層房屋，因有屯門、元朗的新工業和正在籌建中的輕便鐵路，房屋的建設，應沿着屯門、元朗的錦田公路，用帶形的方式去將三個市鎮聯系起來。

如果我們對中英協議是有信心的話，這新界西北地區的規劃，可能在1997年之後為了適應新的政治關係，會重新修改。後海灣的南岸可能提供一些工業發展的機會，當然這個提議是我一個投機性的想法。

總括來說，地產商應該在土地發展公司沒有成立之前，做好準備，聯朋結隊去成立對這方面工作有協助的小組，也應該協調起來，去多做些建設大型市區重建計劃的工程，我想在這裏引用鄔維庸醫生昨天創釋的一個名詞就是"企業精神"entrepreneur spirit，發展商應該憑着"企業精神"去直接向新界發展，向新界進軍。這樣一來，在人口不斷膨脹的香港，對改進生活環境會做出很大的貢獻。

一個香港建築師眼中的現代中國建築

現代中國建築創作研究小組于1987年8月在新疆烏魯木齊舉辦學術研討會上發言

幾年前，我在國內作了幾次報告，並通過國內的建築雜誌例如《建築師》、《建築學報》和《世界建築導報》發表了一些文章。當時我對建築設計及傳統和地方特色的關係所發表的意見，認為要以溫故知新的作法，從研究古代建築的例子中，熟習傳統的構思淵源及深入了解現代生活方式，從而創作出有地方風味和傳統特色的建築設計。

最近的一年來，我對這個題目，又作了進一步的深思，對以前的意見有些不滿，或者說深化了一步。現在則認為，建築設計主要的成果，就是創造一個適當的建築環境，來給人類享用，所以現代的建築物是應當反映出現代人的生活方式及現代化的建築技術，而反映傳統和地方特色，只是一個副產品而已，所以盲目地研究古建築或用古建築的各種特色，例如：紅墻綠瓦、八角形窗框等放在現代的建築外表上，並不能算是反映了傳統和地方特色。

我們不妨看看中國傳統的古建築(1,2)，它們的設計，客觀上是反映當時的生活方式及建築材料與結構的形式。因為當時運輸困難，所以除宮廷及大型公共建築外(3,4)，所有其他建築物多是用本地材料的，而當時的結構及施工方式，是由工匠控制，加上閉關自守的國風，造成沒有外國進口材料的影響，所以清朝以前的建築創作，從我們現代人的眼光看，是豐富了傳統及地方特色的(5)。

清末到民國以來，因中國對外大規模的開放，導致有外國租界等外來的風味，甚至洋風四起，將傳統都拋之戶外。從清末到今天，雖然是有一段很長久的日子，但是，我們在國內的一般建築創作，依然是擺脫不了這股洋風(6,7,8,9)。

從我手上的資料，看國內的現代建築，可分四大類：

1

2

3

4

5

6

7

8

9

第一類是屬於"畫蛇添足"型

以為在建築物的外形上,加了斜瓦屋頂或中國園林式的窗框,就是交待了傳統和地方特色,所有這些,都屬於這一類。難道我們穿上了西洋的衣服和戴上禮帽,就是一個洋人嗎?這一類的創作只是在外表上表達了作者的一番心意,並不能深入反映傳統和地方特色(10,11,12,13)。因為只是利用古建築的特徵,譬如用中軸綫來設計現代醫院,這種佈局,根本是食古不化。我對古建築的理論,了解不深,但我想中軸綫是從當時的宮廷儀式和向南背北的風水要求創造出來的,對現代醫院的操作,並不一定適用,試想病人那裡有心情去欣賞這樣左右相對稱的宏觀建築?
他們只是想能夠盡快躺在病床上要醫生治病。

第二類是"仿古型"

這一類的建築物(14),還是受一般人的喜愛,像仿古的瓷器一樣,很多人喜歡在家裡放一隻唐三彩馬或用仿雍正茶杯來品鐵觀音茶。

所謂"仿古",顧名思義,這一類的建築物是古建築的複製品,而在建築費用上,很不經濟,當然這種手法是對在古建築旁的新建築設計。在未有創造出一個富有傳統及地方特色的新建築形式時是一個交待(15)。我認為貝聿銘的香山飯店(16),雖然屬於這一類的產品,但是他的手法已經開始有擺脫盲目仿制的造法,而反映出一些新現代建築型的精神,可算是仿古派的姣姣者。

第三類是"國際型"

這一類的創作,是由參閱外國建築雜誌或清華大學出版的《世界建築》和從外國考察得來的,而沒有經過消化的產品;大師級的創作,可以媲美國外高水准的設計,例如得獎的北京貿促會國際展覽中心(17),深圳的華聯中心大廈(18),廣州的珠江實業中心(19),台北的中興紡織大樓和美術館(20,21),北京的崑崙飯店和上海的華亭賓館等等(22,23)。
達不到水準的話,就可以造出不太可觀的後果(24)。本人並不完全反對這一類的創作,因為我覺得,只需要一點消化的過程,這一類的創作可以演變成高質的新現代中國型。

第四類就是"新現代中國型"。

在敘述這一類型之前,粗略表達本人對建築創作與傳統及地方特色的淺見。新現代建築設計是,一方面能反映現代生活方式、人生哲學;另一方面也反映現代建築技術、地方環境氣候及用途。這樣看來,因為中國現代的社會,有很多方面的混作及生活方式已經演變到和外國等量齊現的程度,例如上班工作情況(25),醫療服務(26),体育活動等方

面(27)。而在建築材料及技術上，也和外國的相同，眾所周知的飯店及機場等建築(28,29)，根本多數是為外國旅客而設的，所以一般國際型的設計是情有可原的，那麼剩下來只有人生哲學和生活方式同外國有分別。這樣一來，只有在居住的房屋或大型屋村的設計中比較容易反映傳統及地方特色(30,31)，在交待氣候方面是可以有地方特色的，但一般現代中國的創作，是缺乏對這方面的交待。

以上淺見，不是說居住的房屋外，其他的建築物的設計，是很難有傳統及地方特色的表現，只是難於造成好的創作，因為我們只要看看日本方面，就知道是可以辦得到的(32)。日本的現代建築設計，是充滿了先進建築技術和高效率的生活精神，同時，在建築空間中，反映了日本傳統的氣氛。這象設計集團，對設計學校的哲學是值得在此一提的，他們認為學校是一個市鎮，課室是一間房屋，學校是一段回憶(33,34)。

香港在每一方面都很接近國內，它有傳統生活的一面(35,36)，也有蓬勃的商業一面(37,38)，有居屋的各種問題(39)，有富有貧(40,41)，也有人滿之患(42)，但也有調劑生活的方式(43)，整個社會是充滿實事求是、緊張的經濟氣候、繁忙的生活節奏。而所有的建築材料和生活所需物資，都是進口的，所以一般由香港建築師設計的建築物，是表露了這個社會的特點，可算有些地方特色(44,45,46)。但外國建築師所經手的設計，就容易分辨出來是外來貨，因為它同週圍的環境有一點脫節了，對傳統和地方特色沒有交待(47,48,49)。

國內的社會雖然同香港的情況不同，但有很多方面，是一致的，大家很熟習國內的這些情況，我想以第三者的眼光來選一些生活小照，相信看過後，會有許多回味(50,51,52,53,54,55,56)。因怕在急於現代化中，很容易忽略了我們的根底，我覺得人生哲學及生活方式，是對建築設計有大的影響，而建築師的想法，往往是集中於外形及建築技術方面，如果忽略吸收其他比較抽象的事物，忽略了人生哲學及生活方式的根底，在建築設計上，是等於人失去了靈魂，怎能表達傳統及地方特色呢？

如果不理一切，只顧實事求是，以實用及經濟為原則，這樣就會創作出很單調的建築環境，例如唐山重建的民房(57)，及北京的中日友好醫院(58)等，這些建築物對氣候及生活方式置之不理，只有表達到建築技術的一面；另一方面，如果只是集中用傳統建築中的特徵來交待地方特色(59,60)，那麼建築設計就會停頓在後現代派的水平。例如這一台灣雜誌裡的創作一樣(61)。

26

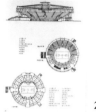
27

28

29

30

31

32

35

37

38

39

40

41

33

34

36

42

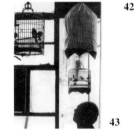
43

44

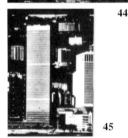
45

46

國內第四類的建築，是"新現代中國"型。

雖然到目前為止，我還沒有找到完全符合這一類的建築創作，但是我想在這裡介紹兩三個頗為成功的建築創作，來表達我的心意，第一個就是北戴河全國政協休養所，雖然在設計中是有斜屋頂，但這斜屋頂並不是傳統的複製品，它的設計是吸收了週圍環境的精華，和達到我以上所講的，在建築創作中，能反映到傳統和地方特色的因素(62, 63)。第二個就是台灣溪頭青年活動中心(圖64)。第三個就是台北的世界貿易中心(65,66)，這個中心跟深圳的世界貿易中心相比，真是有天埌之別(67)。

以上引用不少香港和台灣的例子，是因為香港及台灣在建築設計上，先走了一些路程，而國內正好開始在尋找自己新路的時候，最好是觀察香港及台灣兩地的錯失和成就。這樣，不但可以節省一些開支，而且也可避免走許多彎路，相信會有參考價值的。

我在今年五月底，參加了中國建築學會和香港建築師學會舉辦的研討會，在我的記錄中有兩句四言警語，是為了準備寫這篇文章而記下的，當時不知道是不是張欽楠先生說的? 或是我聽了張先生的報告後有感而寫下的提要? 這兩句就是描寫我們建築師現在的兩種通病"食古不化，食洋不化"。

最後我希望大家"食古早化，食洋早化"，切切實實，共同努力，創造出一個新現代中國的流派。

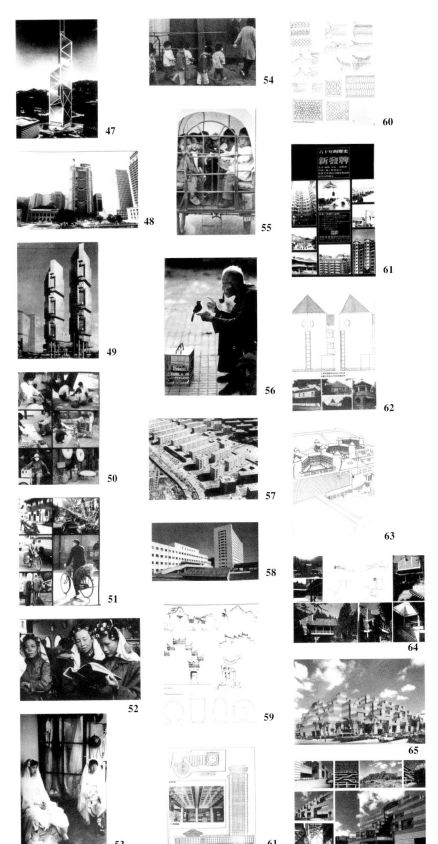

聯繫各界　發展科技

香港北區扶輪社午餐會發言，14/07/1988

香港科技協進會成立於1985年，在短短的三年時光之中，幸賴社會各界人士不斷的鼓勵和支持，加上會友們協力幫助發展會務，現在已經有了一定的規模，奠定了進一步發展的基礎。

我現在想用四個標題來介紹我會發展的經過及將來的展望。第一個標題是成立過程和宗旨；第二是以往的成就；第三是會務發展；第四是今後的展望。

(一) 成立過程及宗旨

1984年秋，時逢新中國成立35週年紀念，香港一批資深的科技界人士，受邀到北京去參加國慶典禮，在觀禮台上，我們大家有一個共同感受：受邀者雖然同是香港科技界的工作人員，彼此也許心儀已久，但各行專業，少有來往，以致隔膜殊多，更遑論學術上的交流了，這豈不是一個損失嗎？有見及此，回港後，大家便倡議在香港組織一個容納各種科技人員的協會。

發起人是陳耀華教授，倡議人是各專業的代表，大家共同奔走，遂在1985年6月19日正式成立"香港科技協進會"，陳耀華教授便是本會的第一屆會長，當時納有會員80多人。

香港以一個彈丸小島，可以成為一個高度現代化的城市，本港科技界的貢獻肯定是它的一個重要支柱。雖然本港科技界各行業和領域已各自有專業學會存在，但都限於各行業的範圍，沒有進一步跨越科技界各行業之間的橫向聯繫。事實上，我們誰都知道世界上每一件科技產品或工作，都是綜合性地互相協作產生的。一個綜合性的"香港科技協進會"的成立，應可在不同專業聯繫上，通過互相溝通，產生橋樑作用。它，可以使本港科技界綫狀的聯繫，跨前一步編織成為綱狀的交流，發出整體運作的功能；通過彼此密切的合作，對本港社會的繁榮發展，必能發揮更大的作用。

香港科技協進會成立的目的，就在於：一方面希望協調科技界各行專業之間的業務關係；另一方面也希望促進彼此間的人際關係，使本港科技界能夠和諧合作，提高科技水平，造福社會。

為了溝通各類型科技工作者的目的，本會會員包括下列各類：

甲：正式會員——凡符合資格，在學術界、專業界或工業界服務的科技工作者，均可申請為正式會員。一般資格要求是：學術界科技工作者，需有講師級或以上的資歷和經驗；專業界需要有認可的專業學會入會條件；工業界應首先是從事科技發展有關的工作和企業。

乙：附屬會員——凡熱心於科技發展，而本身專業與科技無直接關係的專業人士，俱可以申請為附屬會員。

丙：公司會員——有規模而從事與科技有關的公司，可以申請為公司會員。

本會會員人數已從創會時的80人，不斷增加到現在已有約200個正式會員，他們都是本港資深的科技工作者，分別服務於本港各工商界、專業、學術團體中，其中學術界的會員，有港大及中大教授9名，博士40名，專業會員包括醫生、建築師、工程師、測量師等，廠家會員大約有30名。

(二) 以往的成就

香港科技協進會在成立初期，即獲得楊振寧博士、黃麗松前校長和馬臨前校長答允出任名譽顧問；安子介博士及陳壽霖前議員答允出任名譽會長。去年又獲得何添博士、何鴻燊博士、李嘉誠博士及鄭裕彤博士等4位關心科技發展人士慨捐港幣二百萬元，用作購買會址之用。香港科技協進會之所以迅速地在香港奠定基礎，各方面的熱心支持是一個很重要的因素。

在過去幾年內，香港科技協進會不只開啟了本港各科技團體及學會之間的聯繫，更成功地與大陸及國際同類型的組織取得了初步聯繫，得到了認許，建立了聲望。

幾年來我們的活動方面，除了定期性每月會內專題的科技講座、服務社團的科普教育；更曾於1985年9月與中國核學會合辦了一個"核技術展覽會"從科學技術的角度上，全面地、客觀地向香港市民介紹了對於核能的認識。通過這類型的工作，本會更廣泛地和本港和大陸團體接觸，達到溝通的目的。

科學普及化是推動科技的一個重要環節，在這方面，本會除出任"香港中學聯校科展"的顧問，為每一間參展學校提供有關科技指導外，更曾與香港電視廣播有限公司合作，在"青春前綫"的節目中，提供了13集有關本港科技的資訊，叫做"科技點滴"的，有效地向青少年推廣科技知識。還有，去年本會與九龍西區扶輪社合辦了一個"科技發明獎"、與市政局科學館合辦了一個"香港科技攝影比賽"等，都是本會在科普方面綽綽的貢獻。

(三) 會務發展

我會每月不斷舉行有關科技的研討會，從去年11月開始，相續有張延生教授講氣功治療，中國統計訊息咨訊服務中心考察交流，黃康權博士主講有關"高溫超導體之應用"，安子介博士介紹安子介中文寫字機之發明及運作。

同時本會對在青年一代中，推廣科技的工作，甚感興趣，今年本會繼續擔任中學聯校科展的顧問，該科展將於10月23日至30日，在香港展覽中心舉行。此外本會也積極參與社會有關科技活動，例如：大亞灣核電廠事故及維修研討會，由香港政府統籌和本會主辦的世界保護環境日研討會其主題是"環境與文化傳統"。明年在香港舉行國際性電腦研討會，本會同意單獨贊助。同時本會將會透過學術研討會介紹及引進新的科技產品到香港的市場。

有關國際關係方面，本會曾探訪英國的皇家學會，並與皇家學會同意連同中國科學技術協會合辦學術及專業交流。今年較早時候，本會曾探訪北京中國科學技術協會主席錢學森先生，討論今年內的各種交流活動之安排，包括11月舉行之四小龍的國際會議。3月中旬與深圳市科學技術委員會及科技開發交流中心洽談雙方交流事項，我們和中國民辦科技專家協會合作籌辦，在深圳舉行科技交易會，這將會是中國民辦企業家與香港工業家首次的科技接觸，這個交易會原本訂於9月舉行，現因有些技術問題，可能延期到明年。本會有些會員在許多科技研究工作方面，有廣泛經驗及專業知識。今年內本會組織一系列的內部科研小組，研究下列項目：精密工程、基本物理過程、人工智能及專家系統、微電子及集成技術、超級計算、生物工程、未解釋現象或不明現象、工業自動化及機械人，科技與社會等。小組將於9月舉行對外的交流活動，小組的努力將會刺激及促進香港這類罕有的研究工作。

本會今年開始，設立了公司會員，並接納震雄機器廠有限公司為第一家公司會員。本會希望更多公司能參加本會，以求達到與工業界更密切的關係。

今年11月間，本會將主辦一個國際性的會議，主題是亞洲四小龍之新工業發展，會議將於11月14號至16號在香港舉行，並將於18號至27號到北京及西安進行會後考察活動。

這個國際會議是亞洲區首次舉辦，本會希望藉這次會議能增強本會與亞洲區內各會的密切聯繫，並同時能從這次會議鼓勵彼此在新科技發展上之交流。

(四) 今後的展望

香港最近成立了"科技委員會"，是對解決香港科技發展中的種種問題之第一步，本會在4月舉行了一次討論委會今後工作的研討會，委會主席、我會前會長潘宗光議員，也參加討論。眾所周知，香港科技發展是次於亞洲其他三小龍。香港政府在推廣其他事業，有很積極的機構，例如:貿易發展局、生產力促進局、建造業訓練局，但在科技發展方面，就沒有同類的推動機構。我認為，政府應考慮成立科技發展局，來積極推動香港的新科技發展。

我希望，今後我會能表達會員的期望，及與有關當區交換意見，對香港長遠科技發展作出一些可行的建議，協助政府推廣科研及創辦一些實際的科研設施，為香港的科技發展來努力!

A speech delivered at the Rotary Club North luncheon on 14/07/1988

Unite all fronts to develop Science & Technology

by Ronald Poon, President, Hong Kong Association for the Advancement of Science & Technology (HKAAST)

The HKAAST was founded in 1985. Within the short span of three years, the Association has definitely grown in size and in recognition with the support of people from various walks in Society as well as from our members.

I would like to describe to you, in four sections, our Association's development and our future aspirations, Section One deals with the establishment process and our aims, Section Two deals with our past achievement, Section Three our present work and Section Four our aspirations.

1) <u>The Establishment Process and our Aims</u>

In the autumn of 1984, on the occasion of the 35th Anniversary of the founding of the People's Republic of China, a group of well known Hong Kong personalities in the field of Science & Technology was invited to attend the celebration ceremonies in Beijing. It was then that they all felt that there was a distinct lack of communication between the various concerns and professionals related to development of Science & Technology in Hong Kong. On their return, the preparation was underway to form an association consisting of all those whose work relate to the development of Science & Technology. Prof. Chan Yau-Wah spearhead the formation work with a group of representatives from the various professions. Thus on the 19th June 1985 the HKAAST was founded. Prof. Chan became our first President. At the time there was over eighty members.

Hong Kong despite its small size, has become a highly modernized city and it cannot be denied that the science & technology input does provide a fair support in the realization. Although there exist in Hong Kong various professional institutions and associations related to science & technology work but because of their individual limitations, cross fertilization work amongst them has hitherto been dismal. Everyone knows that the creation of a science & technology product or project can only materialize with the co-operation of various parties. A multi facet Association such as ours can act as coordinator in the promotion of cross fertilization amongst the various institutions and associations and make invaluable contribution towards the prosperity of Hong Kong.

Our aim, therefore, is firstly to liase with the various professions and institutions involved in the development of science & technology and to promote amongst their members, personal contacts in the hope that people involved in the development of science & technology in Hong Kong can join together in the promotion of higher standards and co-operation in the work related to science & technology, thus benefiting the society as a whole.

In order to achieve our aims, we have created the following categories of memberships : -

A) Ordinary member - any qualified practitioner engaged in the furtherance of science & technology whether in academic institution research or professional or manufacturing field. Academics should be lecturers or above. Professionals should have qualification admissible by their respective professional institutes and industrialists should be engaged in work related to science & technology.

B) Associate member - any qualified persons otherwise in the field of science & technology who are involved or interested in the advancement of science and technology.

C) Company member - those companies doing business related to science & technology and of significant status.

Our membership has increased from the original eighty when we were founded to nearing two hundred now.

Our members are all highly qualified practitioners in the industry, in the profession and in academic circles. Amongst them, there are about nine professors and forty doctorates in our academic memberships. Thirty industrists and numerous doctors, architects, engineers and surveyors in the professional section.

2) **Past Achievement**

At the founding of our Association, Prof. C.N. Yang, Dr. Rayson Huang and Dr. Ma Lin became our Hon. Advisers, Dr. T.K. Ann and the Hon. S. L. Chen became our Hon. Presidents. Last year Dr. Cheng Yu-Tung, Dr. Stanley Ho, Dr. Ho Tim and Dr. Li Ka-Shing became our Hon. Presidents as well and jointly donated HK$2 ml. for the purchase of our Association premises. The speed in which our Association has established itself is heavily dependent on the warm support we have received throughout.

In the past three years, our Association not only developed a link between the various associations and institutions involved in the development of science and technology but also successfully established contacts with similar associations in the mainland and overseas.

We have in the past three years maintained monthly meetings on subjects related to science and technology with guest speakers, and participated in community projects related to our aims. In September 1985 our Association jointly with the Chinese Nuclear Society organized an exhibition on Nuclear Technology. The exhibition showed general information on nuclear energy and production processes.

Popularize science and technology is an important step in the advancement of science and technology. In this area our Association has been acting as adviser to the Joint Secondary School science exhibition. We have also co-produced with Hong Kong Television and Broadcasting Co. thirteen episode on science and technology information with the 'Youth Frontier' TV programme. Last year together with the Rotary Club Kowloon West organized an inventor's award programme and jointly organized with the Hong Kong Science Museum a Science and Technology Photographic competition. All these projects contributed significantly towards the deepening of knowledge on the subject amongst the youth population.

3) **Present Work**

Since November last year we continued to have monthly meetings. We have Prof. Zhang Yan Sheng on 'Qigong', a visit by China Statistical Information and Consultants Service Centre, Dr. H.K. Wong on 'the use of High Temperature Superconductor' and Dr. T.K. Ann on his Chinese Word Processor.

This year we continue to act as adviser to the 'Joint Secondary School Science Exhibition' which will take place in the Hong Kong Exhibition Centre between 23th and 30th this month (July 1988). We participated in the discussion concerning the remedial measures for the Daiya Bay Nuclear Plants. On 6th June, at the request of Government, we organized a seminar in celebration of World Environment day with a theme on "environment and cultural Heritage" at the Hong Kong University. In addition we have agreed to act as the sole sponsor for an International Conference on Computer scheduled for next year at the new Exhibition Centre. Meanwhile we continue to introduce new products and information to Hong Kong through seminars.

In international relationship, we have visited the Royal Society in London and came to an arrangement together with the China Association for Science and Technology to promote academic and professional exchanges amongst the three Associations. Early this year we visited the China Association for Science and Technology in Beijing and met their President and discussed with them the various joint activities for this year including the international conference on 'the four Little Dragons of Asia' scheduled for November in Hong Kong. In the middle of march we had discussions with the Shenzhen Science and Technology Commission and Science and Technology Development and Exchanges Centre in joint projects. Presently we are organizing a Science and Technology Trade Fair with the China non-Governmental Science and Technology Entrepreneurs Association which was originally scheduled in Shenzhen in September this year. However in view of some technical problems this fair is re-scheduled to next year. We have many science and technology experts amongst our membership and this year we have established a number of Research Study Groups covering Precision Engineering, Fundamental Physical Processes, Artificial Intelligence and Expert System, Microelectronics and IC Technology, Super computing, Biotechnology, Unexplained Phenomena, Industrial Automation and Robotics, Science, Technology and Society. The groups plan to have an inaugural seminar in September this year. No doubt the work of these Study Groups will stimulate further activities in these rare areas of scientific research.

This year we have also established company membership and have accepted Chen Hsong machinery Co. Ltd. as the first company member. We hope more company will join us thus achieving closer relationship with the industry.

In November this year we are hosting probably the first international conference on "the New industrial development in the 'four Little Dragons' of Asia" in Hong Kong. The conference is scheduled to take place between the 14th and 16th of November at the Sheraton Hotel with a post conference technical tour of Beijing and Xian. We hope through this conference our ties with the oversea Associations will be further strengthen and the exchange of new development information will be further encouraged.

4) **Our Aspirations**

Government has just established the Science and Technology Committee. No doubt this is the first much needed step to solve the problems conformity us in the development of Science and Technology in Hong Kong. Our Association organized a seminar on the work of this committee in April this year with Prof. C.K. Poon, its chairman, our past President participating. Everyone is aware that Hong Kong is lacking behind the other three Little Dragons of Asia. However the Government in dealing with other pressing problems in Hong Kong has in the past created effective boards to spearhead development such as the Trade Development Council.

The Productivity Council and the Construction Industry Training Authority etc. Nevertheless in the development of Science and Technology, we do not have a similar organization. I personally think that Government should establish a Science and Technology Council to effectively provide and facilitate Science and Technology development in Hong Kong.

I hope we can consolidate our members' views and consult Government with a view to develop a long term development strategy as well as to assist Government in establishing research development and research facilities in Hong Kong.

《香港建築》序言

萬里書店及中國建築工業出版社出版，1989

1983年9月，中國建設報社副社長、好友楊永生先生來信，提及出版社的三位先生朱象清、彭華亮及韋然將來港籌辦《香港建築》一書的事，到今已有兩載矣！在這兩年內，誠蒙出版社社長周誼先生竭力推動，及港方各好友、同行的支持，總算把這件事辦妥。港方的統籌工作是由前香港建築師學會長徐志樑先生為首負責，多年友好鍾華楠先生、何弢博士、郭彥弘教授及本人擔當顧問，參與籌劃。

本書的內容包羅萬象，從香港現代建築發展的歷史、新市鎮規劃，到各類型的建築物，例如公共建築、商業、教育、醫療，以及住宅、屋村、市政和園林建設，都有專文評述，更有數十幢精選的建築物用圖文作詳細介紹。

雖然開埠期間的香港，在建築設計上是富中國南方的建築風格，例如高度的斜頂、深深的屋檐，店前的騎樓等特徵，但同時因為受英國統治，也有濃厚的19世紀英國殖民地的色彩。現代多數香港人的生活方式，差不多都是洋化了，而所有的建築材料及施工方法，都是外來的，所以香港現代的建築設計是顯現着各國的建築板樣面貌。這種作風也帶着亞洲各大城市建築設計的通病，即忽略了地方氣候及風土人情。香港的現代化建築設計，可以用以下一副十言對來形容：建得快，仿得多，多姿多彩；缺突破，無風味，味如嚼蠟。香港的現代化建築設計，雖然有它的缺點，但也有些優點。很多國內的建築師，往往以為香港的建築好看不好用，建築物的本身似乎說明香港表面上的繁榮和緊張擁擠的現實，對內是不合時宜的，因而不去研究了解香港建築設計方面的一些好因素。雖然香港的建築設計不能說是世界的群首之一，但它們始終是在建築條例、經濟要求和國際文化交流影響下產生的可喜成果。從事物本身的發展來看，這往往是產生建築流派的必經之路。

香港跟內地有密切的關係，有許多共同點，諸如倫理道德觀念、語言、文化背景等，而內地在最近幾年內，幾個特區的發展方針，都是用仿效香港的口號，香港距中國最近，是資本主義社會的一個櫥窗，又有很多國內的機構對中國考察人員來說，香港是深入了解西方的最好的跳板。從國內到國外去考察的建築師，最好先研究香港的建築設計，這樣可使中國現代建築的發展，避免許多浪費和彎路。同時，我希望中國有關人員，能了解問題之癥結，放棄盲目複製和仿效外國的形式，例如玻璃幕牆、旋轉餐廳、一千客房的酒店等，最好能讓建築師們充分發揮他們的智慧，創造一個在生活要求上、對中國更加適合的具有地方風味的新建築。了解吸收香港建築的優點，宜仔細地翻閱這本書裏的文章及實例，此外更要研究香港的地方風土人情、社會制度及運作機制，才可以達到事半功倍的效果。

今後，我們要對我們的文化以及形成地區風格的各種因素，進行研究並提高這方面的興趣。我們應放棄仿效別人的做法，去尋找具有真正的、永久性的、地區建築風格的規律來，建築設計應有地方風格及時代精神，才可以改進社會的環境，增加生活情趣，促進社會的繁榮。

Obituary

Mr Dai nianci 1920-1991

Originally published in HKIA newsletter 12/1991

The passing away of Mr Dai Nianci, President of the Architectural Society of China(ASC), is a great loss to all of us who knew him and a tragedy for the architectural profession in Asia. I first met Mr Dai in 1980 during the 6th congress of the Architectural society of China in Beijing. The deliberation of his approach to modern architecture, with sharp and humorous remarks often illustrated with hand gestures and facial expressions impressed me most during the workshop meetings.

Mr. Dai was one of the driving forces behind the cementing of relationship between the ASC and the Hong Kong Institute of Architects (HKIA) and took part in all of our joint efforts in the promotion of architecture in the past ten years, such as the opening of the first exhibition on Hong Kong Architects' work in Beijing 1982, ASC's official visit to HKIA in 1983, HKIA's official visit to ASC in 1985, the opening of the second exhibition of Hong Kong architects" work in 1987, to name just a few.

When a deadlock appeared on the representation of the architects of Taipei and Beijing in the Architects Regional Council Asia (Arcasia), way back in 1986, it was Mr. Dai who led an ASC delegation to Kuala Lumpur during the second Asian Congress of Architects to meet face to face with a delegation of Taipei architects, to try to resolve the situation. His dedication and sincerity impressed us most during that occasion and his skill of negotiation set the mood for a hopeful solution.

As president of ASC, he was looked upon truly as the father figure of his profession, not only as a leading designer but also as champion in the promotion of the profession's standing in society. During his tenure he managed to establish the respect of the profession amongst politicians and common people. He also managed to maintain unity amongst all the architects throughout China.

In the controversial design of the Queli Guest House, Mr. Dai developed what can be called the "modern Chinese Classic" style, in which he demonstrated a new concept to be used on buildings immediately adjacent to historical buildings. The design provoked admiration as well as criticism but, all in all, it woke us up to the complexity of the problem. It was, of course, design which was dear to Mr. Dai's heart. During his years as Vice Minister of the then Ministry of Urban and Rural Construction and Environmental Protection, he often admitted to me that he felt the strain of the post and that he was not suited to deal with ministerial matters. He longed to get back on to the drawing board. I fear that his years as Vice Minister probably robbed him of many years of his life. After stepping down from his ministerial post, he immediately rushed back to his drawing board and one of the occasion, when I paid him a visit to his office in the Ministry's building in Beijing to discuss matters concerning our two Institutes' joint activities, he started off by showing me the design drawings of his latest project instead. Though we were twenty two years apart, he had the ability to make me feel as if I was his contemporary. This quality of his earned him tremendous respect from the younger generation of architects in China.

Arcasia held its council meeting and forum for the first time in Beijing in late September this year. Unfortunately Mr. Dai had already fallen ill and was unable to attend. I was told that even in his sick bed, he insisted on checking the design drawings of his latest project. His dedication to work was an inspiration for us all.

Though he has left us now I feel that his devotion to work and his enthusiasm over international relationship will serve as guiding light in our pursuit of a better environment in this world.

在中國建築學會第8次全國會員代表大會暨學術年會上的發言

葉理事長、各位前輩、各位老朋友、各位女士及各位先生：

這次我從香港來北京參加第8次全國會員代表大會暨學術年會，我感到非常愉快。從1974年起開始來國內我已經參加了3次的全國代表大會，跟學會已經有10多年的關係。

參加這樣高水準的盛會，聽了各位女士及各位先生的講話，見到許多老朋友，結交了許多的新朋友，我也有很多收穫。

希望今後再進一步，加強大陸與香港建築界同行的交往，迎接1997年的到來。

作為中國人，我們雖身居香港，但對國內建設的每一項成績，對國內同行在建築設計中的每項成果，都感到非常高興。

作為建築師，我們雖然在香港做設計，對國內近10多年來建設規模這麼大，你們有這麼廣闊的天地做設計，我們是很羨慕的。藉此千載難逢的盛世，從事建築設計，你們是很幸運的。

現在我想對怎樣提高建築設計水準，作為補充意見，談談我的一孔之見，敬請各位批評指教。

我聽說大陸幾十年來都提倡群眾路線。我認為，我們建築師也應當採取群眾路線，藉此增強建築師與大眾的交流，取得共識。比如大陸建設很多的住宅，建築師不能太自以為是，要多聽聽居者的見解，他們是上帝，要舉辦設計展覽，讓居者品評，這是提高設計水準的一個辦法。

建築師要從建築"沙龍"中走出去，步向社會，不要只集中去幹量大質小、大體粗作、大而無當的三大設計樣本。要增強建築師的社會意識，增強社會的建築意識和大眾的參與意識。

我希望大陸多舉辦建築設計比賽，從比賽中可以分辨真假高低。收藏古玩的人，都是從比較中發現真偽。官方的和民眾的優秀設計評選，近幾年已經開始做了，這是一件很有意義的事情。我認為，這件事不是做多了，而是做得還不夠。我們學會雖然也舉辦這一類的活動，但是學會因為還有很多其它的事情要做，而且經費與人力不足，所以要推廣設計比賽，就要多依靠其它單位去辦，唯一注意的是要求評判是要有份量的建築師去擔當就可以了。《建築師》雜誌做這方面的事情我是支持的，而且還會積極地參與。

另外一個提高建築師水準的方法，就是說服大家接受有建設性的評論，和鼓勵多些建築師去作評論。大家做完每一件事後，都要檢討得失，何況建築設計？國內建築甚多，評論甚少，評論家更少，大多數介紹作品的文章，全是以導游的方式來寫，舉個例子：一篇介紹一幢新飯店的文章，內文這樣寫："西部為飯店大堂，中部為商店、餐館以及室內花園等……"其實建築師一看平面圖就完全知道的，而且這些文章是寫給建築師看的。大陸的文學、電影、藝術等領域評比多，評論也多，社會也重視，可是與大眾十分密切的建築，在這方面卻遠遠不如。

缺少評論就很難提高創作水準，文藝如此，建築亦然。希望各位先生及女士齊動手和動口來做評論，而且運用各種輿論和媒界(如電視、廣播及報刊等)來做建築評論。

以上意見可能不準確，敬希各位先生女士諒解。

我的話說完了，謝謝各位。

1992年3月8日

HKIA-ASC Linkage

Originally published in HKIA newsletter special issue "China Linkage", 01/1992

1. **Historical development**

The first formal interchange between Hong Kong Architects and their counterparts in China must have occurred in 1973 when the first group of HK Architects toured China. This was followed by second tour in 1974 organized by Mr. S. Liang of the Kui Kwong Development Ltd. The group visited some ten major cities in China & held numerous discussions with local architects.

Some years later, in 1980, Prof. Reginald Kwok, Mr. Chung Wah Nan and Mr. Ronald Poon were invited to attend the 6th Congress of the Architectural Society of China (ASC) in Beijing and took part in their workshop meetings.

It is not until 1981 that HKIA for the first time formally established a relationship with the ASC, under the Presidency of Mr. Ronald Poon. In that year preparation was underway to organize jointly with ASC the first ever overseas Exhibition by the ASC on "Traditional Chinese Architecture", to be staged in Hong Kong in 1982. The exhibition was staged in the New World Complex & the late Mr. Huang Hua Bin, a Vice President of the ASC, was invited together with the Hon. Mr. Donald Liao, then the Secretary for Housing, to officiate at the opening ceremony. At the same time HKIA launched a major exhibition of members' work. A duplicate set of Exhibits was prepared in late 1981 for a touring exhibition which took place in Beijing on the 5th June 1982 with the Hon. Li Ximing, Minister for Urban and Rural Construction and Environmental Protection in China & the late Mr. Yang Tingbao, President of ASC officiating. Mr. Ronald Poon & a delegation from HKIA attending the opening ceremony in Beijing. The Exhibition consisting of some 800 photographs representing the works of about twenty leading member firms toured major cities in China over a period of several years. In the same year the HKIA China Forum was established to facilitate members' exchange of work experience in China.

In 1983 the HKIA cooled off activities with ASC. Ironically in the same year ASC decided to honour three HKIA members as Honorary Council members of the ASC. They were Prof. Reginald Kwok, Mr. Chung Wah Nan and Mr. Ronald Poon. It was only through the personal efforts of the three that exchange activities was resumed in 1984 with the arranged visit to Hong Kong of the President of ASC, Mr. Dai Nianci, the Vice President Mr. Yan Zixiang and the Secretary General Mr. Gong Deshum (who is currently the President of the newly formed Chinese Institute of Architects under the ASC). The visit was organized & financed on a personal basis. This was followed by four professional study visits to HK from various architectural groups from China. In the same year Mr. Ronald Poon again, on a personal basis, organised through the Hangzhou Branch of the ASC, a study tour to Hangzhou and Huang Shan with a number of HKIA members.

In 1985 under the Presidency of Mr. Edwin Wong, the formal relationship with ASC was resumed with a courtesy visit to ASC in Beijing. In the same year a lecture tour of HKIA members took place in Guangzhou at the invitation of the Guangzhou Branch of the ASC. In addition Mandarin courses were organized at HKIA for members and a dictionary was produced for frequently used architectural terms in English and Chinese.

In addition the China Committee was established and a symposium on "HK Architects' Practice in China" was organized by Mr. Leo Zee. This was followed in 1986 with a symposium in Hong Kong on "Modern Building Development in China" with invited Chinese speakers.

In 1987 under the Presidency of Mr. Jim Kinoshita, a delegation of 14 HKIA members attended a symposium held in Beijing organized by the ASC. Concurrently a members' work exhibition was staged in the Exhibition hall of the Ministry of Urban and

Rural Construction and Environmental Protection. However in 1988 again no interchange occurred between HKIA and ASC. Towards the end of 1989, after the 4th of June incident, there was an international conference in Beijing organized by ASC and Mr. Chris Haffner, then the HKIA President, together with Mr. Ronald Poon attended.

The linkage between HKIA & ASC can be described as "hot & cold" between the time of the official establishment of the relationship back in 1981 & now. Unfortunately the whole issue seems to depend on the attitude of the current HKIA President at the time. Occasionally a few individual members who are keen on maintaining the relationship managed to sustain some activities between the two Institutes. In the ten years from 1981 to now, individuals such as Messrs. Chung Wah Nan, Ronald Poon and Leo Zee have been consistently organizing lecture tours and interior design courses as well as attending conferences throughout China. Most of these activities were carried out on a personal basis and at their own expense. Fortunately ASC regards these activities as interchanges with members of the body of the HKIA. Hence as far as ASC is concerned, the relationship has been regularly maintained.

2. Current situation

The current situation is very encouraging as the present HKIA President, Mr. Stephen Poon's enthusiastic over maintaining and broadening the linkage, not only with ASC, but also with other international professional bodies. Hence in the ARCASIA Council and Forum 6 meeting held at Beijing in September this year, we saw the largest contingent (some 19) of Hong Kong architects attending the occasion, while in the past only one to two members attended. HKIA also had some formal interchanges with ASC and the Ministry of Construction. In November this year we witnessed a resumption of tour activities with a HKIA group visit to Tai Yuen, to participate in work sessions on urban projects. This year we also saw the re-surfacing of the China Committee, now re-named "China Affairs Committee". In addition the HKIA Council is planning to invite the President of ASC to Hong Kong. It seems that this year the HKIA is experiencing 'hundred flowers bloom' period as far as the linkage with ASC is concerned & it could not have come more appropriately as we edge towards the 1997 situation.

3. Future development

In the academic aspects Hong Kong is a very good training ground for chinese architects. HKIA can play a leading role in facilitating short and long term training programmes in collaboration with the two local Schools of Architecture to provide mid career as well as fundamental academic course for the architects from China. For convenience of access HKIA can organize such activities in Shenzhen or Guangzhou in association with their local schools and their local branches of the ASC. This would provide a new dimension in our linkage with ASC and ensure that after 1997 any Chinese architects practicing in Hong Kong would have acquired similar training to our local architects. It is inevitable that there will be an influx of architects from China and instead of concentrating only on a stringent practice qualification system, we should also provide a more thorough training programme for the Chinese architects, who would help in balancing the 'exodus' situation of the local profession nearer to 1997. In addition Hong Kong will become an academic centre for the Chinese architects seeking employment in the Western Countries, leading to the possibility of the establishment here of even a third School of Architecture.

In the practice aspect, HKIA should arrange more exchanges with the ASC and all its branches in China such as lecture and study tours and annual official visits not only to Beijing, but also to other major cities in China as well. HKIA should also re-generate exhibition exchanges of members' work which have proved to be popular and influential amongst the architectural profession in China. The latest work tour organized by China Committee to Tai Yuan is a good start. Many more of this kind of exchange will not only benefit the Chinese architects, but also serve as a mid-career course for our members.

The Chinese economy has only slowly bounced back since 1989's June 4 incident, but then China recently experienced a serious natural disaster, which calls for large scale rebuilding work. It is time our Institute actively promotes opportunities for our members to offer their services in China. We should re-establish the China forum for members practicing or interested to practice in China, we need to facilitate cross exchanges of experience, as well as formulating programmes to promote practice opportunities in China.

Our future lies in a close relationship with our Chinese counterparts and our practice can only survive if we take on a regional view, particularly towards China as our nearest neighbour. In the past very few Hong Kong architects took any interest in their profession outside of Hong Kong. Very few participated in international conferences. The time has come for us to step outside our 'cocoon' and spread our wings. The recent surge of interest of the HKIA Council to participate in the international scene is encouraging. let us hope that it is a sustained interest throughout the future presidencies, so that we can truly develop a continuous working relationship with the international arena and, in particular, with our Chinese colleagues.

擺脫束縛　開展評論

原載建築論壇叢書第一冊《建築與評論》，天津科學技術出版社出版，1996年9月

我今天十分高興能夠和大家同聚一堂，討論建築評論。早在上屆中國建築學會的年會上，我已有意在各領導人面前提出國內缺乏有建設性的建築評論，而這類評論是促進提高建築設計的基本條件。但年會之後，當事人好像左耳入右耳出，這個提議就此石沉大海。當我今年三月在京與楊永生先生談到這個問題時，大家不謀而合地提議舉行一個研討會來推動建築評論，而且把研討會的發言稿刊登在《建築師》雜誌上，使更多人能領略到評論對推動建築設計創作的好處。有建設性的建築評論能刺激起創造性的新建築設計，而且能作為人民對建築設計的權衡指標。在近年國內百花齊放的情況中，人民往往被花枝招展的建築物外貌所誤導，尤其看到那些玻璃幕牆及建築物上的仿古亭子及符號，以為就是現代中國建築設計的好榜樣。在缺乏有建設性的評論的情況下，人民和年青的建築師就無從分辨。我們中國是禮儀之邦，中國人不太喜歡對他人作出批評，尤其是對同行或前輩。外國有專業建築評論家，而且多不是建築師，但他們的評論是受多數的建築師尊敬的，因為他們的專業知識深厚，而且是局外人，可以提出中肯的意見。我這個看法，不是因為我崇洋而是以事論事，在這方面外國的例子是值得我們學習的。

大家都知道在有關建築設計的雜誌上，所有對建築物的介紹或評論都是以導遊和稱讚的方式寫出來的，完全沒有批評的含意。對我來說這只是浪費紙墨，徒勞無功。話雖如此，但是我們不應該單方面提倡有建設性的評論。因為如果不培養能接受批評的心態，就不能夠很大方和開心地接受批評，並且會記仇於作出批評的人。這方面一般的中國人的氣量是比較小一點。舉個例子，在香港的外國人開會時無論在會上怎麼爭吵、怎麼樣翻臉，會後多數可以暢談痛飲，完全沒有會後仇，但我們中國人的會就多數辦不到。所以要使評論發揮它的效力，就要培養一個寬大的心胸。我希望從今以後，我們建築師能擺脫這個傳統的束縛，接受有建設性的評論，認真地發揮，創出真真正正符合國情的現代化中國建築設計。

建築風格與古城風貌

原載《建築學報》1995年第2期

提倡"古都風貌"是十分恰當的，也是有積極意義的。"辭海"對風貌一詞的解釋是"風采容貌，亦指事物的面貌格調"，當然，"風貌"主要是涵蓋城市的建築。但又不但是建築，一個城市的風貌還涵蓋着城市的精神文明這樣一個層面，也就是說既指"硬件"，也指"軟件"。

要實現"古都風貌"有兩種方式，一是創造出一個完全仿古的城市；另一種方式是加深理解"古都風貌"的涵意，在實際生活中創造出符合現代化要求的新的建築。

前種方式使全部建築物都以復古手法設計，人要穿古裝，車要有馬拉，這種做法在現在的北京只可以在城市內劃出些小區來實現，有如深圳的"錦鏽中華"或香港荔園的"宋城"，而它在功能上只能是起到游樂場作用，它不可能滿足現代城市運作的需要。

現今北京的一些新建築，為了要表面上滿足現今流行的"古都風貌"的要求，在高層的屋頂擺上一些仿古的亭子或者在立面上生硬地貼上一些古典建築的符號，在一般建築師還在摸索"古都風貌"正確表達的初期，這些方案就變成官方能接受的"古都風貌"典型。試問，如果一位女士穿上時髦的衣服，頭上戴上古裝帽子，這就是"古貌"嗎？一般北京現代的建築物，樓身是現代化的外型，而且多數有玻璃幕牆，而頂上加上綠瓦的古典亭子，這可能是一種表達"古都風貌"最簡單的解決方式。但如北京滿佈這一類的建築物，可能把"古都"演變成迪斯尼樂園的氣氛。

"古都風貌"這種提法多次在新聞媒界中提及，因此人們印象頗深。但許多人都忽視了北京市的領導人在給張開濟先生復信中提到的"首都風貌"這一說法，雖然只是一字之差，但後者是比前者更為準確，更為貼切的提法。"古都"當然是指明清的北京城，今天要想恢復也好，搶救也好都是無論如何也辦不到的。而建設首都風貌，則無疑是現實的，必要的。作為泱泱大國的首都—北京，理應有它自己的獨特風貌，就其大部分街區來說，它既不應該是仿古的、復古的建築的堆砌，也不應該是迪斯尼樂園的翻版或者是紐約的再現，而應該是具有五千年文明的中華民族現代社會的風貌，体現中國的四個現代化，蒸蒸日上的民族心態和社會進步，体現出10億人口大國政治文化中心的風貌。

建築物的風貌重點不在頂部，而是在建築物的下半。因為只有人透過眼睛才領略到其"風貌"，而大多數的人是在馬路上穿梭而過，根本看不見高層頂的風貌。這是現今在屋頂上放古亭做法的一離題地方。傳統的大屋頂是適應當時氣候、室內生活空間的各種要求而產生。而視覺上是人的眼睛在街上可以看到的。最重要的是其造型是完全出於適應功能的要求，而不是像當今高層頂上的大屋頂或小亭子完全是為了外表裝飾而設的。無論如何，建築物的風貌是整体的。首要是發揮建築物的功能從其外形上充分表達出來，無論在外牆上或屋頂上也都要表達到建築物內部的功能，使人們能清清楚楚地分別出建築物的用途，而其外形是配合內部功能的要求而設計的，不單是用來作裝飾之用而已。

"古都風貌"也好，"首都風貌"也好，唯一能適應現代化社會環境的，就是從功能、規模、地方色彩、地方氣候方面去發揮，不能以表面、以假的面具去嘩眾取寵。在"內應外合"的外形設計上能充分發揮古都建築的特點。在具体体現上述原則時就自然會產生出一個合理的"古都風貌"。(見附錄(14))

Speech delivered at the Urban Council annual debate on 06/01/92.

Chairman, after the sumptuous X'mas lunch in the terrace of my home with a vintage port in my glass and my family around me, I decided to read all the 34 speeches from last year's annual debate, the X'mas gift from Roy Spencer. As you know the speeches ranged from mini versions of the state of the nation review to a 6 page detail study of a small but significant area of our Council's work-public toilets.

The purpose of reviewing last years speeches was merely to help me to decide whether or not I should speak at all during this annual conventional debate. You probably are aware that a number of newly appointed Councillors including myself have refrained from speaking at the traditional Council open meetings so far and, instead, spent our time during these meetings trying hard to provide an attentive backdrop for the charade in front of us.

After reviewing the speeches I come to the conclusion that I should make my maiden speech in the Council meeting this time because I realize that the annual conventional debate is not only an opportunity for each of us to show our area of concern in our society but also an opportunity to reveal more of our personalities, our strong and weak points to fellow councillors as well as to the public at large. The speeches are like lines in your palm. People in the know can easily read out your character from one look at your speech. This parade of personalities once a year will help to improve the working relationship amongst ourselves. So I decide to speak.

Knowing that if I give you another "state of the nation" type of speech coupled with a detail research report, say, on the merits of squatting posture as against the seating position in the discharge of our daily waste, I would be regarded as "the man for all seasons". However five minutes after my speech, whatever I had read will be like dust in the wind. The material at best, can only add weight to the paper achives of our Council and will instantly be forgotten.

I am a practical man and I like my proposals to bear fruit as soon as possible. Therefore I shall only speak on two points to-day. Hoping that by the time I rise to speak again next year, what I am going to propose will already have been implemented allowing me another opportunity to raise two new proposals.

My first proposal concerns the traditional monthly open meeting of our Council. This meeting, I gather is designed to give an opportunity for the press and the public to attend our meetings and to witness the Council in the action at the same time allowing selected subject matters to be exposed to the public. Questions and answers are written beforehand like the script of a school play and the quality of some of the deliberations are not far from the standards of a school play either. Most of the questions circle round complaints of inefficiency in the Council's and the Department's daily work matters e.g., dog waste left lying in a lane in the backside of a housing estate. The Chairman of the select committee concern are always able to provide convincing answers. Though challenges to the answer given can be made by way of supplementary questions, this does preclude anyone of us offering any constructive comments. In any event most of the questions concern such "run of the mill" matters that, no wonder, we have a deteriorating audience rating in the gallery. It defeats the original purpose of providing a hunting ground of news from the Council for the press.

As for us Councillors who had not yet acquired the art of appearing attentive while taking a "forty winks", it is a shear torture to sit through two hours of meeting particularly after a few glasses of wine during the monthly Council lunch.

As far as I am concerned all those questions can be dealt with through several more constructive channels.

1) by slipping a note to the relevant Select Committee Chairman who will then consult with the Department to provide the answers at subsequent Select Com mittee meeting.

2) by raising it through any other business at the regular Standing Committee meetings.

3) The question and answers can be shorten to provide a press release, thus, facilitating more effective information to the press.

Taking the above comments and considering that all our meetings are now open to the press and the public, the monthly Council open meeting is, surely, a thing in the past and should be abandoned.

The other point I wish to make to-day concerns how our Councillors can make further contribution towards the work of our Council.

For the seven months I have been with the Council, I feel like a man without the use of his two arms constantly being mouth fed by the Department. Do not get me wrong. I am not complaining about the Department's efforts but I am critical of the limited method of contribution from our own Councillors towards the work of the Council.

We should untie our own hands and help the Department in enlarging the scope of our work rather than just simply being critical or complimentary of the stuff which are fed to us.

Our Chairman has shown us a lead while he was still an ordinary member of the Council last year during his speech at the annual conventional debate. he took a personal interest in a critical area of our work, be it may, a little unsavory and as a result we now have given the Department something new to do i.e., to promote a Hong Kong wide design competition on a proto-type public toilet. If half of us select an area or a project, we shall have at least twenty new projects each year. Elected Councillors can select 'black spot' projects within their district and Appointed Councillors can concentrate on policy issues or Hong Kong wide matters. This will give individual Councillors an opportunity to identify himself or herself with a particular project thus giving further opportunity to acquaint individual councillor with the public. To alienate the Department's worry over added work load, these projects need not necessary lead to creation of more ad-hoc committees. We should aim for projects which one or two Councillors can undertake themselves and report back to the Standing Committee or Select Committees. In conclusion we should be more creative and do something useful with our own hands rather than become dependent on being month fed.

With these remarks, Mr. Chairman, I support the motion.

Speech delivered at the Urban Council annual debate on 12/01/1995

Originally published in Urban Council official record of proceedings, 1995

Mr. Chairman, as you know, I always write my Annual Debate speech in pleasant surroundings in order to lessen the horrific task. This year, it is of no exception. However, this year I started early as I did not want to spoil my Christmas holidays. I wrote this at the poolside of the Westin Phillipine Plaza Hotel in Manila, during the morning break of the 6th Asian Congress of Architects, amidst splashes and laughter produced by some unruly Taiwanese holiday makers. This scene is reminiscent of the scenery we have come to experience in recent times in Hong Kong and come to think of it, in these Chambers occasionally where budding politicians try frantically to make themselves seen and heard.

Very often, we see them exposing their total lack of understanding and ulterior motives unintentionally, periodically making life unbearable, at times creating some excitement in our newspapers and occasionally making a reasonable constructive intrusion.

This scene is inevitable in the formative days of Democracy and particularly unavoidable when the pace is at neck breaking speed. When an organization reaches a certain point of development, new blood is necessary and vital to survival in this ever changing world. Like a human body at old age, occasional blood transfusions help to rejuvenate bodily functions and almost makes a new man out of you if the blood type is correct, but a total transfusion can be quite devastating. It either makes you into a completely different man or it may kill you if the blood type is foreign to yours. Therefore, progress and political changes should be gradual and sympathetic to the local situation.

In the last 2 years, the political scene in Hong Kong is similar to a place next to a volcano. We had the Patten eruption, we had the tidal waves which swept away all appointed seats and now, we are going through the after tremors created by the new breed of politicians. Let's hope that like a volcanic eruption, after a while things usually settle back to peaceful existence. And above all, the produce of volcanic eruption is usually fertile and provides new impetus for future growth.

I hope patience and tolerance will prevail as these theatrical disturbances, though painful at times, are only temporary and Hong Kong shall overcome and prosperity and stability will prevail.

To ensure that our island will not sink as a result of the volcanic eruption, I propose that we form a disaster rescue unit or back benchers club, or better still, call it by a name we have come to accept, amongst ourselves, the Motorola Fellowship, to provide a platform for alternate views to be heard.

This is possibly my last formal speech in these Chambers for the foreseeable future and I would like to keep it shorts as I have probably spoken too much already in the last three and a half years. Anyway, I would like to leave more time to those that follow me. It remains for the thank you, Mr. Chairman and my fellow colleagues of the Urban Council for your tolerance for my occasional outbursts during my term of office. For me, it was a most enjoyable and beneficial vocation. I would like to take the opportunity to wish the Urban Council every success in the future.

With these words, I support the motion.

在建築論壇第二次研討會上的開幕詞

中國建築論壇第二次研討會於1996年8月在天津市舉辦
原載《建築師》1996/12期及建築論壇叢書第二冊《比較與差距》，天津科學技術出版社, 06/1997

舉辦中國建築論壇的設想是在一個很湊巧的場合裡產生的。1995年3月22日，我在北京參加鍾華楠先生為了新任香港建築師學會會長一職到京與中國建築學會及有關單位的領導人進行學術交流活動，晚上在下榻的西苑飯店咖啡廳與好友《建築師》雜誌編委會主任楊永生先生把酒言歡，我與楊先生談及通過學會舉辦學術活動，在籌辦工作上需要很多人力、物力，如果能夠舉辦一些小規模而又有深度的研討會，通過私人贊助，便能對推動中國建築設計起一些實際的作用，如古人的蘭亭會。當時楊先生提意先在深圳舉辦一小型研討會，邀請10至12位建築師、學者出席並提交論文，用兩三天的時間去詳細討論提交的論文，會後可把論文出版，公諸於世，使更多的人受益。結果，《中國建築論壇》就這樣產生了。我認為，建築評論是改進建築設計的原動力，但可惜我們中國人對評論不太好感及接受。我覺得要跟上世界設計水平，我們就要放棄固執的思想，接受批評。所以，我提意把第一個論壇研討會的主題定為"建築評論"。這第一次研討會於1995年10月在深圳舉辦得十分成功，而且全部論文在《建築師》雜誌第67期上刊登出來。今日，承蒙天津科學技術出版社的支持，已經把全部論文印製成書，成為《中國建築論壇》叢書第一冊。天津科學技術出版社還願意以後主辦每年一次的中國建築論壇研討會及出版提交會議的論文。

今次研討會的主題是"比較與差距"，可算是把建築評論再提高到另一台階。通過一些例子深入研究我國建築設計與國外的差距，從而對我國的建築設計作分析並找出缺點或長處。作有建設性的比較首先要有不偏不倚的心態及豐富的設計經驗，對國外的實例要作深入的了解，對國外的設計運作要熟悉，才能對國內的例子作出合理的評論。有比較才有進步，有比較才能把我國的建築設計水平提高。國外有很多國家在建築設計上，已經歷過很多派別演變，而日本卻從傳統建築設計進展到模仿西方階段，而且在短短的時間內創造出自己的獨特風格。我國從民國開始只滯留於模仿階段，直至現今也擺脫不了模仿，談何創造出自己的風格呢？

比較學是一門高深的學問，但沒有比較就沒有進步，我希望今天舉辦的研討會，能激發比較學熱，激發我國建築師的雄心，努力創造出中國新的現代建築風格！

Reflections and the way forward

A Public Lecture for the AA Workshop in Hong Kong on 27//08/97 at the Hong Kong University

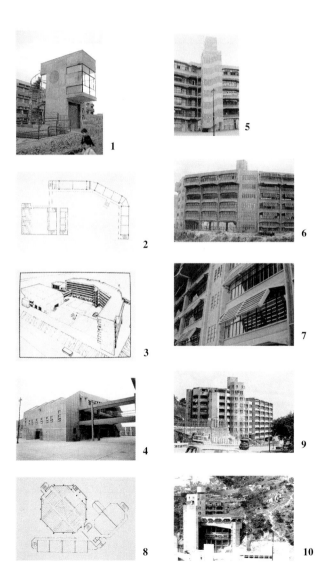

I am a late starter, Architectural life for me began in earnest after I was 50. Unfortunately in the last 5 years I do not have any building design works in Hong Kong. But the workshop theme is Hong Kong therefore when Prof. Patrick Lau spoke to me two weeks ago pressurizing me to give a talk saying that as I am one of the few of the old breed of ex AA left in Hong Kong I must participate. In order that one of the near extinct generation of ex AA graduates in Hong Kong is being heard by the budding generation of architects, I decided to take up my dancing shoes and lead you through 6 of my oldies so to speak. I took this opportunity to look up our office files and I found to my surprise my first constructed work in my professional career in Hong Kong built in 1969. A guard house in the staff dormitory of the Shaw Studio in Clear Water Bay Road.(1) I was then heavily influenced by Brutalism and started to develop a liking to exposing the staircase. It was called the guard house but its main function was to provide a visual entry point to the compound. It has a hairdressing salon on the first floor and a small tuck shop on the ground.

The second project I like to show you is a 24 classroom Secondary School in Kwai Chung called Siu Ming Secondary School situated on top of a little hill. (2-7) Here I utilized the various elements of a school, the classroom block, the laboratory blocks and the assembly hall to provide a sense of enclosure in a very exposed site situation. The external treatment is again very brutal with fair faced concrete, precast sun shades which were somewhat unconventional in the early 70's in Hong Kong. Here my approach to architecture appear to formalize. The different elements in the school are grouped into distinct entities, the classroom, the laboratories, the washrooms block and the assembly hall with the main staircase single out as a focal point.

Another school I did was the Pooi Tun Secondary School in Hammerhill Road, Kln. (8-10) This is a difficult site, building against a steep hill at back. The only bid of near flat area of the site was given to the basketball court, so the assembly hall has to be sited half way up the hill behind the classroom blocks. This not only provided a focal point for the site but also tied all the elements of the School together in a compact configuration. The compact nature of the project to me reflected abstractly a kind of local identity. Only Schools in Hong Kong has such restricted site conditions. The classrooms are all designed with cross ventilation as air-condition was still a luxury in schools in the early seventies. The hexagon shape was used because it fits into the compact grouping of the various blocks and it produces a more interesting block shape as opposed to the usual square shape of most School buildings. Here I have used the exposed staircase again to provide a focal point and a chance for the student to be aware of the greenery on the hill side behind as they go about their classes.

The fourth project is the money printing factory for Thomas de La Rue (HK) Ltd. at Tai Po Industrial Park.(11-14) This is a joint design project with David Russell. We have here purposely separated the various major elements of the factory and expressed them separately e.g. the printing factory, the machine and workshop block, the office and vault area, the staff canteen with staff changing room under etc.. There are two distinct entrances to the building other than the goods delivery bays. One's for the clients of the Company and the other for the staff. We have to make two entrances as the two functions required different security treatment. Externally we reflect the two entrances by the two stair towers in different height configuration. The taller one of course, is the clients' entrance. The printing process and machinery installation are highly technical and precision is the order of the day. Client specified a particular tiling and window opening requirements therefore from the off set of sketch design we have chosen a 1600 x 1600 grid as a base design module which fits all openings and finishes including external tiling. As a result dimensioning is very simple and there are very little cutting of tiles needed in the installation. On the whole both the external and internal walls appeared to be very neat and despite the anticipated after thought additions on the external wall like external lighting etc. the wall still able to look tidy because all installations had to fall into the modular grid. This project was awarded the 1986 HKIA design of merit award. Another project which was jointly design with David Russell is the Home for severely handicapped children in Tai Po.(15-20) The brief is a little demanding as not only the children have to be housed in special environment but the staff who have to be housed on site also require special attention. On top of that we have to retain an old house in the middle of the site. The old house contained the chapel and therefore is the spiritual center of the whole project. The geometry of the layout was developed from the shape of the old house creating three courtyards for the project each with its own character. The entrance courtyard centers round an old tree in the middle, the courtyard on the right is a little open as the staff canteen overlooks it. The courtyard on the left is completely private being surrounded by children's living area. In good weather the children are wheeled out to enjoy the sun in the courtyard. The Children's dormitory opens onto the living area with a sun roof as the children needs warmth all the time in winter. This direct connection between the dormitory, living space and courtyard enable the children also to be wheeled out quickly in case of fire. The slanting roof line was originally design to mount solar panels for the heating of water which was abandoned as it was uneconomic. However it does serve to provide a fragrant silhouette of a group of building in the south of China. This project received the Joint Council for the Physically and Mentally Disable 1994 Design award.

The last project I am going to show you is a 34 storey commercial/residential tower in Caine Road.(21-26) The intention is to design a residential tower with a strong identity admist the surrounding townscape sympathetic to the climatic condition as well as reflecting a high quality of residential accommodation. The residential units are initially

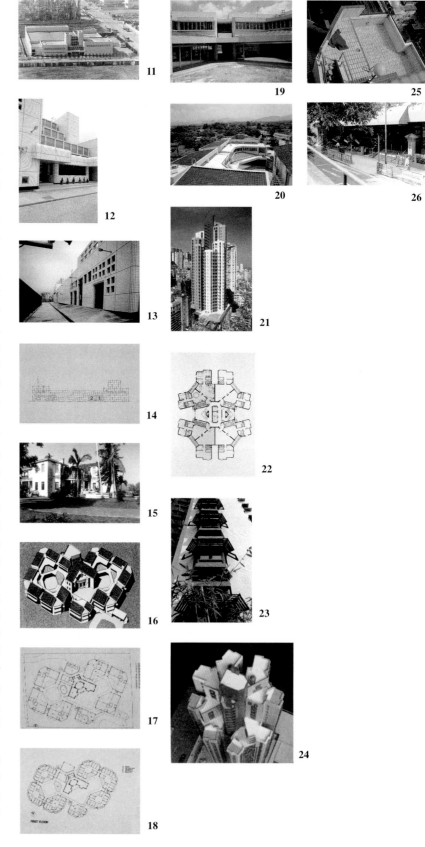

designed with a flexible layout in order to provide flexibility to suit the ever changing market situation in Hong Kong. Each floor can have eight 65m2 units or four 130m2 units. The configuration of the block permits sea view for all units as well as cross ventilation and some aspect of south facing. Each window opening has built in air-condition slot and sun shading device to cater for the particular climatic conditions. The spiral shape of the upper part of the block is the result of a unusual interpretation of the street lighting regulations avoiding the usual set back method. The added advantage of the spiral shape is it provided several additional penthouses with flat roofs each containing a small swimming pool opening directly off the living space.

The lower podium has a club house and office spaces with two split level shopping arcade on the ground and lower ground linked by a half level with the pavement. This split level shopping arcade provided double the amount of shops having street access and developed on intermediate space for shopping and entry into the building, transforming the front of the site into part of the public domain.

External space around buildings is an area of my particular interest and I like to take this opportunity to show you some examples both good and bad around Hong Kong.

Intermediate space

These spaces if well designed can provide an intermediate space between the building and the pavement or the edge of the site giving it a meaningful relationship between private and public domain as well as linking the building with the urban context.

The most impressive example to me is in the Economist Building in London.(27) I came across this building or rather this group of buildings in my student days in the AA. In fact if I remember correctly it was one of the studio programmes for us to study the Economist Buildings and to produce an alternate design. What an ambitious demand to make on students in their second years. To provide an alternate design to the one designed by what must be one of the top architects in London. The beauty of the scheme is the ingenious way Peter Simthson divided the Client's requirement into three independent buildings on an L Shaped Site creating a complete pedestrian throughout fare through the site linking the streets in the rear and front of the site. Thus merging the site and the building in the urban fabric of the City.

In Hong Kong the earliest example of what I am concerned appeared in the overhanging shop buildings in the fifties.(28) This provided weather protection as well as an intimate intermediate cover space in front of the shop which encouraged lingering and browsing activities.

The modern equivalent of the overhanging shop houses can be found in the Leighton Centre designed by Wong & Tung providing covered walkway for the pedestrians.(29) If all the shops alone the street are provided with similar covered walkway then the pedestrians will enjoy their shopping more.

In this project which I show you earlier,(30) I was some what inspired by the design of the Leighton Centre but, the result is somewhat improved I hope. By not having columns on the edge of the site, I have managed to provide a clear visual Linkage between the intermediate space and the pavement.

The usual excuse of maximization of space on the ground level particularly in Hong Kong has produced unimaginative design fronted by curtain walls along the shop fronts making a hard division between the private and public dormain.(31) I can show you that this excuse is not valid but merely show a lack of design consideration on the part of the

27

28

29

30

architect. In Hong Kong by giving up part of the ground floor to the public, the building planning regulation allows additional floor area to be added onto the building to compensate the lost of the ground area.

If you must come straight down to the street with your vertical external walls, I would at least treat the base design like the Entertainment Building design by Palmer & Tumer (Remo Riva) by spraying the comers so that the corners of the site is soften by the extra space between the pavement and the site boundary.(32) Unlike the New Shell Buildings next door by the same firm, perhaps not by the same architect.(33)

A bigger scale example is the space in front of the Time Square project designed by Wong & Ouyang complete with clock and TV screen.(34) The idea is there but unfortunately it lacks an intimate touch. At least the architect should put in some trees and sitting benches. I call this that novel rich example - big empty & without a soul.

The grandest examples of a well considered intermediate space, I belief is the Hong Kong Shanghai Bank design by Norman Foster.(35) Where almost the whole of the ground floor is devoted to public access. It is indeed a sweeping statement.

To me the best example of the external space treatment is the Citibank Plaza in Garden Road by Rocco Yim.(36-38) Like the Economist Buildings, instead of occupying the whole site with a bucky podium design, the architect divided the client's requirement into two blocks rising directly from the ground leaving a good part of the site to serve as an intermediate space merging the site as part of the urban pedestrian fabric. Here the site sits in between the Hong Kong Park, Murray Building and the Bank of China with linkages to what used to be the Hilton Hotel. In this case the architect scored extra marks because his design actually forces the pedestrian through the building in order to gain access to other places beyond. Thereby making the heart of the building a meeting point for the urban pedestrian metric.

Here I cannot resist by showing you my latest work though it is in Beijing.(39-43) We did the master plans design of a commercial district covering 20.85 hectares and was responsible for the design of three major buildings in the district. Sites no. 4 and 5 are two of sites linked by bridge buildings with a total floor area of 310,000m2. In view of time constraint I shall only present the intermediate space of the project. In this project there are three levels of intermediate spaces interlined by escalators and open wells providing ample lingering areas for the pedestrian. The pedestrians route go straight through the main shopping mall on ground and 1st floor to the back of the building which fronts onto a pedestrian shopping street and to the right to building no. 4 and to the left through the 1st floor pedestrian bridge to the Bank of China head quarter building on site no. 6. Transforming the shopping mall into part of the pedestrian system of the district. This project was voted one of "the ten best design for the Capital 95" in an exhibition organised by the Capital City Planning and Construction Committee of China.

I hope my area of interest is also yours and that in your next design in the urban context, you would give a little more thought on the intermediate space.

31

39

32

40

33

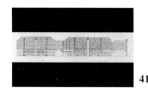
41

34

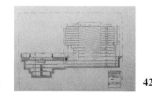
42

35

43

36

37

38

譜寫香港的新歷史

原載《建築學報》1997‧6期

香港回歸祖國，是歷史上的大事件，是洗卻國辱，振奮民族精神的令人興奮時刻，我和香港具有民族自尊心的人士一樣，為此而雀躍而歡呼！

香港被英國統治100多年，由於影響所及，使香港不自願地披上了殖民主義文化特有的色彩，這個由歷史演繹出來的現實問題，相信由於香港重投祖國懷抱，並在民族感情和港人愛港的作用下，不但使顛倒的歷史徹底翻轉過來，而且香港人也會以豪邁的心情，並有能力譜寫香港的新篇章，使香港在過去的基礎上，更加昌盛和繁榮。

回顧歷史，展望將來，使人們充滿信心。香港的回歸也給香港的建築設計帶來新的契機。

國內改革開放以來，自80年代起，香港的建築師與國內的人員、展覽、研討、考察等多方面雖已有過雙向的交流，但，這是在有限的條件上進行的，是十分不夠的。隨着香港的回歸，如今香港和國內的距離更近了，交流也會更為頻繁和順利，這將使香港建築師在廣泛的交流中得益。

香港是一個彈丸之地，在地少人多的環境中，有其特殊的生存規律及特別的建築條例。國內卻有廣闊空間，因有不同的條件，而有不同的建築規範。兩地之隔，自有兩地可取之處，香港建築師在向國內同行學習過程中，這對創立和完善香港地方的建築自我風格上，無疑也會有所啟發和幫助的。

中國的傳統文化，是我們共有的一份豐厚遺產，香港過去的特殊環境，使西方文化大行其道，在不知不覺中，已使許多人數典忘祖。我們應該熟知中國傳統的優秀文化，在匯中西之長，集古今之精華中，努力去尋找適合表現香港風格的設計路向。

相對之下，香港的建築師缺乏創作源泉，缺乏創作靈感，因而在過去的一個階段，盲目模仿和照搬西方的現象也是很普遍的，今後如何沖出局限，如何有實質的突破，這成了向前邁進的一個關鍵。

香港的建築師們，會利用回歸的契機，多到各地吸取養份，開闊視野，增加新知；特別是在中國文化中多加擷取養料，能與國內同行加強溝通和交流，彼此取長補短，共勉共進，不但能解決香港建築師當前存在的問題，使其保持活躍的思維和不絕的創意，同時也可使香港建築設計更上一層樓，在世界的東方繼續放出它耀眼的光輝。

我深信香港的明天會比今天更好，香港人一定能寫好自己的新歷史。

Design for Sites 4 and 5 of the "Xi-Xi Project"

Talk delivered at AA Asia seminar in Beijing on "Contempprary Vernacular" on 28/09/1997 and published in "Contemporary Vernacular" by Christopher Chew Chee Wai, on AA Asia monograph one, published by AA Asia, 1998

I have been in pursuit, consciously and sub-consciously, in the design of contemporary vernacular since 1980. In order to give you a brief background of the approach to the design of Sites no. 4 & 5 in the "Xi-Xi Project" in Beijing. I would like firstly to shown you three of the projects I have designed since the 80s.

1

4

2

5

3

6

The first project is the Home for Loving Faithfulness in Taipo, Hong Kong. (1-6) This is the first project I have came across which contains all the ingredients in the making of contemporary vernacular. The setting of the site is in the New Territories of Hong Kong where it is surrounded by paddy fields and traditional villages. The height restriction on the site produced a maximum two storey development. The function of the project being a home for the severely handicapped children and staff quarters. All these factors can be found in vernacular architecture. On top of that we have to retain an old house in the middle of the site which in style belongs to the modern vernacular of the 50s in Hong Kong. The existing house contains a chapel hence the spiritual centre of the project. The geometry of the new project is developed from the shape of the existing house in relating to the site producing three courtyard environment with different qualities. The front courtyard opens to the road being the formal approach to the Home. The courtyard on the right is semi-private opening onto the entrance courtyard with the staff dining room on one side and the courtyard on the left is completely surrounded, providing a privacy needed for the handicapped children.

The project consists of five blocks of children and staff accommodation, one block of staff dining and accommodation and an administration office block with the existing house consisting a chapel, a reception room and a guest room. The childrens' block has two dormitory rooms on the ground floor for the children linked directly onto a living space with a glass roof providing the much needed sunlight where the children stay most of the day. Severely handicapped children require strict environmental conditions and the glass roofs above all living spaces admit the winter sun and movable sun reflectors in the roofs help to reflect the summer sun. Overall the home is designed as a big family dwelling to create a homely atmosphere for the children. The staff quarters on the first floor has views over the surrounding countryside which provide a relaxing environment after a day's work.

The second and third Project are both small hotel projects in the north western areas of China.(7-16) In an area full of cave dwelling with sharp contract of monotone and colorful renderings in their traditional architecture. In the monastic buildings they managed to reflect a highly sophisticated blending of colours with the landscape. That part of the country is basically very poor and people still dresses in the pre-economic revolution style. The site is in a valley north of the provincial capital Lanzhou.

The hotel has 168 rooms 30% of the rooms are allocated to budget travellers. (17-22) The traditional courtyard concept has been adopted here as it has been time proven to be the most suitable solution for low rise situations for the particular climatical condition in the area. Rooms for the budge travelers are built against the slope taking benefit of the natural ventilation characteristics of the vernacular architecture 4 storey in height with a natural air extract system built at the back of the rooms. We have experimented with two types of roof structures. One with turf on top to further enhance the preservation of energy within the building and the other with solar panels to supplement the heating of water within the building. The construction is mainly by the traditional ram earth method. Therefore the pursuit of modern vernacular is through the regenerated use of traditional space, time tested methods to combat the severe climatic economic conditions of the region and reuse of traditional construction methods.

The third project is a small hotel on a hill slope outside the city of Xian.(23) Here the courtyard is elongated to suit the hill contour. The concept is to recapture the compactness in the traditional hillside village and through the massing and simple elevation treatment to recreate a modern vernacular living environment.

In the case of the design of Sites 4 & 5 in the "Xi-Xi Project" I was faced with a dilemma. The project consisted mainly of shopping malls and multi storey offices. These functions do not exist in the traditional vernacular concepts. What exist however, are the grandeur, the scale, the symmetry and

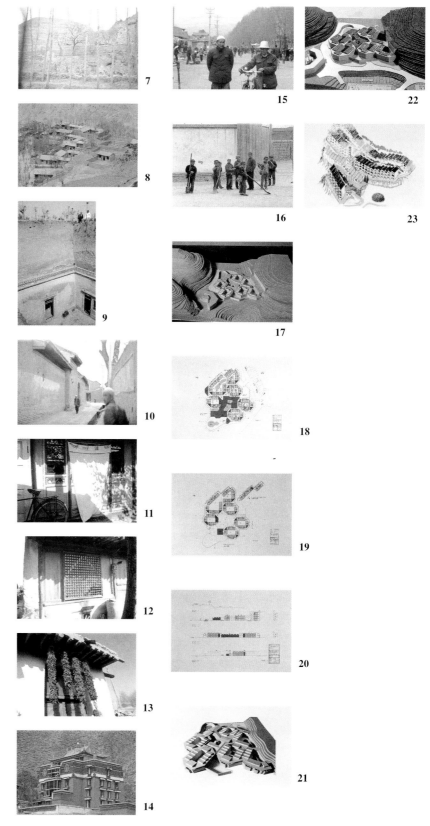

people's habits in the capital city of Beijing.(24-32) I find even amongst the buildings in Beijing whether modern or traditional, there is an air of supremacy and grandeur. In most cities of China people gathers in the pavement very often. In the summer, particularly, as the pavement is cooler at night than their homes. Therefore I find the grandeur in the proportion of buildings, symmetry and people's activities are the most relevant vernacular aspects in the design of buildings in Beijing.

We were asked initially to develop a master plan for Xidan North Street, one of the most busy commercial district in Beijing with a total area of 20.85 hectares.(33-34) Sites 4 and 5 are the most prominent sites in the district with a total floor area of 320,000m2.(35-40) The height limit is 12 storeys (45m). Sites 4 and 5 are designed as two buildings linked by bridge buildings making it into one hugh mass. In order to facilitate the peoples's habit of waiting, mingling on the pavement and to create a social node and entry point to the building, I have designed an intermediate space between the entrance doors and the pavement. This atrium serves as an introductory space to the complex, pooling the people off the pavement and into the shopping mall. The shopping mall is symmetrically laid out to facilitate ease of orientation within each floor. The ground and first floor acting as a cross road in the shopping activities as it is accessible from all directions, with pedestrian ground and upper level links to the pedestrian street at the back and to 1st floors of all the surrounding buildings. This is extending the building into the pedestrian matrix of the city. The bulk of the building is divided by exposed staircases which I am very fond of using on my buildings. They serve to break up the building into more reasonable proportion as well as giving vertical eccentricities to the overall bulk.

The external bulk of the building was developed from the initial sketch concept expressing a degree of grandeur reminiscent of the palace structures. (41-44) This impression is further reinforced by the regimental display of partial sun shading on the office floors on the upper part of the building giving a contract of complexity and simplicity between the upper and lower part of the building not only to express the different functions but also to reflect a traditional treatment method of the place structures.

I have purposely attempted to capture the essence in the discipline of traditional palace design in the external treatment of my building and have recreated an intermediate space to facilitate a time tested habit of the people to enhance the attractiveness of the buildings. These are the two aspects which I can find valid in the Beijing situation as modern vernacular in a building of 12 storey high has very limited application.

This is my attempt of this scale in modern vernacular. I can only treat it as a product of a transition period as I think I still has a long way to go.

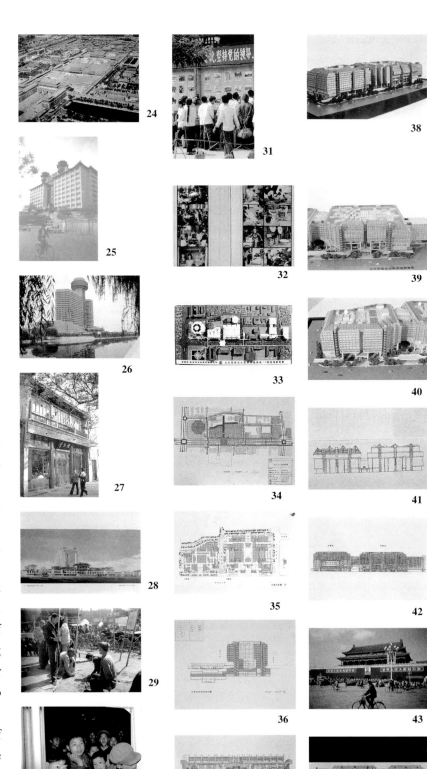

To me modern vernacular architecture in the most successful form is an architecture which when you look at it or when you walk through it you feel a sense of the vernacular without being able to immediately pin point a direct reference to the vernacular vocabulary. To put it another way round. By putting a tradition pavilion on top of a curtain wall building, or using a tiled sloping roof or adding a moon gate at the entrance or having traditional motifs, symbols plastered all over the walls, corners, & floors or using a concrete version a traditional structural modern bracket system, at best can only be classified as a common version of modern vernacular. At worse it can be a bad example of traditional replica in disguise.

If you can captivate on the philosophy of the vernacular design and create spaces to nocture the on going vernacular behavior of the people, you can then create a vernacular spirit amongst your building design. This I call true modern vernacular.

建築設計 —現狀與改革

政協第九屆全國委員會第一次會議大會發言 03/1998
原載楊永生主編《建築百家言》一書,中國建築工業出版社出版 1998

我國現代建築設計的期望,是追上國際先進水平,但眼前所見到的成果,確實有一段的距離。像外形設計,只有小量的實例能仿效到國際水平的模樣。多數的外形是簡陋而缺乏對氣候及本地風格的交代,而平面佈局、內部空間處理以及功能的考慮是遠遠不及國際水平的。以致因為經濟改革的要求,建築師為了滿足建設急需,以急功近利及盲目追求突破的心態,不去面對現實、面對國情,在設計上下苦功,而盲目地抄仿所謂國際水平的設計。這既浪費國家資源,又建成了一批不中不西、華而不實、與國情脫節的實例。在這方面,有關領導也要負起一些責任。因為他們往往希望在任期內建設起一批創舉性的大樓來為本地增添些繁榮的景象,但沒有考慮到長遠的利害。

建築設計水平的起落是受到兩個重要的因素影響,其一是建築教育,其二是執業及管理水平。

在建築教育上,四、五十年代有一批受過外國良好教育的高材生,例如梁思成、劉敦楨、楊廷寶、童寯、陳植等回國執業及教學,所以我國的第二代建築師能夠從他們那裡吸取當時世界上比較先進的經驗及做法。但這批第二代建築師經"文革"的艱苦日子及缺乏繼續與外面交流的機會,從而與世界潮流已脫節,而經過他們教育的第三代建築師因種種原因,與國際水平更加拉開了一段的距離。因此,建築設計的發展就出現了些青黃不接的現像,雖然這批建築師在經濟改革後也多次到外國考察,但走馬觀花,不能深入研究,有時甚至產生反效果。

我認為只有保持長期穩定的中外交流,才能取人之長,補己之短。我舉一個例子:美國在30至40年代有一批德國建築師精英逃亡到美國,在各大學及私營單位內另起爐灶,我指的是格羅庇烏斯(Gropius),門德松(Mendelsonhn),布勞耶(Breuer),密斯‧凡‧德‧羅(Mies van der Rohe)等,對美國近代建築設計起了極大的推動作用,後來又培養了一大批新一代美國建築師。所以,在缺乏大批外國留學生的情況下,適當地利用外來人材及技術,對一個國家的建築設計發展是有極大的促進作用。現今美國新一代的建築師,已經完全是"國產"並能獨當一面,把建築設計水平推進到另一個新台階。

香港是另一例子,在50年代以前香港的建築設計完全由外國人把持。但從50年代開始有一批留學海外的中國建築師遷移到香港,大展鴻圖。同時香港大學建築系也於1950年成立,由著名的英國建築師哥頓白朗當系主任,並聘任了不少外國的導師。60年代以後香港的建築設計已經由外國人包辦轉為港人的天下。另一方面,早在五、六十年代發展商已開始不惜巨資聘請國際著名的建築師在港做設計。而七、八十年代因香港經濟起飛,更加突出,例如貝聿銘(I M Pei),哈瑞‧賽德勒(Harry Sydler),諾曼‧福斯特(Norman Foster)等都在港留下了不少高水平的建築設計。這樣多姿多彩的中外交流,產生了一批土生土長的新一代香港本地建築師。他們大多數是香港大學建築系的畢業生,有些是到過外國深造或執業的,而現今香港的大部分建築設計工作是由這一批新一代的建築師負責,他們的設計水平是有目共睹的。

現今我國的建築教育情況,實在令人擔心。因為在缺乏經驗豐富的建築師的情況下,經濟已經起飛,所以就算有好的教師他們也忙於建築系設計院的工作。由於一般大學教師的收入不多,也難怪他們棄教從業,學生因受了教師及社會風氣的影響,也無心專心學習,急不及待地先學走路而後學站立。

為長遠之計,最好能在經濟條件允許下,聘請一批海外有建築教育經驗的華僑建築師到國內執教。因為他們比較容易了解國情,而且有愛國之心;另一方面,大力支持一批年青建築師到外國深造或執業,並鼓勵他們學成後回國投入服務。這樣,"國產"的建築師才有名揚世界的一天。

在執業管理方面,這幾十年內,在執業手法及管理程序上依然是老一套。有影響力的設計院規模大,運作笨重，再加上眾多退休員工，就像一艘大型油輪一樣，要改變方向要一段漫長的時間和決心。主要的問題，是缺乏與海外同行交流這方面經驗的機會。近年來外國設計單位到國內進行設計工作的已不少，但因大多數的國內建築師的心態尚存有保守的態度，對新的執業方式或設計方法有所抵觸。對新的執業及管理的手法是要經過一段時間或一、兩項與海外建築師合作的工程，才能領略一二。現今國內的情況要跟上國際水平還要走一段漫長的路。但如果把現有的設計院改為集團式經營，把各所改為私有股份制而原設計院把各所的設備售與各所來換取一部分各所的股份。此外每月向各所徵收辦工樓的租錢來支付退休員工及其他總務的開支。當然把現有的設計院私有化並不簡單，要從詳計議，而在這裡因篇幅所限，只是拋磚引玉，希望能引出一個好的解決辦法。

單把制度及辦公方式改革是不夠的，另一方面應舉辦一些學習班，請海外建築師講述執業方式和工程管理程序，或鼓勵一些分拆後的設計所與海外設計單位合辦設計院，從而向海外建築師汲取有關經驗。同時，政府可以開放建築設計市場，引進海外建築師在國內開業。

這樣，我國設計水平趕上國際先進水準是指日可待的！

下真功夫，走向世界

在建築論壇第三次研討會"現狀與出路"發言稿, 1997年9月東南大學,南京
原載建築論壇叢書第三冊《現狀與出路》,天津科學技術出版社, 08/1998

放眼全國各大城市，我們可以看到除了一些傳統民居、古代建築，五、六十年代受蘇聯建築思想影響的建築外，還有在這十數年內受經濟改革推動的不中不西、似仿非仿的當代建築，頂上有大屋頂、小亭子，樓身上有玻璃幕牆等等。

中國因閉關多年，加上"文革"十多年的動盪，改革開放後才大開門戶，引進外來的技術。但對建築設計來說，最近幾年才有一些海外建築師在國內做設計，但還是屬於少數，而且還存在些保護主義，所以海外建築師還未能大展身手，更未能堂堂正正地自立設計院執業。雖然最近幾年有少數國內的年青建築師有機會到外國工作，但回歸的數量不多，引至國內設計的一套手法，只能是50及60年代方式的演變，沒有根本性的改進。大多數由"大師級"建築師起草稿、設計方案，之後交其他建築師去做初設，有時做施工圖的又是另外一套班子，缺乏連慣性，往往做出來的最終設計與大師的原意不合，而大部分的設計工作得不到大師的指導。大師們因"文革"停工多年，而改革開放初期，忽然間工作量劇增，無時間去好好地思考及深造，所以出現一些膚淺的作品。國內缺乏有建設性的評論，以至所有評論多數是稱讚式的，所以誤以為所作為極品，不再作檢討，再接再勵。近年來的建築設計大多數實屬表面功夫，五花百門，多姿多彩，但缺乏真功夫，與外國現代建築設計比較，尚有一段很大的距離。但因經濟改革的推動，建築量大大增加，但建築設計的深度不夠，所以產生了大量浪費資源的後果。各大城市都有大量的大而無當，花枝招展的建築物，實是遺患無窮。

另一方面，經濟改革後，錢的價值突升，帶引起不少不正規的行為，有部分建築師在工餘的時間，甚至在工作的時間內接外間私下委託的項目，因沒有設計院的技術支持，而且個人的經驗所限，設計出來的項目多屬水平不高，始終都因貪圖私利，為害不淺，又因公私兼顧，導致精神緊張，本職的工作也受到不良的影響，結果兩敗俱傷。

四、五十年代因有一批受過外國教育的高材生，例如梁思成、楊廷寶、童寯等回國執業及教學，所以第二代的建築師能夠從他們中吸取當時世界上比較先進的經驗及做法，這批第二代的建築師經"文革"艱苦日子及缺乏繼續與外交流的機會從而國內第三代建築師因種種原因與第二代的是有一段距離，建築設計的發展就出現了些青黃不接的現像。雖然這批建築師在經濟改革後也到過外國考察，但從一般性考察得到的只是走馬看花，不能深入，甚至產生出反效果。

我不是崇洋，但我認為只有長期性的中外交流，才能取人之長，補己之缺。我舉一個例子，美國在30至40年代有一批德國建築師的精英逃亡到美國，在各大學及私人單位另起爐灶，我指的是 Gropius, Mendelsohn, Breuer, Mies van der Rohe 等，對美國近代建築設計起了極大的影響，後來又也栽培了一大批新一代的建築師。所以在缺乏大批外國留學生的情況下，適當地利用外來人材及技術，對一個國家的建築設計發展是有極大的帶頭作用。

經濟改革是成功的，而且會繼續進行，雖然建築量不會像改革初期那樣劇增，但也能維持相當的規模。我提議各大城市有關政府單位支持成立一些特種的"建築設計工作室" "Architectural Design Workshop" 以私人有限公司方式經營，給予適當的設計項目開業。這些特種"建築設計工作室"適量地聘請海外建築師或邀請海外建築設計單位加入合作經營，在本地聘請有水平的建築師加入，利用三分之一的工作時間，開設"再教育"班，討論工作方式、設計思路等有關課題，同時邀請本地及海外有經驗的建築師作報告，從而提高員工的設計水平及改進工作方式。有了設計項目，按現時項目的規模收入，是可以支持聘請海外建築師的，實行邊教育邊執業，可以有效地提高設計水平，彌補建築教育的不足。雖然同現在的建築設計研究院組成時的出發點可能有些相同，但現今大多數的建築設計研究院因業務壓力或種種其他原因，已經變相。在前路茫茫之際，應及早回到邊設計、邊研究、邊教育的路上來。

《香港著名建築師作品選》前言

中國建築工業出版社出版，1999

為響應1999年國際建築師協會在北京舉行大會，並且是有史以來第一次在亞洲區舉辦大會，中國建築工業出版社楊永生先生要求我主編一本《香港著名建築師作品選》一書，我是覺得榮幸之至。誠蒙各香港同行的合作，成功地收集了頗具代表性的20多個香港建築師的作品，希望藉國際建築師協會大會的舉辦，介紹給世界同行香港建築師的成就。

香港的建築設計是組合了經濟條件、建築條例和建築師印像中最先進的西方建築設計的結晶品，缺乏地方氣候及傳統文化精神的內容。

目前，在香港依然聚集和呈現着各國的建築樣板，還沒有形成它自己獨特的風格，毫無疑議，我建築師應以適當的建築設計創造一種既符合香港文化歷史背景和生活方式，又有強烈地方色彩和風格的地方建築。

在亞洲，有很多以本地風格來設計的嘗試。但有些卻很幼稚，只是表面的。也有許多設計，只取引人的外表，起初雖然使普通人覺得新鮮，好像適合口味，但它們被歡迎和接受的程度是有限的，猶如事過境遷一樣，在視覺上、心理上慢慢就產生了一種相反的作用，甚至那些俗不可耐的地方風格樣板，後來竟成了人們的眼中釘。

放眼亞洲，我們可以看到亞洲高速發展的城市裡的一些情況，所有對建築風格有影響的建築，都有着濃厚的西方風格，例如從建築材料(即使是本地出產，也仿效西方同類產品)、施工方法、生活情況等方面也可以看得出來。而且還會發現，雖然各地氣候是不同的，但在高層設計方面，最有影響的是建築條例、結構要求、施工程序、業主要求和實際用途等因素卻是相同的，難怪亞洲的高層建築設計都是千篇一律。

我們建築師無力單獨扭轉和改變以上因素，盡管具体地區的建築條例有所不同，對建築師來說，唯一可以充份表達地區風格的地方，是從建築物以內和以外的空間設計構思中表達出來。

要在建築設計上達到一個優質的空間構思，首要的是徹底了解當地人的傳統觀念、價值觀、文化背景、軼聞趣事，以及他們的生活方式、最新的建築科學發展情況、氣候條件等。這些都是建築設計上悠關問題。我認為，有了對這些方面的了解，就必然胸有成竹。建築的地區風格，會從設計圖中自然表現出來。也就是說，當你熟悉所有客觀條件，頭腦裡形成地方風格的概念，動手設計時，隨之而來的，就會水到渠成，躍然紙上。而單單靠憑空冥思苦想，是不能達到預期目的的。

今後，我們要對我們的文化以及形成地區風格的各種因素進行研究並提高這方面的興趣。我們應放棄仿效別人的做法，同亞洲其他建築師一道，去尋找具有真正的、永久性的地區建築風格的規律來。否則，亞洲全部大城市的建築，就會面目雷同，張李不分，變為西方建築設計的翻版。如果不幸而言中，其後果將會令人觸目驚心。

古為今用 ——現代化傳統建築設計探索

1998年12月建築學術交流會，台北會議發言

如何使傳統建築現代化，這個問題我從沒有專心去研究。因為我覺得在設計上要實現這個精神，只能夠在不知不覺的情況下，才能有效地在設計上有所表現。當然建築師首先要對本地的傳統建築有所認識，及對構成傳統建築的風格和影響設計因素，例如當地居民的生活方式、營造法式及氣候的情況要了解，吸取傳統建築的精華，再對本地現今的情況徹底研究，才能溫故知新，創出有傳統精神的現代化建築設計。回顧我的設計工作，我是從1980年起不知不覺地把我腦袋已消化的中國傳統建築精華，演變於適當的建築設計項目中，使建築整體有傳統建築的觀感。而內部空間及佈局有傳統空間的層次及特色，但採用的設計手法是現代化而且絲毫沒有採用傳統的構造、符號或色彩。因為好的現代化建築設計是要實事求是，配合現今的國情，不可盲目抄襲，要對地方氣候、風土人情、周圍環境及業主的要求有適當的交待。其實，現代化傳統設計只是一個代號，其真正的名堂應該是現代化建築設計。因為傳統建築是根據當時的生活方式、建造材料及技術和地方氣候而產生，現代的建築設計也是根據這些因素來構成。這些因素經過時間的考驗，大部分已脫胎換骨，但小部分例如居民的生活習慣及地方氣候可能仍會留下一些傳統的精神，所以現代建築設計如能夠充分地表達到地方風格，便能反映到傳統的精神。

1

2

第一個項目是我和大衛·羅素合作設計的弱能兒童院，位於香港新界大埔。這是我第一個富有現代化傳統建築的條件的項目。工地位於香港的郊區，周圍是農田及傳統的村落。發展的高度限制於2層高，而項目是一間提供弱能兒童及看護人員的大型住屋。此外，應業主的要求，要保留在工地中部一座50年代的小屋(3)，屋內有一小教室，因此成為項目的宗教精神的核心。項目布局的定位是從要保留的小屋的外形、角度伸展而來，把土地分割成三個不同環境的庭園，來配合三個兒童院的主要功能(4)。前院面對馬路為兒童院的大門，右邊的庭院是半開放式，與入口庭院

相連，庭院兩旁有員工食堂，第三個庭院是完全密封，提供了一個比較清靜的空間給兒童的活動。

項目有五座兒童及員工的住屋(5)，一座員工食堂、一座行政中心及保留下來的小屋，屋內除一小教堂外，還有一接待室及客房。兒童及員工的住屋，首層有2間8人的睡房，及一相連有玻璃屋頂的起居室(6)，使得兒童能在室內也能享受日間的陽光。弱能兒童需要有嚴格環境控制的空間，玻璃頂部能引入冬天的太陽而搖控的遮陽板卻能把夏天的太陽反射。兒童的住屋是以住宅的設計方式，盡量提供大家庭的氣氛，員工的宿舍在二樓(7)，窗外能欣賞到週圍的田園風光，給員工在休息的時候有清靜的環境。

整個項目利用了要保留的小屋的造型，創造出一座擴大現代化的大庭院。屋頂的斜頂不單止是為了配合週圍村屋的造型(8)，而且是為了裝置一些太陽能板的需要。因為弱能兒童在冬天的時候需要暖氣，而夏天就要大量的熱水沖涼，所以提供太陽能設施能節省能源。這個項目因為在功能上及業主的要求加上週圍環境的影響，不知不覺地創造一個比較有傳統精神的現代大庭院。

第二及第三項目是在大陸西北部地區的小型飯店。大陸的西北部是滿佈窰洞式的住屋(9,10,11,12)，建築的色彩上有黃沙色的生土建築，也有多姿多彩傳統裝飾(13)(14)。一般宗教的建築有與週圍景色協調的色彩(15)，這個地區基本上非常窮困，居民還穿上經濟改革前的服裝(16)。1985年開始正式進軍大陸，第一個項目在西安市郊，臨潼縣的小型飯店牡丹山莊(17)(18)，工地位於近華清池的一個山坡上。自從1974年我第一次到大陸參觀後至1985年間，大陸的單位領導及建築師們都忙於興建滿佈玻璃幕牆的高層飯店，經過多次到大陸參觀後，我覺得在西安地區當時最能適應外國遊客及符合地方經濟的飯店建設，就是以盡量利用本地建材及富有地方特色的現代化設計風格最為理想。所以在牡丹山莊的設計中，我簡單地用半圓型入口及單面斜頂的黑灰瓦屋頂，使得在視覺上有地方風味的感覺，而電梯塔的外形是吸收了傳統磚窰及50年代的工業煙囪的特色演變而成。整個飯店是沿山形建造，務求創造出一個有傳統氣氛的山村，通過佈局及簡樸的立面，造成一個現代化的傳統生活環境。第三個項目鵰鷹山莊的工地位於甘肅省省會蘭州市北面的山谷(19)(20)，飯店有168間房，百分之三十的房間是提供給低消費的旅客。我採用了傳統格式的四合院佈局，因為這種佈局在大陸西北地區經過長時間的考驗(21)(22)。低消費的房間採用了背山的做法，用傳統窰洞通風的設施，

3

4

5

6

7

9

10

11

8

12

13

14

16

15

17

18

把自然風透過管道提供到4層高的每一間房間內。屋頂有兩種造法(23)(24)，一種方法是以綠化的方式，增加保溫及隔熱的功能，另外一種方法是以太陽能的發電版，提供熱水供應。項目的內外牆是用傳統的生土方法建造。這個項目採用了傳統空間及對抗嚴寒氣候和貧窮環境的建造手法，去創造出一個既現代化又富有傳統精神的設計。

第四個項目是位於香港半山的豐樂閣。香港有中國傳統建築(25)(26)、也有英國殖民地式的傳統建築(27)。但香港成為國際性的大都市已過百年，而居民的生活、建築材料及技術已演變為西方模式，只有氣候沒有大的變化。所以豐樂閣的構思是受到現今建築條例、通風採光的要求及生活方式的影響為重(28)。在公寓的平面布局上，根據香港特殊的市場需求，以靈活的住宅設計去適應市場對大小單元需求的變化(29)(30)。每層單元分佈可由8個65平方米的改為4個130平方米的單元。樓身造型是根據南北通風和採光的要求設計，使每個單元都有三面外牆，這樣不僅起居室有良好的自然通風條件，而且從每一個單元都既能遠眺海景又能近觀山色。外窗頂部裝有鋁質遮陽板(31)，這個設施在香港大部分建築物都沒有考慮的。

大樓中部的電梯組合防火梯(32)，採用外塗鮮紅色的中筒，防火梯旋轉其間直插雲端。週圍的四幢樓身外塗灰、白兩色，高低錯落有緻，增加了大樓的雄觀，從而創造出樓身頂部螺旋形的外貌，似盛開的花朵(33)。由於其特殊的造型，由九龍隔海相望，千百座淡色的高樓大廈之中豐樂閣頗有萬綠叢中一點紅的意境(34)。總括來說，整個設計是提供了多項符合香港氣候的設施(35)(36)，整幢大廈的色彩及造型，都充分反映了香港多姿多彩的城市生活方式(37)(38)，是既有時代精神又有地方風格的一個嘗試。

第五個項目是北京"西西工程"的4及5號樓。在這個項目的設計工作中，我面對了一個難題。項目的內容主要是大型商場及辦公樓，這些功能與傳統建築設計是拉不上什麼關係的。但在北京市有富有傳統精神的雄偉建築(39)、大規模的建設(40,41,42,43)，中軸綫及居民的生活方式，無論是傳統或現代的建築都是帶有王者味道及雄偉的色彩。一般城市的居民都有在路旁乘涼的習慣(44,45,46,47,48)，尤其是在夏天，因為路旁總比大部分民居住地涼快，還有地道特產的色彩(49,50)。總括來說，我認為北京的建築傳統精神是它的建築尺度的雄偉、平衡對比的佈局和居民的生活方式。

我的原來任務是在北京西單北大街西側的商業區提拱一個總規劃的設計(51,52)。西單北大街是北京最繁華的商業街之一，項目的總面積為20.85公頃。4及5號工地是區內最主

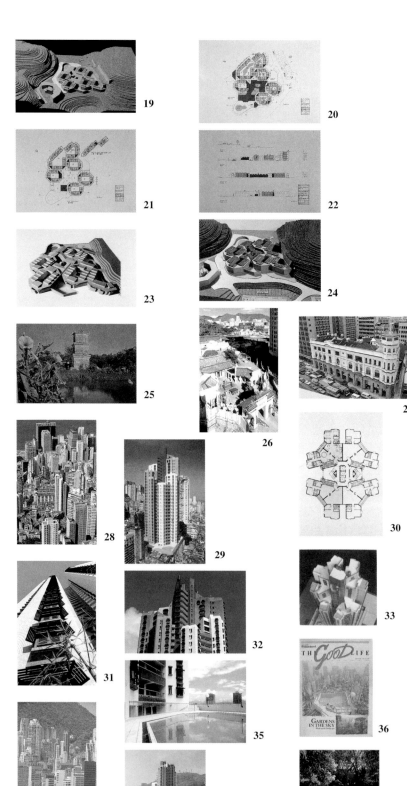

要和最大的工地(53,54)，總建築面積有320,000平方米。建築物的高度限制於12層(45米)。4及5號地是把兩幢大樓以過街樓相連為一體，為了適應居民徘徊路旁的習慣，及創造一個社交中心點和大樓的主要入口(55,56)，我在入口與行人路之間提供了一個相連的空間，這個空間把行人帶引到大樓的入口，每層的商場是從一中軸綫伸展，使每層的商場都有明確的佈局。首層及2層的商場成為室內的十字路口(57,58,59)，把樓外的行人通過首層的四面入口及2層的步廊入口引進商場，這是把這兩層的商場與過街的步廊相連，變為城市行人路網的一部分。

大樓的體形是用室外防火梯分割，減低體量並在立面上提供一些垂直綫條，外形是從構思的初稿深化而來(60)。造型仿彿有傳統宮庭建築的雄偉(61)，樓身頂層的遮陽板與樓身下部簡單的外牆處理，不只表達了不同的功能，更創造出一個豐富和潔簡的對比，與傳統宮庭建築的外貌有異曲同工之感(62)。我有意無意之間把傳統建築的精華融化到外形的設計內，並提供了一個相聯空間去適應居民的傳統休憩習慣，增加了大樓的吸引力。我發覺在12層高的建築物能夠伸展現代化傳統設計是有限的，但這兩個因素在北京地區是與現代化設計有密切的關係。我在這一大規模項目中，嘗試現代化傳統的設計手法，但我認為這只能作為一個過渡式的產品，成功還待一段日子。

最近我在北京龍潭湖公園內有一俱樂部及別墅的項目(63, 64)，因為別墅的日向及視野要求及工地正好是方型，所以基本上我採用了傳統中軸綫的佈局，無形中形成多個富有傳統意味的庭園。

現代化傳統的建築設計最成功的地方，就是當您置身其中，您會感覺到傳統的精神，但找不到一處與傳統建築的實例有直接關係的地方。換句話說，將傳統亭子放在玻璃幕牆的樓頂上或在入口處放一傳統的拱門，或把傳統裝飾符號放在外牆樓角及磚上，或以混凝土仿做斗栱，這些做法只可算為仿古設計的例子。如果您能消化及吸收傳統建築的精華，而且對項目設計有關的事物有徹底的了解，兩者便不謀而合，在不知不覺中表現出傳統的精神。我認為這樣才是現代化傳統的設計。

有關文章及主編書籍目錄

Related articles & edited publications

1. "Guard house" & pedestrian bridge, Far East Builder, Hong Kong, 02/1970

2. Shaw's residence, Far East Builder, Hong Kong, 05/1971

3. Pooi Tun secondary school, Asian Architect & Builder, 03/1972

4. Village shop, Building Record 1974-76 published by Thomson Press, 1976

5. Cactus mansion, Asian Architect & Builder, 07/1979

6. 弱能兒童院,《世界建築導報》03/04/1985,《建築師》06/1997
   Home for severely handicapped children, Asian Architect & Builder, 03/1980 & 12/1981

7. "Moves to improve HK architecture", an interiview, Building Journal Hong Kong, 03/1981

8. 中港建築業學術交流前景探討,專訪,《屋宇建築通訊》第二輯第七期, 1981

9. 揉合中西建設並非抄襲,中港建築師須努力創作,專訪,《建築業導報》第七輯第十一期

10. 本地建築你要知道,專訪, 號外, 10/1981

11. 香港建築師與內地加強交流經驗, 梁華,《經濟導報》30/08/1982 第 34 期

12. "Hope for a new concept in architecture", Asian Architect and Contractor, 12/1982

13. Albron court, Asian Architect & Builder, 01/1981; Building Journal Hong Kong, 10/1981; Building Review, 04/1985; Arcasia newsletter, 7-10/1986; Hong Kong weekend guide published by Hong Kong Standard 18/04/1987; "Contempory Architecture in Hong Kong" by W.N. Chung, published by Joint Publishing (HK) Co. Ltd., 1989; Asian Architects & Builder, 01/1981

14. "Innovation in Architecture" selected papers of Arcasia Forum 1, edited by Ronald Poon & published by Architect Asia Publication Ltd. 1984

15. 豐樂閣,《建築業導報》04/1985;《世界建築導報》03/04/1985,《黃金時代》11/1988;《香港建築》,中國建築工業出版社、中國海外建築工程有限公司合編, 香港萬里書店, 中國建築工業出版社出版, 1989;《良友》12/1992;《建築業導報》03/1995;《世界建築》03/1997;《建築師》06/1997;《世界建築導報》3-4/1985 期, 香港特刊, 潘祖堯客座編輯

16. 紙幣廠,《香港建設》no.78, 1986;《黃金時代》11/1988;《香港建築》,中國建築工業出版社, 中國海外建築工程有限公司合編, 香港萬里書店, 中國建築工業出版社出版, 1989;《良友》12/1992;《建築業導報》03/1995;《建築師》06/1997

17. "Asian Identity" selected papers of Arcasia Forum 2, edited by Ronald Poon & published by Architects Asia Publication Ltd., 1986

18. Printing factory for Thomas De la Rue Ltd, Asian Architect and Contractor, 01/1987; Building Journal, 01/1987; Hong Kong Development no. 78; "Comtemporary Architecture in Hong Kong" by W.N.

Chung, published by Joint Publishing (HK) Co. Ltd., 1989; "Guide to Architecture in Hong Kong" published by Pace Publishing Ltd., 1998

19. "Design direction in Asian Architecture", selected papers of Arcasia Forum 3, edited by Ronald Poon & published by Architects Asia Publication Ltd., 1988

20. "My architecture", selected papers of Arcasia Forum 4, edited by Ronald Poon & published by Architects Asia Publication Ltd., 1988

21. 《馬伯樂畫集》，潘祖堯主編，亞築軒有限公司出版，09/1989
"Paintings of Ma Bole" edited by Ronald Poon, published by Architect Asia Publication Ltd., 09/1989

22. 《歷代文物萃珍》，潘祖堯主編，敏求精舍出版，1990
"Selected Treasures' of Chinese Art", edited by Ronald Poon, published by Min Chiu Society, 1990.

23. 雕鷹山莊，《中國建設發展》，1995年第三期;《建築業導報》03/1995;《良友》12/1992;《建築師》06/1997

24. 建築名師潘祖堯志存高遠，阿鑽著,《良友》12/1992

25. "西西工程"總規劃,《中國建設發展》，1995年第三期;《建築業導報》03/1995;《建築師》06/1997

26. 芭提雅度假公寓大樓,《建築業導報》03/1995;《建築師》06/1997

27. 西西工程，4號及5號商辦大樓,《建築業導報》03/1995;《首都建築設計方案選》首都建築藝術委員會編，中國建築工業出版社出版1996; 建築師 06/1997

28. 《建築與評論》，建築論壇叢書，潘祖堯,楊永生主編，天津科學技術出版社出版 1996

29. 《比較與差距》，建築論壇叢書，潘祖堯,楊永生主編，天津科學技術出版社出版 1997

30. 一本讓人驚嘆與思索的書, 讀《比較與差距》, 顧孟潮,《建築師》10/1997

31. "西西工程 "2號商辦大樓,《建築師》06/1997

32. 牡丹山莊,《建築師》06/1997

33. 根植國土, 志在八方──潘祖堯建築師其人其事, 楊永生、黎朗著,《南方建築》01/1998

34. 《現狀與出路》，建築論壇叢書，潘祖堯,楊永生主編，天津科學技術出版社出版 1998

35. 人民大會堂香港廳,《建築師》06/1997; "香港廳", 人民大會堂香港廳籌建小組編及出版 03/1998

36. 《香港著名建築師作品選》,潘祖堯主編, 中國建築工業出版社出版 1999

個人檔案及作品年表
Biographical chronology

學生時代作品 1968 Student work		"空中庭園"住宅小區方案,英國倫敦建築協會建築學院畢業作品,1968(見32頁) "Sky Courtyards" public housing AA thesis project at Kellett Bay, Hong Kong.(see page 32)
潘衍壽土木工程師事務所作品 1968~1972 Projects executed at Peter Y S Pun & Associates	 	邵氏制片廠警衛屋及行人天橋,1968(見34頁) "Guard House" & pedestrian bridge at Shaw's Studio (see page 34)
	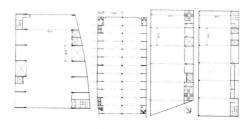	標準工業廠房構思,1968 Proto-type concept for multi-storey industrial projects
		邵氏大屋,1968~1970(見36頁) Shaw's Residence (see page 36)
		兩間電影院及商場,香港九龍馬頭圍道,1968 Twin cinema/department store complex, Ma Tau Wei Road, Kowloon, Hong Kong.
	 	蜆殼電油站和汽車陳列室及維修部,香港英皇道,1968(見41頁) Petrol filling station/motor car showroom & car repair complex, King's Road, Hong Kong. (see page 41)

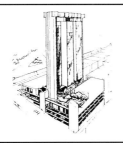
商住大廈, 香港堅道, 1968
Commercial/residential building, Caine Road, Hong Kong.

蕭明天主教中學, 1968~1972 (見 42 頁)
Siu Ming Catholic Secondary School (see page 42)

永勝大廈, 香港德輔道西, 1969~1972
Winsing Building, Des Voeux Road West, Hong Kong.

培敦中學, 1969~1972 (見 44 頁)
Pooi Tun Secondary School (see page 44)

筆架山道 4 間排屋, 1969~1972 (見 48 頁)
4 Terrace Houses, Beacon Hill Road (see page 48)

翡翠明珠戲院, 1969~1972 (見 40 頁)
Jade and Pearl Cinemas (see page 40)

蜆殼預制油站, 1970~1971 (見 41 頁)
Shell's precast petrol station (see page 41)

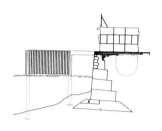
海上油站, 香港九龍, 1971~1972 (見 41 頁)
Marine filling station, Kowloon, Hong Kong. (see page 41)

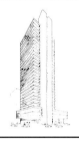
海外信託銀行大廈, 1970 (見 50 頁)
OTB Bank Building, (see page 50)

蜆殼油站, 香港麥當努道, 1971 (見 41 頁)
Shell's petrol station, Macdonell Road, Hong Kong (see page 41)

蜆殼油站, 香港山頂道, 1971 (見 41 頁)
Shell's petrol station, Peak Road, Hong Kong (see page 41)

輕工業廠房, 香港電器道, ML277RP 和 ML281 地段, 1972
Light Industrial Building, ML277RP & ML281, Electric Road, HongKong.

潘祖堯則師事務所 1973~1986 Ronald Poon Associates		石澳村屋, 1973~1974 (見52頁) Village Shop, Shek-o (see page 52)
羅素／潘建築師集團 1976~1986 Russell/Poon Group Partnership		又一村公寓樓, 1973 (見55頁) Apartment complex, Yau Yat Chuen. (see page 55)
		蒲台島度假村, 1973 (見56頁) Po Toi Island resort (see page 56)
	 	商業／飯店大廈, 香港九龍金巴利道, 1973 Commercial/hotel building, Kimberley Road, Kowloon, Hong Kong.
		公寓大樓, 香港高陞街, 1973 Apartment complex, Ko Shing Street, Hong Kong.
	 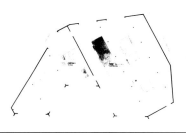	嘉頓麵包廠, 香港九龍深井, 1973 Biscuit Factory for Garden Bakery, Shum Tsang, Kowloon, Hong Kong.

	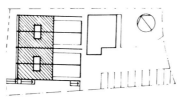 	排屋, 香港壽山村道, 1973 Terrace housing, Shouson Hill Road, Hong Kong.
		安居小區總規劃, 馬尼拉國際設計比賽方案, 1975 Master plan for low cost housing estate, international competition entry in Manila.
香港大學建築系半職講師1978及1979 Part-time lecturer, School of Architecture, University of Hong Kong.		南豐路公寓大樓, 1979 (見58頁) Apartment complex, Nan Fung Road (see page 58)
		山光道公寓大樓, 1979 (見62頁) Apartment complex Shan Kwong Road, (see page 62)
		銀禧體育中心, 1979 (見64頁) Silver Jubilee Sports Centre (see page 64)
香港政府土地供應特別委員會委員 1982~1985 Member, HK Government Special Committee of Land Supply		加達樓, 1979 (見66頁) Cactus Mansion (see page 66)

亞洲建築師學會第一任會長1981~1982 First Chairman, Architects Regional Council(ARCAISA)		弱能兒童院, 1980 (見68頁) Home for the Severely Handicapped Children (see page 68) 獲得1994年香港復康聯會的設計優異獎 Received the Joint Council for the Physically and Mentally Disabled 1994 Design Award.
香港建築師學會會長1981及1982 President, Hong Kong Institute of Architects 創立亞洲建築師學會論壇 1982~1991 Founder, ARCASIA Forum		豐樂閣, 1981~1985 (見72頁) Albron Court (see page 72)
英聯邦建築師學會副會長1981~1983 Vice President, Commonwealth Association of Architects(CAA)		紙幣廠, 1982~1985 (見78頁) Printing factory for Thomas De la Rue Ltd. (see page 78) 獲得香港建築師學會1986優異設計獎 Received the HKIA Design Merit Award 1986
香港政府建築設計上訴法庭委員會委員1982~1997 Member, HK Government's Building Appeal Tribunal		城市理工大學, 1983 (見80頁) City Polytechnic University (see page 80)
潘祖堯顧問有限公司 1986~ **Ronald Poon Consultants Ltd.**		牡丹山莊, 1985 (見84頁) Xian Mountain Inn (see page 84)
中國建築學會名譽理事 1983~ Honorary council member of the Architectural Society of China 創立中建學社1986 Founded Zhong Jian Society	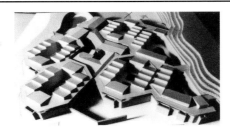	鵰鷹山莊, 1986 (見86頁) Lanzhou Mountain Inn (see page 86)

香港中華文化促進中心顧問 1984
Adviser, HK Institute for Promotion of Chinese Culture

亞洲建築學會顧問 1984~1996
Adviser, Arcasia

青山藝術館, 中國北京后海畔, 1987
Museum/guesthouse, Hou Hai, Beijing, China.

香港科技協進會會長1998
President, Hong Kong Association for the Advancement of Science & Technology

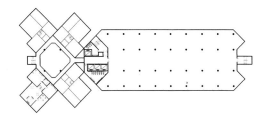

商住大樓, 泰國曼谷市, 1988
Residential/commencing complex, Asoke Dundaeng, Bangkok.

聽葉山房, 1989 (見90頁)
House, Iean (see page 90)

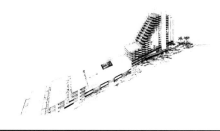

華獻市度假公寓大樓, 1989 (見92頁)
Residential condominium, Huahin (see page 92)

北芭堤雅度假公寓大樓, 1989 (見94頁)
Holiday condominium, North Pattaya (see page 94)

清邁山莊度假村, 1989 (見96頁)
Misty Hill Holiday Homes, Chiang Mai (see page 96)

中國蘭州市城關區人民政府技術顧問 1986 Adviser, ChengQuan District Government, Langzhou City, China.		蘇榮特皇路住宅小區, 1990 (見48頁) Housing estate, Survintawongs Road (see page 48)
香港房屋協會主席 1991~1993 Chairman, Hong Kong Housing Society 香港市政局議員 1991~1995 Councillor, Hong Kong Urban Council	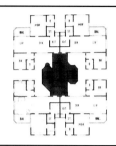	商住大樓, 泰國曼谷市, 1990 Commercial/residential complex, Soi 53/55 Sukhumvit Road, Bangkok.
敏求精舍主席 1991~1993 Chairman, Min Chiu Society.		曼谷市42街公寓大樓, 1991 (見102頁) Residential condominium, Soi42 (see page 102)
香港市政局博物館委員會副主席 1992~1995 Vice Chairman, Hong Kong Urban Council Museum Select Committee.	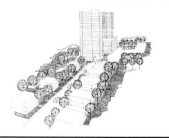	拉曼9街商住大樓, 1991(見100頁) Commercial/residential complex, Rama 9(see page 100)
香港市政局重建大會堂工作小組主席 1992 Chairman, HK Urban Council working group on redevelopment of City Hall	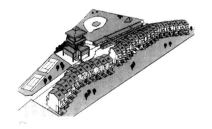	曼谷市郊河畔小築, 1992 (見104頁) Riverside housing, near Bangkok (see page 104)
第八屆中國人民政治協商會議全國委員會委員 1993~1998 Member, the 8th National Committee of the Chinese People's Political Consultative Conference(CPPCC)		肖山市金家浜花園, 1993 (見106頁) Riverside garden, Xiao Shan (see page 106)

香港律師紀律委員會委員 1993~ Member, Solicitor's Disciplinary Tribunal Panel 華南理工大學顧問教授 1993~ Advisory Professor, South China University of Technology	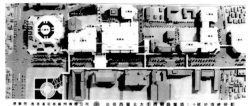	"西西工程"總規劃, 1994 (見110頁) Master plan for "Xi-Xi Project", (see page 110)
與楊永生創辦中國建築論壇 1995~ Co-founded the China Architectural Forum with Mr. Yang Yong Sang 中國河北省易縣基礎設施和規劃首席顧問 1996 Senior Adviser, Infrasturcture and Planning, Yi County, Hebei Province, China.	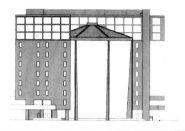	二號地商辦大樓, 1995~1997 (見114頁) Office/commercial complex, No2 site (see page 114)
中國香港特別行政區第一屆政府推選委員會委員 1996 Member, China-Hong Kong Special Administrative Region First Government Selection Committee		四及五號地商辦大樓, 1995~1997 (見116頁) office/commercial complex, No 4 & 5 Site (see page 116) 獲得95首都建築設計匯報展"首都十佳建築設計方案"獎, 1995 Received the "Ten Best Design in the Capital" Award.
中國香港特別行政區第九屆人大代表選舉會議成員 1997 Member, China-Hong Kong SAR Electoral College for 9th Peoples' Congress Deputes		維他奶廠房入口庭, 中國上海, 1997 Entrance at Vitasoy Factory, Shanghai, China.
		武康路公寓大廈, 1997 (見122頁) Apartment complex, Wu Kang Road (see page 122)
中國桂林市秀峰區發展顧問 1998 Development Consultant, Xiu Fang District, Guilin, China		龍潭湖渡假村, 1998 (見124頁) Holiday Villa complex, Lun Tam Lake (see page 124)

第九屆中國人民政治協商會議全國委員會委員 1998~2003
Menber, the 9th National Committee of the Chinese People's Political Consultative Conference(CPPCC)

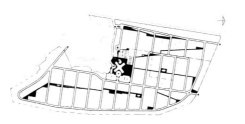

都江堰別墅山莊, 1998 (見126頁)
Holiday villas, Dujiangyen (see page 126)

深圳少年宮, 1998 (見128頁)
Youth Centre, Shenzhen (see page 128)

附錄
Appendix

(1) 1974~1982 在香港建築師學會內參與的委員會名單　360

(2) 1981~1982 報章報導　361

(3) 歷年參與國際會議及學術會議記錄　373

(4) 給英國建築師學會雜誌的信和有關報導　377

(5) 國際建築師協會1987年大會發言　380

(6) "建築報"有關中國建築論壇報導　381

(7) 亞洲建築師協會報導　382

(8) 現代中國建築創作研究小組杭州會議報導　407

(9) 香港科技協進會報導　408

(10) 題字及畫　411

(11) 市政局陳詞報刊報導　413

(12) 政協工作的報刊報導　414

(13) 房協工作的報刊報導　415

(14) 其他工作的報刊報導　419

(1) HKIA Committees participation record 1974~1982　360

(2) Press cutting 1981-1982　361

(3) Attendance record of international conferences and academic symposium　373

(4) RIBA Journal press cutting and letter to the RIBA journal　377

(5) Speech delivered at the XVII UIA Assembly, Dublin, 1987　380

(6) Press cutting on the work of the China Architectural Forum　381

(7) Article on ARCASIA　382

(8) Press cutting on the Hangzhou meeting of the Research Group on Modern Chinese Architecture　407

(9) Press cutting on the work of the Hong Kong Association for the Advancement of Science & Technology　408

(10) Gifts of calligraphy and painting　411

(11) Press cutting on the speech for Urban Council　413

(12) Press cutting on the work of CPPCC　414

(13) Press cutting on the work of the Housing Society　415

(14) Press cutting on other work　419

附錄(1)
Appendix(1)

1974~1982 在香港建築師學會內參與的委員會名單
HKIA Committees participation record 1974~1989

(1) 香港建築師學會執業討論組創辦人及主席 (1974~1978)
(2) 香港建築師學會銀牌獎評判 (1974 及 1981)
(3) 香港建築師學會執業委會副主席 (1975~1976)
(4) 香港建築師學會建築師執業法案委會主席 (1976~1982)
(5) 香港建築師學會，英聯邦及英英國註冊建築師委會香港大學建築系驗察委員會會員 (1977~1981)
(6) 香港政府修改消防條例委員會會員 (1977~1980)
(7) 香港政府消防員及註冊建築師執業委員會會員 (1977 和 1980)
(8) 香港建築師學會保險組主席 (1977~1978)
(9) 香港建築師學會會員及註冊委會副主席 (1977~1978)
(10) 香港建築師學會建築教育會委員 (1977~1978)
(11) 香港建築師學會專業考試委會委員 (1977~1978)
(12) 香港建築師學會有限公司經營委會委員 (1977~1978)
(13) 香港建築師學會公制化委會委員 (1977~1978)
(14) 香港建築師學會公關委會委員 (1977~1978)
(15) 香港建築師學會／香港大學顧問委會委員 (1977~1978)
(16) 香港政府註冊建築師及工程師執業委員會會員 (1978~1981)
(17) 英聯邦建築師協會大會籌委會主席 (1978~1979)
(18) 香港建築師學會對外事務委員會主席 (1978)
(19) 香港建築師學會國際事務委員會主席 (1979~1980)
(20) 建築師、工程師、測量師法例小組聯絡人 (1981 和 1982)
(21) 香港建築師學會執業問題工作小組委員 (1981)
(22) 國際"山頂"建築設計比賽評判 (1982)
(23) 香港演藝學院建築比賽評判 (1982)

(1) Chairman, & Founder, HKIA Discussion Group (1974~1978)
(2) Juror, HKIA Silver Medal Awards (1974 & 1981)
(3) Deputy Chairman, HKIA Board of Architectural Practice (1975~76)
(4) Chairman, HKIA Ordinance Committee (1976~82)
(5) Member, HKIA/Commonwealth Association of Architects (CAA/Architects Registration Council UK Visitation, Board to the School of Architecture Hong Kong University (1977 & 1981)
(6) Member, Government's Revision of the Fire Code Committee (1977~1980)
(7) Chairman, Fire Services Department/HKIA Liaison Group (1977~1980)
(8) Chairman, HKIA Insurance Committee (1977 & 1978)
(9) Deputy Chairman, HKIA Board of Membership and Registration (1977~1978)
(10) Member, HKIA Board of Education (1977~1978)
(11) Member, HKIA Professional Examination Committee (1977~1978)
(12) Member, HKIA Limited Liability Committee (1977~1978)
(13) Member, HKIA Metrication Committee (1977~1978)
(14) Member, HKIA Publicity and Press Committee (1977~1978)
(15) Member, HKIA/HK University Advisory Committee (1977~1978)
(16) Member, Government's Authorized Person & Registered Structural Engineers Registration Committee (1978~1981)
(17) Chairman, Organizing Committee of the CAA Triennial Conference in HK (1978~1979)
(18) Chairman, Board of External Affairs HKIA (1978)
(19) Chairman, Committee on International Institution HKIA (1979~1980)
(20) Convenor, Architect/Engineers/Surveyor Joint Ordinance Committee (1981 & 1982)
(21) Member, HKIA Working Party on Practice Problems (Government Department) (1981)
(22) Juror, International Architectural Competition "Peak" 1982
(23) Member, Advisory Panel for the Selection of Architects for the Academy of Performing Arts (1982)

華僑日報 YAT PO

中華民國七十年公曆一九八一年八月廿九日 星期六

建築師感有心無力
改善社會環境難作貢獻
地產發展商港府有責任

建築學會會長潘祖堯解釋原因表示遺憾

（特訊）香港建築師學會會長潘祖堯，昨日出席香港西區扶輪社午餐會，演講「建築師與香港環境問題」。午餐會由社長陳有生主持。

潘祖堯會長稱：香港是世界首屈一指的商業經濟中心之一，它的普通民居的生活環境，卻是世界最差的之一。

香港缺乏現代化的城市規劃，住屋供應，教育及交通系統都不足，令到中下層階級的市民對此社會甚為不滿，而這些市民佔本港人口卻有百分之八十以上，所以可以講是危機四伏。

建築師雖有心有力，但因兩個主要因素所阻而不能對香港的社會環境有甚麼貢獻：

（一）業主；
（二）政府。

業主有很多種，當然，則師亦有很多種去服侍他們，這個情形有醫學及法律有發生。但很少病人會教醫生怎樣開刀，亦很少有客人干預律師的工作。

但多數業主會有自己的主見，不合乎常理的要求，堅持要減收則費，若不答應的話，就提議削減我們的服務，而達至減收費用的目的。很多時，建築師往往當作繪圖員被業主要求抄襲某雜誌或另一個則師的設計。很多香港業主及本港建築師是坐井觀天，所以設計比先進的國家落後十至十五年，一方面很多香港人已經接受一分錢一分貨的觀念，買花錢去買中名牌貨如打火機，買這些奢侈品的人很少會討價還價，但多數業主會對的建築設計並非多餘而是必須的，及建築物的設計及施工討價還價，不接受一分錢一分貨的原理。當然，好的建築師亦不會接受業主這種苛求的寒酸要求，而達至今日來說，業主的寒酸苛求的惡習，不接受建築設計並非多餘的。

一般點就全部由建築師負起。

如果有很多實例如廖創興大廈及信和旭和道山泥傾瀉而樓的事件發生，社會人士便會更加醒覺。可笑的是，建築師費用在整個物業發展費用當中是佔很少的比例。一個一億港元的工程來說，建築師的費用是佔總投資額1.3%，因建築師的費用並非是根據建築工程的費用，而包括地價。而建築費到今日來說，佔總投資額三分之一。這類的工程多數需時三年，而每年建築費用等於全部投資額之百分之0.43，但息口已超過全部投資額百分之二十以上，或者算每日17萬師的負担。建築師如收費低於建築師學會訂下之最

低收費率，我們有時稱他們為簽字則師，每一個星期做60小時工作，他們往往不能細心處理每一工程，同一寫字樓可有60至80個工程，亦可有20至30單是在施工中。就算他們能一個星期做60小時工作，也不能在每一個星期內每一工程花費一小時。這情形下錯誤百出，而以每天息口17萬來算，不起這樣的花費。

然而想在車胎費用方面省金錢，而改用雞牌車胎，因為各車胎公司出品不標準，以致發生交通意外，省下之錢亦不夠付修理費十分之一。

另一因素是歸咎政府，不合潮流的建築條例是一障碍，但最主要的問題是官方的見解及處理問題的方法，在今年年初則師會員投訴，搜集有問題的工程，本會組成短短兩個月內向會員機構作出詳細的報告。這報告包括了八十多宗實例有不公平處理的情況。雖然最近兩方面已將報告呈上新界政務司及工務司，了解及提出某些折衷辦法，但主要的問題就是公務員處理時的態度問題。

兩三年前公務員是很有合作精神，在照高層指示下盡力協助建築師處理批則問題。但近兩三年內，情況大變，合作精神全失，官方的籍口是人手缺乏，但依我們看來，就算人手充足，他們的態度亦不會更改，而在人手不足之時，我們要花的費用更多時間與經驗不足的官員作更多無謂爭論，實屬費時失事，對我們專業人士來說是一個侮辱。

這種情況有人也會提議是廉署成功的後果，但如果道屬事實的話，本人情願貪污活動繼續發生。建築圖則是很有責任的，如反會員專業人士當當與官方接頭，不會敬章批准，這對香港的發展會有深入研究某一性戀問題，我覺得以上見解的情況，更值得優先處理。

總括來說，業主方面的問題是比較容易解決的，只要他們用醫生或律師的態度來雇用建築師，便事半功倍。當然，日本會有嚴格的會規，詳細的僱用合約，及提供最新的執業資料，對政府方面我要借用國內一句常用語，「問題不大，但技術上有困難。」（天）

SOUTH CHINA MORNING POST

MONDAY, MARCH 2, 1981

New bank will set example for all

IN reference to an article by Stephanie Williams concerning the Hongkong and Shanghai Bank redevelopment (SCM Post, February 16), I concur with her enthusiasm over the design of the proposed bank building and would add my own congratulations to the bank over such an enlightened appointment.

This article gives me the opportunity to highlight a few problems facing the profession in Hongkong.

It is undeniable that the standard of Hongkong's architecture is as bad as Ms Williams implied but the developers and the Government, and to a certain extent architects, should together take the blame for the state of affairs here.

We are working virtually in a desert as far as architectural appreciation is concerned in Hongkong, where cost of land is so astronomical that the need to develop as quickly as possible often overrides all other considerations, including the quality of architecture.

Therefore we architects cannot be blamed solely for this "arid" situation and God forbid if we ever gave her the impression that we architects in Hongkong indulge in complacency.

Therefore, given the same cost and time parameters (for cost alone, the proposed bank building is 2½ times the normal standard — Admiralty, for instance — and 18 months spent for schematic design is a rare luxury) and the same enlightened client, we could perhaps avoid being "knocked into a tin hat."

Very few local developers realise that with more time and more money spent on the design, more saving would subsequently be made in the maintenance and running of the building and I hope the new building venture will set a precedent for many to follow.

It would appear that new technologies are being proposed in the redevelopment. This breath of fresh air from outside will, we hope, prompt the Government to reappraise its outdated building regulations, and open the way for us to design a better and more creative environment.

RONALD POON
President
Hongkong Institute of Architects

盼與港建築界加強聯繫
楊廷寶和戴念慈在北京參觀香港建築圖片展覽

中國新聞社記者 徐泓

香港建築圖片展覽揭幕式結束後，專程從南京趕來致開幕詞的中國建築學會理事長楊廷寶教授和副理事長戴念慈總建築師，向記者暢述了他們參觀這次展覽的感受。

楊老和戴總不僅同為中國建築界高齡，他與已故的清華大學建築系教授梁思成齊名，被譽為中國建築界的「南楊北梁」，是建築界的元老，桃李滿天下。戴總就是楊老主持設計的北京中國美術館的得意門生之一。戴總近來在設計的北京中國國際會議廳在國內外頗有影響。

戴念慈總建築師講一口略帶浙江口音的普通話，平易近人。他對記者說：「我一九六二年、七三年、八零年三次到過香港。二十年來香港的地理位置，廣泛接觸到世界各地的先進技術，刺激香港建築業向現代化發展，許多最新的施工技術、最新的建築材料和設備在香港幾乎都採用了。從某種意義上可以說

我們的視野，豐富設計思想，促進中國內地建築業向現代化發展很有益處。」楊老談起九月將在香港舉辦中國建築圖片展覽，他說，「中國上下五千年的歷史，創造了輝煌燦爛的建築，其中包括豐富多采的宮殿、精巧的住宅和有歷朝歷代留下不同風格的建築，還有幾十個少數民族獨具特色的建築，內容之宏大豐富，世界罕見。我想通過展覽和學術交流，使港澳同胞、海外僑胞了解並領略中國古建築。

香港是世界建築的一個展覽櫥窗，我們要吸取先進技術，就要向香港同行學習。」戴總還說，「我一直很注意香港高層高密度住宅的發展。內地一些百萬人口的特大城市也有用地緊張的問題，市中心住宅是向空中發展的趨勢，這方面香港的實踐對我們很有參考價值，某些處理方式可以取法，當然不能照搬，而要設計方案和我們的實際結合。」戴總在談到中國內地與香港學術交流活動時，讚揚香港建築師學會會長潘祖堯在其中起了很好的作用。戴總說：「中國建築學會和香港建築師學會的正式交流是從潘祖堯上任後開始的。兩三年中潘先生還是亞洲建築師學會主席，他為中國大陸與台灣建築界的學術交流也做了不少工作。楊老最後告訴記者：「中國建築學會在香港澳和台灣同胞開展了理事名額，我們希望通過各種學術交流活動，彼此加強了解，互相學習，互相勉勵，共同為繁榮中華民族的建築業。」

【中國新聞社北京六月七日電】

潘祖堯談內地建築
認為有不少先進處
面臨保存傳統與吸收西方藝術矛盾

【本報消息】據新華社香港北京八日電：香港建築圖片在北京展出之際，本社記者訪問了前來參加開幕式的香港建築師代表團團長、香港建築師學會會長潘祖堯。

潘祖堯一行是六月三日下午抵達北京的。他說，這次香港建築圖片展覽是香港建築師學會在外地舉辦的第一個規模最大的展覽，是香港和內地整個建築交流計劃的一部分。

他對這兩個學會之間已經開始的學術交流活動的前景充滿信心。他說，我們之間的學術交流已經有了一個良好的開端，相信今後一定會持續下去。

展覽會期間，潘祖堯在清華大學作了題為「淺談現建築藝術的矛盾」的報告，還要去上海、成都、昆明、南京等地的建築院校訪問，並配合香港建築圖片展覽在鄭州、西安、上海的展出，向內地建築界人士介紹近代建築潮流和趨勢。

在談到香港建築今後的打算時，潘祖堯透露，明年年中還將在香港舉行另一個大型建築展覽。這個

展覽將展出目前在法國巴黎展出的「中國人民生活與建築」的部分展品。身兼英聯邦建築師學會副會長（亞洲區）和亞洲建築師學會主席的潘祖堯這計劃將這一展覽在亞洲區巡迴展出。

潘祖堯說，近幾年來，他幾次到過內地，他都能看到各地建設的新進展和新變化，都有新的感受。

在的建築方面很有進的東西。他對內地的「預製式」建築物很欣賞。

他在進一步談到需要研究的問題時指出，現時中國內地建築需要面臨著保存中國傳統建築藝術及吸收西方建築藝術的矛盾。香港亦存在著如何使建築物反映出地方特色和民族風格的問題。因此，香港和內地建築界需要互相交流資料，共同研究和解決這個問題。

潘祖堯說，香港建築設計師近年香港已有不少設計人才到深圳特區為內地作樓宇設計。香港建築設計師也到深圳特區最為集中作品。香港的建築設計師們願為祖國的四化建設貢獻力量。

Architects plan reform fight

HONGKONG architects are fighting hard to press for procedural changes in the Government's examination process for architectural designs.

Several architects said yesterday that their intended targets are the Public Works Department's Building Ordinance Office and the New Territories Administration.

Hongkong Institute of Architects president Ronald Poon yesterday said the institute would hold talks with PWD officials soon after supporting materials for a main report on architects' practice problems were ready.

Mr Poon said the report contained some proposals to improve the submission procedures of building designs.

Mr Poon refused to disclose details of the proposals.

Earlier at a luncheon meeting of the institute, Mr Poon said 112 valid complaints from members had been collected, alleging mishandling of their designs by Government officials.

Several architects — who asked not to be named — said that New Territories architectural projects, which are subject to the approval of district office officials and finally the Secretary for the New Territories, have created the most conern among architects.

The architects said it was doubtful whether the New Territories administration office had sufficient number of qualified officials to handle professional architectural designs.

"Also, their basic criteria for approving designs are still ambiguous to us, despite increasing construction projects in the New Territories, especially in new towns," one of the architects said.

The architects have been persistently clamouring for drastic changes in the way the Government handles their designs, for which approvals often comes, they say, after great delays.

Normally, approvals for designs come at least six months after submission, but period up to nine months had been recorded.

"We are often criticised of having a lack of imagination and creativity in building designs but we are in fact limited by many obsolete and undesirable regulations," one architect said.

Some architects have to return to the Building Ordinance Office for very minor amendments for as many as three times, the architects said.

Any innovative or creative designs are drawn back, because of fears that non-acceptance from officials, will result in delays, they claimed.

The architects explained that for projects in the New Territories, besides letter of compliance and the occupation required same as for urban projects, a further design disposition has to be obtained from the Secretary for the New Territories.

WEDNESDAY NOVEMBER 4, 198

South China Morning Post

'Appalling system' must be changed, says architect

Red tape and the attitude of Government officials has brought chaos to the local building industry, says a leading architect.

And he has called for changes in the building laws to end the delays and frustrations of getting building plans passed by the Building Ordinance Office.

Mr Ronald Poon, president of the Hongkong Institute of Architects, delivered a withering attack on both the Government and cost-cutting developers in a speech at a lunch meeting of the Rotary Club of Hongkong Northwest yesterday.

"The environment in Hongkong is appalling," he said.

"There is a lack of proper planning, be it in housing provision, education or transport."

He said architects were prevented from contributing to improvements in the environment by the twin stumbling blocks of their clients and the Government.

His institute collected 100 cases of member firms complaining about problems with Government departments, and submitted these in reports to the Secretary of the New Territories and the Director of Public Works in May.

He said the PWD "gave us the impression that in a majority of cases complaints against the Building Ordinance Office were not justified, and only 10 cases 'appear to have some justification'."

He described this as "a typical bureaucratic reply. We are naturally disheartened."

He added that the New territories Administration offered "constructive remedial measures."

A second report will be ready next year, giving suggested changes in the existing process for building plan approval.

But Mr Poon said the institute is still working on the second report and the other two architects' associations are also interested in the issue.

Mr Poon

They had not yet decided whether to compile their views into a single report or submit their own reports.

He complained at discrimination against the private architect, pointing out that Government architects need not submit their plans to the scrutiny of others.

"Surely academically and in practical experience, Government architects are the same as private practitioners."

He said the present building legislation was originally designed to be executed with a good deal of discretion as sites or plans were rarely the same.

"During the last two years the attitude of Government officials has changed for the worse. . . The sense of co-operation has gone.

"It can safely be said that not a single active architect's firm in Hongkong can claim it has not experienced some 'unreasonableness' in its dealings with Government departments in the last two years."

He said fear of the ICAC may have slowed the procedures, but added:

"If anti-corruption encourages work-to-rule, disrespect to professional people and a careless attitude then I think there is definitely something wrong somewhere."

He said if it was not enough that architects had possible prison sentences hanging over them, the Government should establish an Architects' Ordinance to regulate the proper behaviour of the profession.

"With that achieved, surely the Government need not apply double standards for their own architects and private practitioners."

Mr Poon also attacked clients whose main consideration was cutting back on architects' fees.

"Most of our clients come with their ill-informed preferences and preconceived and outdated ideas and insist on cut fees. . ."

He said most clients here — and some architects — were unaware of developments in the rest of the world.

"Therefore the majority of the architectural design in Hongkong is about 10 to 15 years behind the times."

He said most developers were only concerned with quick profit, with no social conscience, and were not concerned with the quality of their buildings.

"Little do they realise that with the shortage of land supply, a chance lost in capitalising on the maximum potential of a site is lost forever."

He compared saving money on the architect's fees — a very small proportion of most projects — to buying a Rolls-Royce then running it on the cheapest replacement tyres.

Central waterfront tunnel plan backed by experts

By PAUL BARAN

The Government's proposed elevated road along the Central waterfront will be a waste of taxpayers' money, environmentally dangerous and a blight on the landscape.

And although the alternative proposal presented last week to the Town Planning Board by the property owners' lobby will cost more to build, it will be better for Hongkong in the long run.

These were among the views expressed on both proposals by local urban planning experts, architects and environmentalists interviewed yesterday by the SCM Post.

According to the president of the Hongkong Institute of Architects, Mr Ronald Poon, the Government needs help in developing an overall planning strategy.

The alternative proposal, he said, "looks more realistic and concerned with the environment," while the Government's proposal "looks very questionable."

The lobby wants a $1.7 billion underwater road tunnel running parallel to the Central waterfront instead of the Government's proposed $750 million elevated flyover over Connaught Road.

The controversial flyover, which would run from the Macau ferry pier to the Cotton Tree Drive interchange, would pass directly in front of the third or fourth floor windows of several Central hotels.

Since the Government seems incapable of coping with today's increasingly complex planning needs, Mr Poon suggested that it consider setting up a panel of international experts working with local participation to develop an overall strategy.

"With a population of well over six million," he said, "this community deserves to be looked at by proper planners."

Although he was quick to point out that the Government had good planners, he explained that they "sit under various departments which inevitably make them very ineffective."

It was unfortunate, said Mr Poon, that most planning decisions were made "by policy makers who have no idea of what planning is all about."

He also criticised the Government for not seeking public opinion on its plans before unveiling the project.

The Government, he said, was "pretty naive" in thinking that it could "plunk this kind of mega-structure on to a major waterfront and get away with it."

"I think it's wasting public money to indulge in this kind of exercise," he said, "when any planning expert would tell you right away that this is not the best way to do it."

At Hongkong University, the director of the Centre of Urban Studies and Urban Planning, Professor Kwok Yin-wang, also said the alternative proposal looked better.

Even when considering that the property owners' lobby is working "in its own interests," he said, the Central parkway proposal idea "has a much better argument in terms of noise pollution and aesthetics."

However, he said he was not entirely convinced the alternative proposal would solve the air pollution problem, explaining that the tunnel would simply move the fumes further north.

While agreeing with the need to solve the area's chronic traffic congestion, he said the question was not where to put another road, but how to build it.

The tunnel concept runs into snags, however, said Prof Kwok, if the Government plans to reclaim more land north of the proposed alignment.

"What happens in the long-term if the Government decides to reclaim further?" he asked. "What do you do with the tunnel then?"

But he said "the lobby's points seem pretty reasonable on the surface."

Environmentalists, meanwhile, heartily endorse the Central parkway idea.

One source said that since the alternative proposal calls for about 10,000-20,000 sq metres of parkland on top of the tunnel, "obviously from an environmental point of view, putting in a tunnel is better."

Although he pointed out that the disposal of marine mud from the tunnel's construction could be a problem, he said it was likely to be a short-term headache when considering the overall advantages of the concept.

Environmentalists point to the experience of other cities in attempting to solve traffic congestion along their waterfront areas.

Many cities, including several in Britain, they said, have had to scrap elevated road plans after strong objections from special interest groups.

Over the past decade, they said, most cities have been treating their waterfront areas as recreational resources instead of cutting them off from the rest of the community.

Moreover, they said, the flyover would ruin the Hongkong Island skyline, which is partially responsible for bringing about 2½ million tourists to the territory annually.

As Mr Poon put it: "Other governments are trying to bring their people to the waterfront, and here you have one that is trying to do the opposite."

SOUTH CHINA MORNING POST   SUNDAY, FEBRUARY 14, 1982

Jockey Club 'yields' in design row

By PAUL BARAN

The Hongkong Institute of Architects has won its fight in convincing the Royal Hongkong Jockey Club to hold a limited design competition for its proposed performing arts academy in Wanchai.

According to a Jockey Club spokesman, Mr David Yau, designs from six local architectural firms have been submitted to the academy's project board, which is now considering each proposal.

But when the successful design will be announced is still unclear, said Mr Yau, explaining that the board is under no obligation to choose one of the six submitted designs.

The project is scheduled to be built over the next three years on a site which is currently open ground between HMS Tamar and the Arts Centre.

News of the limited competition signals a victory for the institute, which was growing alarmed at how the Jockey Club had been handling design plans since announcing its intention to build the academy in September.

Architectural sources said at the time that the institute was upset over the Jockey Club's "breach of local practice" by approaching the Public Works Department for plans before consulting the institute.

They pointed out that there was nothing illegal about the Jockey Club's move, but it was considered "unethical and unprofessional" as it was a project which would wind up as a cultural landmark for Hongkong.

Despite the Government's involvement in the scheme — by providing the land — the Jockey Club, they said, should have consulted the institute as a matter of courtesy.

Also bothering the institute was the Jockey Club's approach to 10 local architects without consulting it first for an opinion or price levels.

Sources now say that it took a lot of work on the part of the HKIA president, Mr Ronald Poon, to convince the Jockey Club to agree to the limited competition.

But when asked about what it took to get the project board to change its mind, Mr Poon declined to comment, explaining: "I'm walking on a tightrope. You don't expect me to aggravate them further do you?"

The Jockey Club's Mr Yau, meanwhile, said that the six architects were selected "after a series of consultations" between the institute, the project board, the club stewards and assorted Government departments.

Initial estimates of the cost ranged from $200 million to $1 billion.

One architect source, commenting on the competition, said: "The Jockey Club finally decided to do it right."

However, it was pointed out that design plans were originally going to be ready by the end of last year, which means that the competition, although a good step, will delay things further.

華僑日報　一九八一年三月十五日星期日

影响本港建築條例落後
亞洲建築師學會及國際建築師學會等，其實乃歐美十年前作品中區新型玻璃牆幕設計

·李龍生·

國際建築師學會係一個世界性的機構。每三年召開一次大會。最近一次大會將於今年六月在波蘭華沙召開。亞洲建築師學會隸屬於英聯邦建築師學會之下，於一九七九年九月在印尼雅加達成立。由印尼、印度、菲律賓、斯里蘭卡、巴基斯坦、孟加拉及泰國等十國組成。首任會長由現任香港建築師學會長潘祖堯擔任。每年舉行一次大會。總部在本港。港督麥理浩爵士為贊助人。現有成員五百餘人。

現任會長潘祖堯說，該學會的主要工作是：（一）是提高本港建築師設計水平，為此常有邀請外國同業進行學術交流，舉行比賽會、演講或學術研討等（二）代表本港建築師在本港及亞洲發言，（三）與大合辦新建築師，每年本港大學畢業並通過考試的新建築師有數十名。（四）改善本港市民的生活環境，研究進行各省能源的經濟設計問題，（五）進行專題研究，例如與水電工程師等合作成立一個小組，研究改善本港市民生活環境及市容的願望及能力，但由於多種原因，未能在短時間內作出來。

潘氏說，而建築師好做者往往離不開市民之需求，例如藥王殿建成後，建築師多新材料，新工藝及新設計保守縮待態度，而在歐美早在十年前便風起的玻璃牆幕設計，而在歐美早在十年前便盛行。

潘氏認為政府過時的條例，亦是一個重要原因。他說，如條例十六G，要求新建樓宇維持原貌要高昂，這妨礙了地方不斷的發展需要。又如外牆的全鋼結構已不少，但香港仍的規定要包混凝土作防火保護層等。另外，政府對外國請來的建築師與本港的建築師即使同樣到外國回來的，對其設計審核的態度也不同。頗不公平。潘氏反復強調，本港的建築條例，已跟不上世界發展的步伐，妨礙本港建築設計水平的提高。因此，重新修訂建築條例，已是當務之急了。

建築師的組織，世界各地均有。如香港建築師學會等。

（生）

Outline for Architects' Ordinance

SOUTH CHINA MORNING POST — WEDNESDAY, MARCH 31

IN response to the invitation from Mr Rick Tsui Hin-fai (SCM Post, March 13) I shall try to explain the Hongkong Institute of Architects' proposals for an Architects Ordinance.

Several of the Hongkong professions are already regulated by ordinance — the medical, legal, accountancy and engineering professions for example. The purpose of an ordinance is to maintain professional standards, to safeguard the public from charlatans and malpractice and to create and maintain codes of practice and ethics.

Much of the above is beyond the control of the present statutes controlling the listing of authorised people. After all, existing legislation was designed to ensure public safety and hygiene only. Environmental, professional and other considerations were not included.

The Hongkong Institute of Architects can only exert its influence on those who are its members (approximately 30 per cent of the architectural professionals in Hongkong). A separate controlling body created by an ordinance and to which all those who wish to practise as an architect must register would seem to be called for.

Basically, our proposed ordinance will resemble the Architects Registration Act 1931 in the United Kingdom and the resulting Architects' Board will be equal to the Architects' Registration Council in the United Kingdom (ARCUK).

The Government, the profession, the academy are all represented on the board of ARCUK. It is proposed that the Hongkong Board of Architects be similarly constituted and, like the existing Medical and Bar Councils, the board will be an independent body appointed by the Governor.

In effect, all architects who wish to practise as an architect and to use the title of the profession, whether they be in the Government or in private practice, will be required to be registered with the board. Registration will be required annually and a registered architect in private practice and offering his services to the public will be required to take out a Practising Certificate.

Use of any description containing the word "architect" will also be restricted to those properly qualified and registered. Misuse will, as in the UK, be an offence punishable by fine and/or imprisonment.

In an environment so lacking in consumer protection laws, this measure is perhaps, even more vital than in the UK. It is further complicated by the fact that in Hongkong the Chinese custom is to address anyone from the design and supervisory offices as "architect" be it may interior designers, structural engineers, mechanical and electrical consultants.

Amendments to the Building Ordinance and its regulations will be suggested but they are kept to the minimum. Basically, they will ensure that registration be a prerequisite to appearing on the Architects List (List I).

It is of course our ultimate aim that the Building Ordinance should be amended to ensure that "architectural" plans can only be submitted by those properly qualified as architects (as in the United States, Europe and much of Southeast Asia). After all, most reasonable people would, I am sure, agree that only a properly qualified surgeon should perform brain surgery and a gynaecologist would not be considered so qualified.

However this will take time as there are still a number of people submitting architectural plans to BOO in Lists II and III and it is not considered desirable to destroy their livelihood in "one foul swoop."

Properly managed, the effects of this ordinance will enable registered architects to assume much more responsibility in the performance of their existing statutory duties under the Buildings' Ordinance. It could well be that the very much understaffed Buildings Ordinance Office could be relieved of some of its more onerous chores such as the checking of plans, issuing of occupation permits etc, enabling it to concentrate on the control and enforcement side of the ordinance, thus providing an effective deterrent to the construction of illegal structures and unauthorised additions and alterations to buildings.

We believe that only with the establishment of such an ordinance will the public be properly provided with an effective, legally constituted body able to control the profession. This will ultimately result in a greater understanding by the layman, a greater discipline within the profession and an improvement in our environment.

RONALD POON
President
Hongkong Institute of Architects

明報晚報
1982年6月1日

對於高層建築物
國內建築專業人士
希望借助本港經驗

建築師學會再呈建議書

確保則師專業水平
港宜修例籌組委會

執業者皆須入會註冊受資格檢定

（本報專訊）香港建築師學會將於未來數月內，積極促請政府訂定「建築師法例」，並成立一個「建築師註冊委員會」，以確保專業水準。

香港建築師學會會長潘祖堯，昨日接受本報訪問時，作上述的透露。

潘祖堯指出，目前建築師會員之專業道德標準，但是卻沒有建築法例，只注重安全方面，但是若由私人之法例有關之事由建築師立案，則甚為麻煩，故應由政府訂定有關之法例，並附以法例之有關細則。

潘祖堯稱，該會已制定了一份草稿，並曾經向政府諮詢過一次，當時遭到有關方面的反對，目前認為已將有關之草稿加以修定，相信可獲得接納。

潘氏指出，醫生和律師等專業人士，均有所屬之公會，而獨建築師卻缺乏，由香港建築師學會、大學、政府及非會方建築師代表組成，以審核會內執業人士之資格。至於何時能成立，潘氏稱，現時很難說。

潘氏解釋謂，香港建築師學會只是一個學術的機構，並非一個執業人士的委員會。而會員的成員數目約五百名，佔行內專業人士三至四成，由於在政府工作的則師不入會，所以缺乏一個機構對此加以控制。

潘氏並謂，現時如果會員收費過低，違反會方之規定，會方至多驅逐他們出會，但該人士仍可繼續在本港執業。此舉有助改善目前「一盤散沙」的狀況。

築師，都要向此委員會註冊，成為有關之會員。

潘氏指出，如果收費過低，則肯定有「偷工減料」的現象，而巡視地盤之工作亦會減少，進而影響到興建樓宇之結構安全，這是很嚴重的問題。

潘祖堯以個人身份向當局呼籲，改善目前之入則程序，容許則師決定模字是否可以興建，而不用工務局之批准。

潘氏指出，因為則師既是具有專業資格的人士，自然有這方面的能力，工務局可作為顧問的地位。

辦第三間大學可解決則師荒
或資助留學生學成後回港服務
用電腦繪圖建築師學會研究中

為了長遠解決本港建築師人手不足的問題，香港建築師學會建議本港應增設一所大學以增加收容的學位。

事實上，非官守議員王澤長今年三月在立法局財政預算辯論中，亦促請政府盡快考慮興建第三間大學，但政府對此至今仍未作出決定。

香港建築師學會會長潘祖堯，對記者跟說，本港建築師人才跟居民的比例遠低於鄰近地區，以新加坡為例，新加坡人口只有二百多萬，但建築師學會人數跟本港差不多，相對來說，本港的有關比例則太低了。

潘氏稱，現時香港建築師學會會員中，第一類註冊工程師人數約三百餘人，而第二類註冊工程師則有二百餘人。

潘氏表示，即使現時建築業陷於不景氣的時期，仍感到請人有困難；一旦建築業復甦，則情況更難樂觀。

潘氏認為，長遠解決建築師人手不足的方法，就是增建一間大學，以擴充有關之學位。另一個方法是仿效現時馬來西亞的制度，潘氏指出，馬來西亞政府出錢資助學員到海外深造，回來後要規定為本港服務三至五年。

潘氏透露，香港建築師學會已成立一個專門研究業務委員會，研究業務上所需的工具，及施工的有關專業資助。潘氏稱，委員會屬下小組的其中一項計劃，是研究電腦圖設計的可行性。潘氏稱，新加坡有三間寫字樓，合資購回一個價值一百五十萬元的電腦，估計在三年內就可以收回成本；若在香港推行此概念，則相信不用三年就可以收回成本。潘氏稱，電腦的的圖設計是順世界性的潮流，不但可增加工作的效率，並且可以解決人手不足的問題。

潘氏認為，現時本港在這方面只處於一個「初步階段」，除了缺乏專業知識的電腦人才，亦缺乏儲藏資料的磁帶一般行）。

China Focus

HK-China architectural exchange

China's Minister for Urban and Rural Construction and Environmental Protection, Mr Li Xi Ming, officiating at the opening of the exhibition in Beijing (above). Viewing the exhibits (below) are Mr Li and Ronald Poon (second from left).

An exhibition of more than 800 pieces of work by Hong Kong's architects in Beijing in June marked the start of the first ever architectural exchange programme between China and Hong Kong.

The 10-day exhibition, sponsored by the Hong Kong Institute of Architects, will be followed by a similar demonstration by Chinese architects in Hong Kong in early September.

From Beijing, the Hong Kong exhibits will be moved to other Chinese cities, including Zhengzhou, Shanghai, Chengdu, Kunming, Nanjing and Guangzhou.

Taking part in the exhibition are 22 of Hong Kong's professional firms, the Hong Kong Housing Authority and the School of Architecture of the University of Hong Kong.

Speaking at the opening ceremony in Beijing, the president of the HKIA, Mr Ronald Poon, said the aim of the exchange is to provide members of the Institute an opportunity to display their works overseas.

The president of the Architectural Society of China and deputy governor of Jiangsu Province, Professor T P Yang, said the exhibition heralded the start of a useful and regular exchange programme between Hong Kong and Chinese architects.

In Hong Kong, the Chinese exhibition will open at the New World Centre on September 3. Photos of modern and traditional Chinese architecture will be mounted. There will also be some rare architectural relics, such as the earliest architectural plan engraved during the Han Dynasty, as well as rubbings of ridge tiles and gates throughout the centuries.

房屋司廖讚口「多姿多采內容豐富」
中國建築圖片展覽
昨日開幕今起展出
廖本懷潘祖堯主持剪綵儀式

【本報訊】房屋司廖本懷昨晚在「中國傳統建築圖片展覽」剪綵儀式上稱讚這是一項「多姿多采及內容豐富」的展覽。

「中國傳統建築圖片展覽」昨晚正式展出，由中國建築學會代表團、香港建築師學會及房屋司廖本懷等主持開幕儀式，展覽將由今天起一直至二十五日止，在九龍新世界中心展出。

廖本懷表示，這個展覽會對中國及香港來說，都具有一個非常重大的意義，因為這是中國建築學會首次在亞洲舉辦最大規模的中國傳統建築藝術展覽會。

他說：「這項展覽使我們得以從早期的建築設計方面，體會到中國古時建築的風格和技巧。此外，這次展覽會中有些圖片是考古新發現，而首次在香港展出。」

他評論香港在這次交流展覽中所展出的作品時，形容它們「富有時代性及想像力，足以反映今日香港在建築藝術方面的成就。」

中國建築學會代表團團長王華彬，昨日在展覽預展的記者招待會上表示，藉着今次展覽，希望能將中國傳統建築的特色介紹給香港。

同時，亦期望能吸收本港建築的發展經驗，相信透過兩地的交流，必能探索出一條既融合現代化的建築，而又保留傳統建築藝術精華的路向。他相信這方面的工作是十分艱鉅的。

代表團的另一位成員，中國建築學會副秘書長曾堅在會上透露，中國建築科學研究院在二十個月前派了七名建築師來港，接建築工程，學習掌握本港的最新建築技術。他們更組成華生工程師及結構公司，是隸屬於中國海外建築公司的子公司。該公司目前已接受六項工程的顧問工作，其中最大的兩項為深圳的連城及福田「的東方明珠旅遊區」的擴展業務，向東南亞設計。他們更有計劃方面進軍。

王華彬指出，本港建築近年發展快速，多採用西方最新建築技術，且全都走向高空發展，多層大廈林立。中國方面，高樓大廈的建築物已日漸矗立...

（上圖）中國傳統建築圖片展覽會上展出之太和門廣場及內金水橋。（下圖）中國傳統建築圖片展覽預展會上，左起：香港建築師學會會長潘祖堯，中國建築學會副理事長王華彬及副秘書長曾堅。

文匯

公元一九八二年九月四日

SOUTH CHINA MORNING POST — SATURDAY, SEPTEMBER 4, 1982

Give Hongkong its own look, urges top architect

By CAMILLA SUGDEN

"Glass houses in a desert" was how Mr Ronald Poon described Hongkong's architecture yesterday.

As president of the Hongkong Institute of Architects, and one who is making his living out of the said "glass houses," Mr Poon was having a jolly session throwing stones at our glass Goliaths.

He was speaking at a press conference held to introduce a pictorial exhibition of traditional Chinese architecture, currently on show at the New World Centre.

The exhibition has been organised by the Hongkong Institute of Architects with the help of the Architectural Society of China, and is part of a plan to exchange ideas between the places.

Mr Poon made a lively plea for architects in Hongkong to look to Chinese traditions and incorporate them in their work, saying that such efforts would be well-received here, where 98 per cent of the population is Chinese.

The vice-president of the Architectural Society of China, Mr Huapin Wang, was also present at the press conference, and he too went to great pains to expound the virtues of classical Chinese architecture.

With a great show of patriotic bravado, he stressed that the ideas which were the mainspring of the West's first "modern" architects' work — **Le Corbusier, Frank Lloyd Wright**, Gropius and the rest — actually originated in **China**. A fact which the West

Mr Poon (left) chatting with Mr Wang.

comfortably chooses to ignore.

Indeed, while making the point that both Orient and Occident have many ideas in common in this field, he was quick to add that even before the modern period, Western architects were far behind China in their establishment of certain principles.

For example, he stated that the first generation of modern Western architects, who introduced the concept of flexibility in space inside, buildings, were 2,500 years behind the times.

The idea was originally expounded by a Chinese architect, in a book called *Classical Morality*.

He went on to say that the idea of landscaping buildings also has its roots in Chinese architecture, as does the concept that the structure of a building should remain uncovered.

David Piano and Richard Renzo epitomised this idea in the design of the Pompidou Centre, in Paris, but it is one which has long been adhered to by Chinese architects.

Indeed, even the notion of pre-fabrication and standardisation of components, which many would claim as a Western innovation, has long been a key factor in Chinese architecture, Mr Wang asserted.

By making these points, Mr Wang not only put Western architects in their place; he also demonstrated how much the architecture of East and West have in common.

Like Mr Poon, he was much in favour of the idea that Hongkong should assimilate some of the classic principles of Chinese architecture.

And he expressed the hope that the exhibition would raise Hongkong's awareness of the Chinese style.

But he was equally keen that China should learn from Hongkong.

There is in fact an exhibition which is currently touring major cities in China, and has been mounted by various member firms from the Hongkong Institute of Architects.

The Chinese exhibition over here includes some material which has never been seen outside China before.

For example, there is a picture of the earliest known architectural drawing extant in China.

It is on a bronze plate excavated from the remains of a mausoleum from the Warring States Period, (circa 310 BC).

Mr Poon expressed the view that a "culturally thirsty place like Hongkong" should show a keen interest in the exhibition.

He also tentatively predicted that Hongkong's architecture may soon move off in a new direction, drawing on the classical traditions of China for its inspiration.

He cited Japan as an example of where the blending of ancient and modern has worked successfully, saying:

"Japan has used modern materials and technology, but you can still feel a traditional Japanese atmosphere.

"I want architecture in Hongkong to have a Hongkong feeling."

新報人

香港建築師學會會長潘祖堯認為

深圳城市發展設計
有待重新切實規劃

香港建築師學會會長潘祖堯建議深圳市人民政府應聘用一間經驗豐富、又瞭解中國情況的本港或外國城市規劃公司，來制定深圳經濟特區的發展計劃，代替國內一貫寄賴各間大學建築系提出建設方案的辦法。

潘祖堯表示，深圳的城市發展計劃，歷來都是由國內各所大學的建築系設計及規劃。如最近三次修訂完畢的〈深圳經濟特區社會經濟發展規劃大綱〉，就是由同濟大學建築系負責草擬及修訂。

他指出，中國雖然擁有不少城市規劃的專業人才，但他們只有豐富的理論，而欠缺實際規劃大型城市的經驗，來全盤負責這個十多年來中國最龐大的城市發展計劃。

他認為，如果深圳聘用有經驗的外資城市規劃公司，以一兩年的時間，全面研究整個深圳經濟特區的發展計劃，那麼所的城市設計報告會遠比現時的切實。

雖然聘用外資研究公司需要龐大的費用，但潘祖堯認為由於深圳發展計劃是屬於大型發展，如果設計完善，將會吸引大量的外來商家投資，因此深圳市政府應仔細考慮這個方案。

另一方面，因深圳是近十多年來中國第一個有計劃地設計及發展的城市，甚至可能成為中國未來新城市的樣本，因此有關方面應該多花時間來設計，否則對中國的未來城市建設可能帶來不良的影響。

至於經過三次修訂後的〈深圳經濟特區社會經濟發展規劃大綱〉，於九月下旬曾到深圳參看該大綱並提出批評意見的潘祖堯表示，這份報告太理想化，甚至有點不切實際。

他指出，這份大綱對於一切可能障礙發展計劃的因素，都以最樂觀的態度來處理或預測，而不是以「一切都作最壞打算」的原則來設計，因此整個城市的費用和維修費用都異常龐大的國際機場。

他認為深圳是個很複雜的城市。以人口來說，除當地居民外，更有本港的短途旅客，外籍商人或由國內不同地方往深圳當短期工作者，因此需要一間富有經驗的規劃公司來訂定該特區至公元二千年的工業、商業、農業、旅遊業、交通運輸、對外貿易等各方面的發展計劃。如果深圳市政府仍只是倚賴不切實的計劃，深圳城市發展前景將難期樂觀。

他說，在規劃大綱內，工商業區及住宅區的分佈非常零散，水電供應問題沒有妥善計劃；交通及運輸更有很多錯誤的設計，例如在計劃中蛇口區的一條主要交通幹線，竟然有一個九十度的彎角，證明在設計時沒有考慮到蛇口區的人口相差不遠。但是廣東省政府卻建議於公元二千年在深圳興建一個國際大型貨櫃車的駕駛問題。但未來將成為南海油田後勤中心及南中國海唯一出口港的蛇口，卻肯定會有不少的貨櫃車經常行

區的面積僅三百二十七點五平方公里，現時人口僅約二十萬人。即使到公元二千年，人口預算也只有八十萬人，與現在荃灣區的人口相差不遠。此點是與經濟原則相違，因爲以八十萬的人口，就算單是興建輕鐵鐵路也不一定必要，更何況是

柳瑞雄

歷年參與國際會議及學術會議記錄
Attendance record of international conferences and academic symposiums

附 錄 (3)
Appendix (3)

(1) 英聯邦建築師協會香港會議 (04.1979)
(2) 國際建築師學會印尼區住居研究組雅加達會議，香港建築師學會代表 (09.1979)
(3) 亞洲建築師學會常委會在雅加達之創立會議，香港建築師學會代表 (09.1979)
(4) 英聯邦建築師協會建築教育委會哥倫坡會議，香港建築師學會代表 (03.1980)
(5) 英聯邦建築師協會亞洲區曼谷會議，香港建築師學會代表 (08.1980)
(6) 亞洲建築師學會常委會曼谷首屆會議，香港建築師學會代表 (08.1980)
(7) 英聯邦建築師會常委會，溫哥華會議 (09.1980)
(8) 中國建築學會在北京舉行第六次全國會員會代表大會 (1980)
(9) 英聯邦建築師協會亞洲地區哥倫坡會議，香港建築師代表 (02.1981)
(10) 亞洲建築師學會工作小組馬尼拉會議 (03.1981)
(11) 國際建築師學會華沙會議，香港建築師學會代表 (06.1981)
(12) 亞洲建築師學會香港第二屆會議主席 (08.1981)
(13) 亞洲建築師學會教育委會香港會議主席 (08.1981)
(14) 亞洲建築師學會工作小組香港會議主席 (08.1981)
(15) 英聯邦建築師協會肯雅會議，香港建築師學會代表 (10.1981)
(16) 英聯邦規劃師協會亞洲區工作小組香港會議 (12.1981)
(17) 亞太區民居之發展地區性會議於橫濱 (06.1982)
(18) 星加坡/澳洲建築師學會大會 (06.1982)
(19) 亞洲建築師學會第三屆科倫坡會議主席 (10.1982)
(20) 第一屆亞洲論壇創辦人及理事 (10.1982)
(21) 亞洲建築師學會第四屆巴基斯坦會議，香港建築師學會代表 (08.1983)
(22) 英聯邦建築師協會悉尼會議，香港建築師學會代表 (06.1983)
(23) 亞洲建築師學會第五屆馬尼拉會議，香港建築師學會代表 (10.1984)
(24) 第二屆亞洲論壇主席 (10.1984)
(25) 亞洲建築師學會首次大會，顧問、主題主席及記錄總長 (10.1984)
(26) 亞洲建築師學會第六屆孟買會議顧問 (11.1985)
(27) 紀念梁思成教授誕辰八十五週年紀念大會及創辦清華大學建築系四十週年大會 (1985)
(28) 中國建築學會生土學術北京會議 (11.1985)
(29) 亞洲建築師學會第二次吉隆坡大會主題主席 (10.1986)
(30) 第三屆亞洲論壇主席 (10.1986)
(31) 亞洲建築師學會第七屆吉隆坡會議顧問 (10.1986)
(32) 現代中國建築創作研究小組第二屆烏魯木齊會議 (07.1987)
(33) 亞洲建築師學會第八屆芭里島會議顧問 (10.1987)
(34) 第四屆亞洲論壇主席 (10.1987)
(35) 中國建築學會泉州學術會議 (09.1988)
(36) 亞洲建築師學會第九屆漢城大會 (09.1988)
(37) 中國建築學會第七次全國會員代表大會暨學術年會杭州會議名譽理事 (1988)
(38) 香港科技協進會第一次國際亞洲四小龍新工業發展香港會議籌委會主席 (11.1988)
(39) 亞洲建築師學會第十屆曼谷會議顧問 (10.1989)
(40) 第五屆亞洲論壇主席 (10.1989)

(41) 英國建築師協會亞洲會香港第二次會議　(09.1990)
(42) 亞洲建築師協會第十一屆雅加達會議顧問(10.1990)
(43) 現代中國建築創作研究小組第五次深圳會議
　　 (11.1990)
(44) 英國建築師協會亞洲會吉隆坡第三次會議(06.1991)
(45) 亞洲建築師學會第十二屆北京會議顧問　(10.1991)
(46) 第六屆亞洲論壇主席 (10.1991)
(47) 英國建築師協會亞洲會濱城第四次會議　(11.1991)
(48) 中國建築學會第八次全國會員暨學術年會，名譽理事
　　 (03.1992)
(49) 亞洲建築師學會第十三屆拉合爾會議顧問(10.1992)
(50) 英國建築師協會亞洲會新加坡第五次會議(10.1992)
(51) 華南理工大學在廣州舉辦的第一次中國民居國際會議
　　 籌委會委員 (08.1993)
(52) 第七屆亞洲論壇籌委顧問 (09.1993)
(53) 第十七屆國際城市開發協會香港全球會議及全球房地
　　 產展覽籌委會委員 (09.1993)
(54) 亞洲建築師學會第十四屆香港會議顧問 (09.1993)
(55) 兩岸學術交流杭州會議 (10.1994)
(56) 華南理工大學在廣州舉辦的執行研討會小組主席
　　 (11.1994)
(57) 亞洲建築師學會第十五屆馬尼拉會議顧問(12.1994)
(58) 英國建築師協會亞洲會新加坡會議 (04.1995)
(59) 主持中國建築論壇第一次研討會，深圳 (10.1995)
(60) 亞洲建築師學會十六屆新加坡會議 (11.1995)
(61) 第八屆亞洲論壇(11.1995)
(62) 英國建築師協會亞洲會新加坡會議 (05.1996)
(63) 香港房屋協會新加坡考察 (06.1996)
(64) 主持中國建築論壇第二次研討會，天津(08.1996)
(65) 中國建築學會第九次全國會員暨學術年會，名譽理事
　　 (11.1996)
(66) 亞洲建築師學會第十七屆雅加達會議　　(09.1996)
(67) 中國傳統民居第二次國際會議，香港 (08.1997)
(68) 中國建築論壇第三次研討會，南京 (09.1997)
(69) 亞洲建築師學會第十八屆東京會議 (09.1997)
(70) 英國建築師協會亞洲會北京國際會議　　　(09.1997)
(71) 亞洲建築師學會第十九屆科倫坡會議　　(11.1998)
(72) 主持中國建築論壇第四次研討會，重慶(10.1998)

(1) Commonwealth Association of Architects (CAA) Triennial Conference & General Assembly in HK (April 1979)
(2) International Union of Architects (UIA), Regional Conference on Urban Village in Jakarta - representing HKIA (Sept. 1979)
(3) Architects Regional Council Asia (ARCASIA), Inaugural Meeting in Jakarta - representing HKIA (Sept. 1979)
(4) CAA Board of Architectural Education Meeting in Colombo (March 1980)
(5) CAA Asian Regional Meeting in Bangkok - representing HKIA (Aug. 1980)
(6) ARCASIA 1st Meeting in Bangkok - representing HKIA (Aug. 1980)
(7) CAA Council Meeting in Vancouver - as alternate Asian Regional Representative (Sept. 1980)
(8) ASC 6th Annual Conference in Beijing (1980)
(9) CAA Asian Regional Meeting in Colombo - representing HKIA (Feb. 1981)
(10) ARCASIA Workshop Meeting in Manila - as Chairman of the Meeting (Mar. 1981)
(11) UIA Congress and General Assembly in Warsaw - representing HKIA (Jun. 1981)
(12) ARCASIA 2nd Meeting in Hong Kong - as Chairman of the Meeting (Aug. 1981)
(13) ARCASIA Board of Architectural Education Meeting in Hong Kong - as Chairman of ARCASIA (Aug. 1981)
(14) ARCASIA Workshop on Practice in Hong Kong - as Chairman of ARCASIA (Aug. 1981)
(15) CAA Triennial Conference & General Assembly in Nairobi - representing HKIA (Oct. 1981)

(16) Commonwealth Association of Planners Asian Regional Conference & Workshop in Hong Kong (Dec. 1981)
(17) Regional Congress of Local Authorities for Development of Human Settlements in Asia & the Pacific in Yokohama (Jun. 1982)
(18) SIA/RAIA Convention (Jun. 1982)
(19) ARCASIA 3rd Meeting in Colombo - as Chairman of the Meeting (Oct. 1982)
(20) ARCASIA FORUM 1 on Innovations in Architecture - as Founder and Director (Oct. 1982)
(21) ARCASIA 4th Meeting in Lahore, Pakistan - representing HKIA (Aug. 1983)
(22) CAA Triennial Conference & General Assembly in Sydney - representing HKIA (Jun. 1983)
(23) ARCASIA 5th Meeting in Manila, Philippines - representing HKIA (Oct. 1984)
(24) ARCASIA FORUM 2 on Asian Identity in Manila - as Chairman (Oct. 1984)
(25) 1st Asian Congress of Architects in Manila - as Adviser, Theme Chairman, Rapporteur General and Session Chairman (Oct. 1984)
(26) ARCASIA 6th Meeting in Lonavla, Bombay, India - as Adviser (Nov. 1985)
(27) Conference to commemorate Prof. Liang Sze Chengs 85th Anniversary and the 40th Anniversary of the Dept. of Architecture Tsing Hua University (1985)
(28) ASC Earth Conference in Beijing (Nov. 1985)
(29) 2nd Asian Congress of Architects in Kuala Lumpur - as Session Chairman (Oct. 1986)
(30) ARCASIA FORUM 3 on Design Direction & State of the Art - as Chairman (Oct. 1986)
(31) ARCASIA 7th Meeting in Kuala Lumpur, Malaysia - as Adviser (Oct. 1986)
(32) 2nd Conference of the Sub-committee on the Research of Modern Chinese Architecture in Wulumuqi, China (Jul. 1987)
(33) ARCASIA 8th Meeting in Bali, Indonesia - as Adviser (Oct. 1987)
(34) ARCASIA FORUM 4 on My Architecture (Oct. 1987)
(35) ASC Chuenzhou Conference (Sept. 1988)
(36) ARCASIA 9th Meeting and 3rd Asian Congress of Architects in Seoul, Korea (Sept. 1988)
(37) ASC 7th National Conference in Hangzhou - as Hon. Council member (1988)
(38) HK Association for the Advancement of Science & Technology - An International Conference on the New Industrial Development in the "Four Little Dragons" of Asia - as Chairman (Nov. 1988)
(39) ARCASIA 10th Meeting in Bangkok, Thailand (Oct. 1989)
(40) ARCASIA FORUM 5 on Food for thought in Asian Architecture (Oct. 1989)
(41) AA Asia 2nd Conference in Hong Kong (Sept. 1990)
(42) ARCASIA 11th Meeting and 4th Asian Congress of Architects in Delhi, India (Oct. 1990)
(43) 5th Conference of the Sub-committee on the Research of Modern Chinese Architecture in Shenzhen, China (Nov. 1990)
(44) AA Asia 3rd Conference in Kuala Lumpur, Malaysia (Jun. 1991)
(45) ARCASIA 12th Meeting in Beijing, China (Oct. 1991)
(46) ARCASIA FORUM 6 on New thoughts in Mass Housing in the Asian Context (Oct. 1991)
(47) AA Asia 4th Conference in Penang, Malaysia (Nov. 1991)
(48) ASC 8th National Conference in Beijing, China - as Hon. Council member (Mar. 1992)
(49) ARCASIA 13th Meeting and 5th Asian Congress of Architect in Lahore, Pakistan as session Chairman (Oct. 1992)
(50) AA Asia 5th Conference in Singapore (Oct. 1992)

(51) Member of the Organising Committee for the first International Conference on Chinese Traditional Houses in Guangzhou, South China University of Technology, China (Aug. 1993)

(52) ARCASIA FORUM 7 on the Vanishing Asian City - as Adviser (Sept. 1993)

(53) INTA 17th - New Town Experience joint Conference - as Organising Committee member (Sept. 1993)

(54) ARCASIA 14th Meeting in Hong Kong - as Adviser (Sept. 1993)

(55) Academic Exchange Conference for Chinese Architects from Beijing and Taipei, 5th meeting, Hangzhou (Oct. 1994)

(56) Architectural Practice Seminar in the South China University of Technology, Guangzhou, China as session Chairman (Nov. 1994)

(57) ARCASIA 15th Meeting and 6th Asian Congress of Architects in Manila (Dec. 1994)

(58) AA Asia Conference in Singapore (April 1995)

(59) 1st Symposium of the China Architectural Forum on " Architectural Criticism" Shenzhen, China as founder and sponsor (Oct. 1995)

(60) ARCASIA 16th Meeting in Singapore (Nov. 1995)

(61) ARCASIA FORUM 8 on "Asian Cities in Asian Century" (Nov. 1995)

(62) AA Asia conference in Singapore (May. 1996)

(63) Hong Kong Housing Society Study Tour to Singapore (6 June. 1996)

(64) 2nd symposium of the China Architectural Forum on " Comparison & Difference" China as co-Chairman and sponsor (Aug 1996)

(65) ASC 9th National Conference in Beijing, China as Hon Council Member (Nov. 1996)

(66) Arcasia 17th Meeting 7th Asian Congress of Architects in Jakarta, Indonesia as Adviser (Sept. 1996)

(67) 2nd International Conference on Chinese Traditional Houses in Hong Kong (Aug 1997)

(68) 3rd symposium of the China Architectural Forum on "Now and the Future" as Chairman and sponsor (Sept. 1997)

(69) Arcasia 18th Meeting and Forum 9 in Tokyo on "Architecture of the Future in Asia" (Sept. 1997)

(70) AA Asia Conference on "Contemporary vernacular, modernizing architectural traditions" (Sept. 1997)

(71) Arcasia 19th Meeting and ACA 8 in Colombo, Sri Lanka as convenor Arcasia Fellowship (Nov. 1998)

(72) 4th Symposium of the China Architectural Forum on " Architectural & Society" (Oct. 1998)

附錄(4)
Appendix(4)

RECORD

Commonwealth Conference: Regional Grants: Scottish Summer School: RIBA Task Forces: CPD Swap Shop.

PETER Murray reports from Nairobi: As the Prime Ministers flew back from Melbourne and Margaret Thatcher and Ted Heath prepared to battle it out at Blackpool, delegates to the Commonwealth Association of Architects gathered at the National Museum in Nairobi for their biennial assembly which, in comparison, made the Tories' tussles seem like the proverbial Vicarage tea party.

There were two strands to this particular farrago of architectural politics which were inextricably intertwined and at times seemed close to breaking point.

Strand one involves the "breakdown in personal relations" between the retiring chairman of the Commonwealth Board of Architectural Education (CBAE) Lim Chong Keat and the Secretary of the CAA Graham McCullough. This led to McCullough's sudden resignation earlier this year. Since the CBAE and the Council of the CAA are independent of each other, the Council had been unable to keep in check Keat's running of his Board.

The financial problems of the Association have been caused partly by the CBAE situation but also by the elimination of all grants for administration from the Commonwealth Foundation. Up until 1976 the Foundation paid for 80 per cent of the running costs, while subscriptions covered the rest. Now subs have to pay for 100 per cent and the management structure is just too extravagant.

An RIBA motion to put to the assembly proposed that the CBAE should be absorbed into the Executive Committee and that a slimmer and more efficient organisation be created. The RIBA, because it has by far the largest number of members, pays the lion's share of CAA income. Its £25,000 per annum constitutes 87 per cent of the association's subscriptions. July Council agreed that unless changes were made, the RIBA would reconsider its membership of the association.

Strand two concerns the election of a new CAA president.

There were two candidates from the Asian bloc, Hisham Albakri and Visra Selvaratnam, standing against Fred Rounthwaite, the incumbent President.

So as to discuss more easily the changes proposed by the RIBA, it was agreed that certain clauses in the constitution of the CAA should be suspended during the Assembly. This worked well for the discussions on the constitution. The RIBA's case was proposed by Owen Luder and following discussion and an

Top: Geoffrey Rowe, RIBA vice-president responsible for overseas members, in Nairobi. He is the new Honorary Secretary/Treasurer of CAA. Centre left: CAA President Fred Rounthwaite. Centre right: Lim Chong Keat. Bottom: Rounthwaite congratulates Jai Bhalla, following the presentation of a CAA medal in recognition of Bhalla's many years of service to the association.

amendment prepared by Prof Allan Wild of New Zealand (without whom the assembly may well have foundered on more than one occasion) the bases of the Institute's reforms were accepted unanimously. The organisation is to be slimmed down and the Board of Education become a committee responsible to the CAA Council.

The fun and games began when Ronald Poon (Hong Kong) proposed that the constitution be reinstated for the election of the President. It had already become clear that the powerful African caucus was not attracted to either of the Asian candidates and felt that Rounthwaite represented the old guard. It had suggested that all three candidates should stand down and one who would hold the support of the whole assembly be put forward. This could not be done within the constitution, countered the Asians, because two months' notice of nomination must be given.

The choice was either to reinstate the constitution and allow the three candidates to stand, or to ignore the constitution and put forward a new candidate. The problem of the latter was that a new method of nomination and election would have to be drafted and agreed in one afternoon; an impossible task and one that the Asians would challenge almost certainly.

A formula was contrived whereby Rounthwaite would win, then resign in favour of the Vice-President, Peter Johnson of Australia, the candidate the Africans had wanted in the first place.

Rounthwaite did win (12 votes to the Asians' 4) but has postponed his resignation for a month or so in order not to upset the delicate equilibrium achieved at the end of the assembly.

The Assembly united behind one motion: the criticism of the increase in fees for overseas students in Britain.

The meeting called on all member institutes to make representations to the Thatcher Government to impress on it the effects of the high fees: that students must abandon their studies, that political philosophies alien to the Commonwealth are made more accessible and that today's pattern of trading loyalties will be lost.

Sanity was again restored to the Assembly when a motion from Tom Howarth of Canada was unanimously accepted which said that only a restricted part of the timetable of the next assembly should discuss constitutional matters.

RIBA Task Forces

The three RIBA Task Forces outlined by Owen Luder in his Presidential Address have now been established. One group will continue the work of the APPA report on the role of the architectural profession in the work of public authorities. It will examine in further detail the damage already inflicted by indiscriminate cuts in staff and the extent to which they have jeopardised the ability of architects' departments to respond to the needs of the communities they serve. Another group will examine the way that bureaucracy has distorted and inhibited design and development and it will monitor the effects of recent changes relating to development and it will monitor the effects of recent changes relating to development and building con-

The Editor
RIBA Journal
66 Portland Place
London W1N 4AD
ENGLAND.

Dear Sir,

In reference to your Record column of the November 1981 issue of your Journal on the 1981 CAA Conference at Nairobi, I would like to pose a question to you. If in a General Assembly of an International professional association, a very important part of its constitution is suspended without serving prior notice to all its members and obtaining prior approval from them and if that part concerns the qualification of its President for the purpose of electing one, then do you think that association is worth the paper its name is printed on?

I regret that my sanity has started what you called "fun and games" but if the move to suspend the part of the constitution for the election of the President succeeded at the General Assembly then we would have behaved in a very irresponsible manner as very few of us have a carte blanche mandate from our Institutes for such drastic unconstitutional behavior which if executed would have set a very undesirable precedent for future Assemblies. Turning to the RIBA motion and the Alan Wild amendment. True it is fairly constructive but very few people realized that it differs somewhat from what the Policy Steering Committee proposed in their report which was not given its due respect and discussion during the General Assembly, though a lot of work has been done to produce it. I suspect that this is because the Policy Steering Committee was probably created to take the CAA over a rough period and seem now to have served its purpose. However both the RIBA motion and Alan Wild amendment did not produce a slimmer and more efficient organization as such because the CAA hierarchy still contains a Council consisting of a President, five Vice-presidents (regional representatives), Chairman CBAE and a new additions of an Hon Secretary/Treasurer while the old Council consists of a President, a vice-president/Regional representative, one Hon Treasurer/Regional representative, three other Regional representatives, Chairman CBAE, one less than the new Council. What was achieved by these two motions is a cosmetic reshuffle in that the CBAE is now under the direct control of Council (an act which the last two Councils and the two Secretaries failed to pull of over the last six years) while the Policy Steering Committee recommends that maximum autonomy be given to the CBAE. Architectural Education seems to be the only effective product of the CAA since its formation and I suspect it will remain the only useful product for the future. To place the Council consisting of mostly people not involved in the education field in control over the CBAE is questionable and in any case it creates a cumbersome machinery. I think if the CAA is to move with the times it should throw away its "imperial cloak" and concentrate on being the Commonwealth Board of Architectural Education. In the present system the CAA falls between two stools. It cannot compete with the regional body of Architects such as the Architects Regional Council Asia (ARCASIA) nor would it be in a position to produce the friendly localized activities and projects that are going on in the regional bodies. At best it just provides another conference facility but the success of such is very much dependent on the input from the CAA Secretariat (which has not yet established itself with any permanency) and from the host Institute. I would like to propose that the CAA be renamed CBAE, the present Council disbanded and the President CAA becomes the Chairman CBAE and the fire vice-presidents (regional representatives) remain as liaison officers within each region and report to the General Assembly. The Board members shall remain as before i.e., one representative from each region nominated by the regional Boards of Architectural Education where they exist or by the majority of the member Institutes of the region: there shall be a Hon Secretary/Treasurer to be nominated by the President and elected at the General Assembly. The headquarters presently in London be moved to where the Chairman CBAE is residing. The Chairman together with the Hon Secretary/Treasurer will be responsible to organize the tri-annual conference and general assembly which should in future concentrate on stimulating new architectural directions and exchanges of individual experiences. The future Chairman and Hon Secretary/Treasurer should be prepared to contribute greater efforts and involve themselves more with the running of the organization rather that in the past where the salaried Secretary seems to act for them and run the Association from the Secretariat.

Contrary to your opinion the financial problems of the CAA have not been caused by the CBAE at all. On the contrary because of the financial difficulties in the last three years and the failure to secure other funding facilities, the work of the CBAE was virtually stopped by Council's decision to minimize any financial assistance to the CBAE between 1978 and 1981.

The RIBA has always maintained that it pays the lion's share of the CAA income and this attitude is well reflected in their consisted effort to have a major say in the running of the CAA. However it keeps on forgetting that they are merely putting into the CAA roughly ten per cent of the income they receive annually as subscriptions from oversea members whose total number amounts to one fifth of the total membership of the RIBA. It is a well known fact that all the overseas members effectively get in return from the RIBA is the Journal by airmail (by airmail only in last year or so). The work of the CAA is another indirect benefit, however, don't forget that all oversea members pay what would amount to a duplicate subscription to the CAA via their local Institute. Therefore a fair arrangement would be for the RIBA to divert at least 75% (roughly sterling 200,000) of the oversea subscriptions to the CAA who in turn should channel the money into the CBAE as well as regional bodies such as ARCASIA for direct benefit to its oversea members. It is high time that oversea members get some real benefit and stop subsidizing activities and facilities in U.K.

Though at times in CAA General Assemblies opinions seemed to represent certain regions like Africa, Asia, the Americas etc. and this is evidently reflected in your article, however do not forget that the CAA is an assembly of architects and should not be read as an United Nation Assembly because international politics should have no place within the CAA. I would only like to quote from part of a letter I received from my Nigerian colleague who took part in the Nairobi Assembly as follows :

"at the later part of that conference, I discovered the dirty politics of personalities quite apart from the bold principles we were following. I became annoyed and left. In anyway, it is an experience. Ronald, like you and would sincerely cherish the hope of your friendship and association. I love your dogged determination to speak out your mind inspite of what people think or say."

I hope you can find a way of printing this letter in your next issue.

Yours sincerely,

Ronald Poon
CAA regional representative for Asia
President, Hong Kong Institute of Architects
Chairman, Architects Regional Council Asia
Overseas Members, RIBA

附 錄 (5)
Appendix (5)

Speech delivered at the XVII UIA Assembly Dublin 19-21/07/87 on UAP issue.

If we are to move forward as a world body, we must forsake youthful display of sentiment for proper procedure at our Assembly. We must observe and adher to established rules for the conduct of the assembly and not interlace our proceedings with unorthodox interruptions which makes mockery of serious efforts by others, done in accordance with the statues of our Union.

While I was somewhat mesmerized by the sentimental outburst of our President yesterday and while I have to admit for all those who appear to know little about the hardwork of Kington Loo, John Davidson, the Bureau and the Council done in this matter concerning the Philippines, the public display of unity was heartwarming, but for all those who know well the background of this saga and who know that asking PIA to merge with UAP is like asking the north pole to meet the south pole, I can only offer two means of consolation. I can easily use a number of rules from this world authority on rules and procedures for Assemblies, Robert's Rule of Order given to me in Warsaw 1981 by Mr. David Meeker, past executive Vice President of AIA right after the first sentimental meddling on this Philippine Saga at the Katowise Assembly to quash yesterday's proceedings on this issue on the ground of improper conduct of assembly proceedings, but it will split the Assembly and disrupt the nice romantic solidarity generated by those kisses across the rostrum. Instead I would make the following motion:-

The UAP and PIA be given six months from to-day to produce satisfactory evidence of a true merger organization acceptable to Council for the national section of the Philippines, failing which Council is given the authority to act on this matter, on behalf of this assembly, as it thinks fit."

附錄 (6)
Appendix (6)

关于建筑创作的对话

● 杨永生

[编者按] 1997年9月12至14日，由全国政协委员、香港建筑师学会前任会长潘祖尧先生赞助的"建筑论坛"第三次研讨会在南京东南大学举办。这次会的主题是"现状与出路"。收到陈志华(清华大学教授)、彭一刚(中科院院士、天津大学教授)、钟训正(东南大学教授)、张钦楠(中国建筑学会副理事长)、戴志中(重庆建筑大学教授)、仲德崑(东南大学建筑系主任)、邹德侬(天津大学教授)、潘祖尧(深圳左思忠建筑设计事务所总经理、总建筑师)、齐康(中科院院士、东南大学教授)、程泰宁(杭州市建筑设计院顾问、前任院长、总建筑师)、戴复东(同济大学教授)、罗小未(同济大学教授)等14人的论文(以论文提交先后为序)。这些论文将于明年初由天津科技出版社出版发行。

这次会议是由东南大学建筑研究所、建筑杂志和天津科技出版社三家主办。

本报特请研讨会主持人杨永生(《建筑师》杂志编委会主任、编审)根据会上的发言及各位专家、教授的论文，整理出4.篇对话，陆续发表，同建筑界人士交流。

杨永生："建筑论坛"已经开了两次研讨会，第一次会的主题是"建筑评论"，第二次会的主题是"比较与差距"。这两次会的论文都已由天津科技出版社出版了专辑。

这次会的主题是"现状与出路"。我想，还是请大家着重谈谈建筑创作的现状并讨论一下出路何在的问题。不言而喻，"出路"这两个字即已包涵了对现状有不满意的意思，确切一点说，所谓出路无非是如何进一步提高的问题。

张开济大师说过，现在各种奇形怪状的建筑体形都出现了，明日张胆地大搞形式主义。程泰宁把建筑创作的现状概括为"亦喜亦忧的现状"。周凝粹认为"目前我国建筑创新的热情被困扰在形式主义之中，设计中新科技含量低、建筑技术发展滞后已经成为阻碍建筑业发展的瓶颈。"现在，请大家谈谈你们的看法。

彭一刚： 当前建筑创作中确实存在着十分严重的形式主义倾向。过去，在计划经济条件下，建筑形式单调无味，千篇一律。现在，改革开放，经济发展迅速，广开眼界向西方学习，这是一大进步，十分可嘉。但是，市场经济又给建筑创作带来了剧烈的冲击。

建筑行业说到底是一种服务性行业。侍之于人，必取悦于人。建筑师要想取悦于项目，就必须取悦于业主，而他们的文化素养，审美情趣又千差万别。(杨永生：应该承认，在某种历史的原因，我们的业主往往文化素养不高)建筑师要想揽到项目，就要迎合他们的口味，首先是以建筑物的外形去取悦。因而，在评选、决策过程中，往往忽视功能，忽视经济、技术的科学性和合理性，忽视与周围环境的和谐统一。最终，多以形式的"奇特"、"新颖"而一举中的。于是，便出现了张开济大师所指出的夸富媲贫、崇洋媚外、相互攀比、追求豪华等风气。

杨永生： 张老指出的这点，非常中肯。中央号召反对铺张浪费，我看，近几年，建筑上的侈者浪费也不小，建筑上的那些无谓的装饰物既非功能所需，又非美观所要，画蛇添足，统统加起来，浪费多少钱财!还有，本来很有特征的建筑，重新装修，花了不少钱，弄得不伦不类的实例也不少，我们建筑师看来也要在可能条件下说服业主，注意节约人民的财富。

陈志华： 十几年来，多亏改革开放的好政策，建筑创作真个是繁花似锦，好一派阳春三月景象。与世界先进水平差距固然还有，只要观念更新，赶上的日子也不会太远，有半个世纪总差不离了罢。

张钦楠： 现在的问题不是复古主义、形式主义的问题，而是"文抄公"，抄的又不像。形式还是需要的。50年代反对复古主义、形式主义，结果弄得大家在大帽子底下不能创作。

今日中国建筑学所面临的矛盾和问题比过去任何时候都更多、更复杂，好象处在好几个十字路口的交叉点上。这些矛盾，举其要者来谈，有：市场经济与"吃大锅饭"、勤俭节约与追求豪华、持久发展与"破坏性建设"、功能要求与哗众取宠、创新探索与墨守成规、社会效益与唯利是图、城市现状与各自为政、行政干预与创作自由、职业道德与损人利己等等。

在这些矛盾和问题面前，可以有几种态度。一种是无视社会已经或正在发生的变化，我行我素，当然会碰壁。另一种是以取媚适应或者说以带投机性的行为去适应变化。由于建筑是一种长期存在的产品，短期行为只能带来更多的问题。再一种是正确理解和阐释社会的变化趋势，区别不变和可变、长期和短期、积极和消极的因素，不断通过实践去检验自己的认识，以便取得主动权。这最后一种是最有作为的途径。

再谈到十字路口。昂首阔步穿越，难免有遭车祸之险。有段时间，某城市对某种"风貌"开绿灯，效果如何，人们已有定论。老老实实等待绿灯，未尝不为有效。当今不为人不为人所知所顾有盼，看准了，插空子过街，最为有效。

潘祖尧： 改革开放后大开门户，引进外来技术，但还属少数，还未能大显身手，更未能堂堂正正地自立门户。近年来的建筑设计大多属表面功夫，五花八门，多姿多彩，但缺乏真功夫，与国外比，尚有很大一段距离，主要是深度不够，产生大量浪费。各大城市都有大量的大而无当，花枝招展的建筑物，实不堪设想，无疑将起到重大的推动作用。

另一方面，经济改革后，钱的价值突升，引起不少不正规行为。有的建筑师在业余时间甚至工作时间内接外间私人委托的项目，因没有设计院的技术支持，个人经验所限，多数成果水平不高，贪图私利，危害不浅。又因公私兼顾，导致精神紧张，本职工作也受到影响，结果是两败俱伤。

左肖思： 引进境外名家、大师设计或国内外合作设计，几个大城市不乏成功实例。但也有名目为国外著名事务所的方案，实则由在国外打工的中国建筑师为主搞的，水平也不怎么高。

潘祖尧： 设计费太低，请不到高手。我以为宁可多花点钱，也要请高手，这对提高国内的设计水平有好处。

杨永生： 我们建筑设计如同足球，不肯花钱请国外优秀教练，请来的是三、四

关于建筑创作的对话

○ 杨永生

左肖思： 根据党的十五大的精神，在总结前一阶段设计体制改革试点的基础上，加强改革的力度和速度，在全国各省市较普遍地提倡(可以多种经济成份或形式)以一级注册建筑师为主持人的建筑设计事务所。这样，对改善设计环境、引导公平竞争、发挥建筑师的积极性和创造性，和提高建筑设计水平，都会有积极的推动作用。

潘祖尧： 我还建议，各大城市有关政府部门支持成立一些特种的"建筑设计工作室"(Architectural Design Workshop)，以私人的固定方式经营。这种工作室可以适量聘请海外高水平的建筑师或邀请海内外建筑单位加入合作经营，不仅做设计，还要对建筑师进行"再教育"，利用三分之一的工作时间来对建筑师作"再教育"，三分之一做业余创作，三分之一做设计费也可以精神紧张。这样边跑边教育。虽然，这与建筑设计研究院组建的设计发点可能有些相同，但现今大多数设计研究院已经更新。在前茫茫之际，及早到这设计、这研究、这教育的路上来。

杨永生： 现在的设计研究院实际上是只研究或者是很少投入力量去做研究工作。

齐康： 社会现象是复杂的，反映在建筑现象上也是纷繁多样的。也只有这种多样才构成着建筑文化。在运转、滚动、扩散的过程中，增长我们的经验和知识，加强实践和理论相结合的研究。我们要持之以恒地探索，多出精品。

我国这次取得了前所未有的成绩，也产生一定的负效应。我们所做的工作有的是良性循环，有的是负面效应，有成绩，也有缺点，甚至有错误。今天我们贫穷的，也许到了下个世纪某个时期看，是不可避免的，是错误的。建设中的人口问题、土地问题、环境问题、资源问题，在城市化建设中分量越重，将更加频繁。资源的保护、人才的保护、文化的保护等等都应给予重视，同时注意可持续发展的标准不可超越时间、空间及具体地的经济承受能力。

杨永生： 面对近年来城乡面貌的巨大变化，不禁产生一种担扰和忧虑。现在，在"现代化"和"世界文化的同化"大旗下，到处可见那种蹩脚的西方建筑的制品。一些城市特色在消失，地域特征在弱化，建筑的整体文化素质体在较低层次上。忽视自然条件以及政治经济文化发展阶段的不同，无视地域性对建筑文化发展的巨大影响，以致造成当前建筑风格的千篇一律和城市面貌的千篇一律。地域性是一个外延不大的概念。不同的建筑师都应有自己的理解和诠释，建筑师应在作品中突出"自我"。一个建筑师、一件作品，只要能在某一个方向突出自己的特色，就是成功，而众多建筑师从不同方面探索、不断追求，就使建筑风格的地域性体现得更为丰富和完善。

钟训正： 我要谈的主要是给城市多一点绿地和公共活动空间的问题。在发言稿里都写了，这里不再多谈。只想谈谈建筑高层化是否等于城市现代化的问题。现代城市，在某些人心目中就是摩天楼大厦，什么市场环线要客客气气否等于城市的问题。现代城市，在某些人心目中就是摩天楼大厦。所以城市现代要客客气气否等于城市的问题。现在要持续百米建100幢高层的计划。一般主干道上的高层，多是综合体、下层是办公楼、上边是下层宾馆餐饮，主楼不是写字楼就是旅馆，在空中一般不介在建筑材料上。这就带来几个严重的问题：交通问题、能源问题、维修和出租问题。现在，这些问题没有得到解决时，正在造成难于克服的矛盾。首先不是否决它，但我们不能把非常壮观、不言不语。我们在攻大大力挖掘。尤其是景色美、地少人多，资源不足等国情，移景多观。如果大家注意到近年新建的一些政府大楼(包括县长的)及厂房，就会发现其气派是非凡的，浪费也是惊人的。

(续一)

关于建筑评论及其他的对话

● 杨永生

编者按：本文是根据建筑论坛第三次研讨会上的发言及论文，由主持人杨永生整理出来的第四篇对话。前三篇关于建筑创作、建筑学术研究、建筑教育的对话，已分别刊载于本报10月7日及10月24日第二版上。

杨永生： 我们的建筑评论工作，近几年来由于有关部一再强调，由于新闻传播媒体的支持，虽然仍不活跃，但也应该说，有了前所未有的起步。除了专业报发表了不少评论文章之外，一些大众媒体也开始注意发表了这方面稿，例如，中央电视台播发了一些关于建筑评论的节目，《人民日报》、一些文化性报刊的晚报都一些大报发了如《明日报》、一些文化性报刊的《读书》杂志也出了一些建筑界和文化界人士的关键性文章，这是几十年来少见的，令人兴奋。我是那句话，建筑要走百万里[建筑主管部门]、中国筑学会都设在北京百万里]，也就是说，要走出建筑学的殿堂，走向社会，争取全社会都关心、评论建筑的方方面面，让全社会都关心建筑，这无疑是重要的。人们的建筑意识，环境意识、审美意识的重要途径，从而也是尽快地提高建筑设计水平的途径。

潘祖尧(全国政协委员、香港建筑师学会前任会长)： 内地报刊但缺乏有建设性的评论文章，以至大多数的评论都是称赞的，所以常常使读者读以为是推广。人们也不再有作这类的，一见到乎没有人像西方的评论媒介那一吹，要成了好医药，下能到处解忧，结果是一片哀声。

张钦楠(全国政协委员)： 建筑评论也是一条出路，建筑师有义务向全社会、向领导、向业主普及建筑知识，这是个大责任!

杨永生： 建筑评论也好，建筑科普也好，几年了，建筑界有关学术团体费下一点功夫，年年哪怕是办那么几件实事也好，光是号召，还不够，看来，没有一支常年活跃的建筑评论队伍，没有一支科普队伍，也不行，我是想加强同传播媒介的沟通，创造机会使他们更多地、更深地了解建筑学上的种种探讨。

齐康： 关于建筑科普问题，我在去年第二次讨论会上说过，假如我们社会摆脱目前的紧张状态，在真心地愿意多做些科普工作。

现在，我想谈谈市化的问题，一是住宅和住宅群的规划设计，我认为，至关重要。房地产的兴起促进了城市建设，但片面追求高档高利润，低层高密度的组团环境，那么低团高密度的组团环境，那么些低团高密度的组团环境，那些低团高密度的组团环境，那些低团高密度的组团环境，那些低团高密度的组团环境，那些低团高密度的组团环境，那些低团高密度。住在这些住宅中的一定的灵活和可变。我们还要去探求公寓式的公共活动空间，和大住城、一些文化性等，一些文化性的精神生活，这种"广场"就要适应中国人自身的行为特点。

左肖思(深圳左思忠建筑事务所总经理、总建筑师)： 我们对开发商建筑架构在不断发展、却仍然林立、且日益严重。随着采取搞集商式中心商业中心。

我曾提出过国外流行的Down Town模式，我们难道不可以寻求其他的城市模式？

美国的DownTown集中于中心区，而中国的则在城市中分散的利益，大饼一样，周围不断地"摊"。城市中心地段的地位仍然。

刘曾平(华南理工大教授)： 温饱根已过的人居环境，许多地方大大地恶化了生存条件，甚至找不到来复甦的安然的信息空间。

1997年5月，在广州被征占的公共绿地达29万平方米，人均绿地到1995年下降为4.69平方米，这与联合国世界卫生组织提出的最佳居住环境人均绿地不应低于20平方米，差距太大！

在粤东山区，有的为此在改革开放之后也大批太被，裸土地露着许多地方。由于没有规划，各县建筑都往上挤，原来前瞻后山的风景见不了、留下的色谱不但窄小，还奇形怪状，连排水沟也没有，种树的地方更没有，挑电条不都得十分小心，有时甚至真的要求的、也许是发展速度过快，在发展中带来的问题也比较集中的、充分地暴露出来了。

也是在伊斯坦布尔召开的联合国二次人居会上，我国佛山市被评为"人类古迹优秀范例"，这也不仅说明的，主要是由当地领导重视经济建设、兼顾得力，使城市建设得到较化的投入，真正做到持续发展。

这两年，关于建筑的问题不少。除了建筑师本身的一些因素，体力因环境条件方面能否让建筑师在更多方面关系的不同？可见，向全社会普及建筑知识，是多么重要。

**建筑设计与室内(装修)设计被分开为两个各自独立的行业，室内设计很少委托建筑设计把门，通常是由一些附属于装修施工的装修公司，因为他们不另收设计费，你要名家甲级设计单位亲自做设计审图，但由于没有装修施工队伍，又是基本无中不材产，对国外对建筑设计缺之个统一而明确的标准或定义。

建筑创作不单是环境设计，建筑设计包括建筑本身，综合性，内外融合，有整体风格的构思和雕凿，怎么可能会有成功的建筑创作？

附 錄 (7)
Appendix (7)

ARCHITECTS REGIONAL COUNCIL ASIA (ARCASIA)

CONTENTS

1. Introduction by Mr. Ronald Poon 3
2. Architects Regional Council Asia (ARCASIA) by Mr Ronald Poon 4-23
3. Arcasia Committee for architectural Education (ACAE) by Mr. P. Kasi 24-25
4. Snapshots of Arcasia 26-30

EDITORIAL ON SPECIAL ISSUE

INTRODUCTION
BY MR. RONALD POON

The Architects Regional Council Asia was inaugurated some fourteen years' ago at Jakarta, but the formation activities were started some ten years before that in December 1969 at Hong Kong when the first Foundation Council met for the first time. The last ARCASIA meeting which took place in Hong Kong was the second meeting of the Council in August 1981 and it is fitting that Hong Kong will be host again for the 14th Council meeting together with Forum 7 in September this year.

In order to commemorate the occasion, the Hong Kong Institute of Architects (HKIA) decided to produce a written record of ARCASIA for the occasion. Although ARCASIA has many publications concerning its various activities at different times during the last fourteen years, but, to-date it does not have one which gives a reasonably complete coverage of the details of the organisation and its work.

I have been asked by the HKIA to produce such a volume. I realize that though I have been involved with ARCASIA since its inauguration and has kept all the records, there will be inaccuracies and omissions in this volume. There were additional problems encountered during the preparation of this edition which only illustrate the typical frustrations and difficulties associated with international institution work. All the four people I invited originally to prepare papers for this edition did not respond at all initially for months despite several reminders, letters, faxes etc.. At the end only one responded passed the eleventh hour and I had to revise the contents to accommodate him. I wrote to some twenty people who were regulars at ARCASIA meetings for snapshots of our past activities but only three responded. Therefore I have no alternative but to extensively use photos from my own album for the bulk of the illustrations in this edition. So perhaps you can forgive me, if I appear too often amongst the photos. I would like to thank Ipe Mendoza, Rusi Khambatta and Jeong Keun Lee for sending me photographs. However if we do not put pen to paper, ARCASIA and its work over the last twenty four years will only be talks in the air. This volume will serve as a starter for later refinement and I hope that it will also serve as a useful tool for the members who had taken part in ARCASIA activities to reminisce by as well as being a helpful introduction for the new comers and other organisations which are interested in the work of ARCASIA.

ARCHITECTS REGIONAL COUNCIL ASIA (ARCASIA)

BY MR. RONALD POON

1) INTRODUCTION

ARCASIA is an international council of presidents of national institute of architects in the asian region. Its membership is restricted to Institutes of Architects which have the majority of the architects in its country or territory as their members. It is an unique professional institution representing architects in some fifteen countries and territories in Asia. ARCASIA meets annually in a member country to deliberate and to give collective direction and representations to matters that affect the architectural profession in the region.

2) OBJECTS

The objects of ARCASIA are :

To unite national institutes of architects on a democratic basis throughout the Asian region to foster friendly intellectual, artistic, educational and scientific activities.

To foster and maintain professional contacts, mutual cooperation and assistance among member institutes.

To represent architects of the member institutes at national and international levels.

To promote the recognition of the architect's role in society.

To promote the development and education of architects to serve the society.

To promote research and technical advancement in the field of the built environment.

3) THE ORGANISATION

ARCASIA is administered by a Council of Presidents of its member institutes. Each member Institute is entitled to nominate two persons to represent it in the annual Council meeting, one of whom must be its President or his representative. The other is normally the Chairman of its Board of Architectural Education as the ARCASIA Committee of Architectural Education meets concurrently with the Council.

The Council elects its Chairman and three deputy Chairmen, Hon. Secretary and Hon. Treasurer each year at its annual meeting. However all officials, including the Chairman, is expected to serve two terms.

In 1985 the Council created the position for two advisers to the Council to be nominated by the Chairman. Mr. Ronald Poon of the HKIA has served as one of the Advisers from 1985 to-date, Datuk Kington Loo (PAM) 1985-1990 and Mr. Felipe Mendoza (UAP) from 1990 to-date.

Under the Council there is an Executive Committee composed of the Chairman, the three Deputy Chairmen and the immediate Past Chairman. This Committee is responsible for the preparation of the annual programme of activities and initiate projects for the approval of Council, thereafter be responsible for implementing them.

There are three regular committees, namely, the ARCASIA Committee of Architectural Education, the membership Committee and the Awards Committee. Two major events of ARCASIA, namely, the Asian Congress of Architects (ACA) and the ARCASIA Forum.

The ARCASIA territories are divided into three zones for administrative convenience. Zone A consisting of Bangladesh, India, Pakistan and Sri Lanka. Zone B consisting of Indonesia, Malaysia, Singapore and Thailand. Zone C consisting of China, Hong Kong, Korea, Philippines, Japan, Mongolia & Macau. Each Deputy Chairman is responsible for liaison with member institutes in one of the three zones.

4) MEMBERSHIP

Institute of Architects, Bangladesh (IAB)(1971)
Architectural Society of China (ASC)(1989)
Hong Kong Institute of Architects (HKIA)(founding member)
Indian Institute of Architects (IIA)(founding member)
Ikatan Arsitek Indonesia (IAI)(1977)
Japan Institute of Architects (JIA)(1991)
Korea Institute of Registered Architects (KIRA)(1985)
Association of Architects of Macau (AAM)(1991)
Union of Mongolian Architects (UMA)(1991)
Pertubuhan Akitek Malaysia (PAM)(founding member)
Institute of Architects Pakistan (IAP)(founding member)
United Architects of the Philippines (UAP)(1977)
Association of Siamese Architects (ASA)(1972)
Singapore Institute of Architects (SIA)(founding member)
Sri Lanka Institute of Architects (SLIA)(founding member)
(for addresses see appendix 2)

5) THE BEGINNING

ARCASIA was initiated during the 1967 Commonwealth Association of Architects (CAA) New Delhi Conference, for the establishment of a regional centre that was concerned with environmental design and unity among the six Asian member institutes of the CAA. The six Asian institutes (the founding members) were India, Sri Lanka, Pakistan, Malaysia, Singapore and Hong Kong. The proposal was for a Centre for Environmental and Technical Advancement (CETA) and for the establishment of an ARCASIA Foundation Council.

6) THE FOUNDATION COUNCIL (1969-1974)

The 1st Foundation Council meeting in Hong Kong, Dec, 1969. Jon Prescott (HKIA), Leslie Ouyang (HKIA), Jai Bhalla (IIA), David Mcdonald (Director of Public works HK), Zahir-un-Deen Khwaja (IAP).

At the Foundation Council meeting held in Hong Kong on 8th December 1969, Mr. Lim Chong Keat from Singapore was elected the Chairman of Foundation Council. An update report on "a centre for environmental and technical advancement" was produced jointly by Dr. Thomas Howarth and Jon Prescott who was President of HKIA at the time, Mr. Kington Loo, Malaysia and Mr. R.A. Callow, Hong Kong, were appointed Advisers to the Foundation Council. The institute representatives attending the meeting were as follows:-

Lim Chong Keat
Singapore Institute of Architects (SIA)
Jon Prescott
Hong Kong Society of Architects (HKSA)
J.A. Bhalla
Indian Institute of Architects (IIA)
V. Selveratnam
Ceylon Institute of Architects (CIA)
Zahir-ud-DeenKhwaja
Pakistan Institute of Architects (IAP)
Fong Ying Leong
Pertubuhan Akitek Malaysia (PAM)
Observers:-
Dr. Thomas Howarth   Canada
C.A. Callow   HKSA
Kington Loo   PAM
Sim Hong Boon   SIA

The Foundation Council further appointed Dr. Howarth to be its Consultant and Mr. Sim Hong Boon its Pro-tem Secretary.

The Foundation Council held its second meeting on the 12th September 1970 at Singapore. There were new institute representatives joining the Council at this meeting and they were as follows :-

Chan Voon Fee   PAM
Sim Hong Boon   SIA
S. Molligoda   CIA
Michael Munday   HKSA
(alternate member Leslie Ouyang)
Ajmal Hayat Ahmad   IAP

At the same meeting Mr. Suhartono Susilo, President of the Indonesia Institute of Architects attended as an observer.

Two committees were established, the ARCASIA Constitution Committee consisting of Mr. Lim Chong Keat, Mr. Sim Hong Boon and Mr. Kington Loo and the CETA Committee consisting of Mr. Lim Chong Keat, the Hon. Adviser, Mr. Kington Loo, Mr. Selveratnam (CETA stands for Centre for Environmental and Technical Advancement).

The Foundation Council in September 1970 resolved that future assemblies of the national institutes of the Asian region of the CAA and ARCASIA be formally called the Architects Regional Council for Asia (ARCASIA). While recognising the need for worldwide associations, the organization was formed to enable closer working relationships between the individual member institute within the region to deal with matters that are of more immediate importance to the region. Many of the current member institutes of ARCASIA already belong to the Union Internationale des Architects (UIA) being an international organisation and integral with UNESCO.

An interim meeting of the Foundation Council was held on 19th May 1971 at Canberra and the following representatives attended :-
Chairman   Lim Chong Keat   SIA
Hon. Adviser   Kington Loo   PAM
Hon. Secretary   Sim Hong Boon   SIA
J.R. Bhalla   IIA
Mistri (observer)   IIA
Zahir-ud-Deen Khwaja   PIA
Visva Selveratnam   CIA
Fong Ying Leong   PAM
Jon Prescott   HKSA
Ho Pak Toe (observer)   SIA

The third meeting of the Foundation Council was held in the premises of the Singapore Institute of Architects on the 4th and 5th July 1972. It was resolved that the Foundation Council of the Architects Regional Council Asia (ARCASIA) be the Executive Committee of ARCASIA and that future meetings of the CAA Asian Regional members would be called "Conference of the ARCASIA Group". It further resolved that the national institutes of Thailand, the Philippines and Indonesia be invited to become full members. The meeting also agreed to establish a regional Board of Architectural Education. At the same meeting a modified report on the CETA proposal by Dr. Tom Howarth was presented by Jon Prescott and Michael Munday. The meeting then appointed Jon Prescott and Dr. Tom Howarth to draft a revised proposal for CETA which could be used for application for international funds.

Those attended the third meeting were as follows :-

Chairman   Lim Chong Keat   SIA
Hon. Aviser   Kington Loo   PAM
Hon. Secretary   Sim Hong Boon   SIA
Dr. Thomas Howarth   Canada
J.R. Bhalla   IIA
Charles Correa   IIA
S.H. Wandrekar   IIA
Abdul Razak Kazi   PIA
Visva Selveratnam   CIA
Jon Prescott   HKIA
Jason Yuen   HKIA
Sirichai Narumit   ASA
Lai Lok Kun   PAM
K.C. Leong   PAM

Observers:
Rex Koh Kum Chuen   SIA
Lim Soon Chye   SIA
Lawrence Lau   SIA
Albert H.K. Hong   SIA
Prof. Bill Lim   SIA
Prof. Rony Fonseca   SIA

The Fourth meeting of the Executive Committee was held at Ahmedabad,

India on the 24th February, 1973. During the meeting the matter of the regional hand book which was previously endorsed for preparation was brought up and was agreed that due to problems in communication, the work had fallen behind. The ARCASIA CETA Howarth - Prescott report was presented and the meeting agreed on minor revisions before deciding to defer the matter but at the same time to proceed on the basis of national centres whereby each national institute could initiate direct action. The constitution committee was asked to proceed with a draft proposal for the establishment of national centres for Environmental and Technical Advancement.

Those attended the Fourth meeting were as follows :-

Chairman	Lim Chong Keat	SIA
Hon. Adviser	Kington Loo	PAM
Albert Hong		SIA
(acting for Ho Pak Toe)		
J.R. Bhalla		IIA
Visva Selveratnam		SLIA
Barry Will		HKIA
(in place of Jason Yuen)		
K. Arunvongse		ASA
(in place of Sirichai Narumit)		
Observers:		
T. Karim		BIA
D. Choudbury		BIA
J. Parkinson		Australia

The fifth meeting of the ARCASIA Executive Committee was held in Singapore on the 22nd and 23rd July 1974. The meeting decided to expedite the drafting of the Constitution and appointed Mr. Kington Loo and the Secretary to produce the first draft. Letters of resignation from the Chairman Mr. Lim Chong Keat and the Hon. Secretary Mr. Sim Hong Boon were tabled and accepted. The meeting proceeded to elect Mr. Kington Loo as the new Chairman and discussed matters concerning membership of ARCASIA , national subcentres, relocation of ARCASIA Secretariat, Institute representation and liason amongst member institutes. It was decided that the Secretariat be moved to Kuala Lumpur.

Those attended the fifth meeting were as follows :-

Albert H.K. Hong	SIA
Ho Pak Toe	SIA
Sim Hong Boon	SIA
Visva Selveratnam	SLIA
Smath Wickremasinghe	SLIA
A.P. Kaimde	IIA
Charles Correa	IIA
J.R. Bhalla	IIA
Kington Loo	PAM
I Hisham Albakri	PAM
Szeto Koh Yim	PAM
Jon Prescott	HKIA
Ananta Krukeo	ASA
Chotechai Attavipach	ASA
Dr. Thomas Howarth	Canada

Observers:
Darmawan Prawirohardjo IAI
Constantino Agbayani UAP

7) THE ARCASIA EXECUTIVE COMMITTEE (1974-1980)

An interim meeting of the Executive Committee was held in Colombo on the 10th October 1975. The first draft of the constitution was discussed as well as the draft for the Regional Board of Architectural Education prepared by Mr. Lai Lok Kun. Mr. Sim Hong Boon reported that the ARCASIA Handbook was still being prepared by SIA.

Those attended the interim meeting were as follows :-

Chairman	Kington Loo	PAM
Hon. Secretary	Lai Lok Kun	PAM
Visva Selveratnam		SLIA
Surath Wickremasinghe		SLIA
Szeto Koh Yin		PAM
I Hisham Albakri		PAM
Sim Hong Boon		SIA

Observers:
Justin Samarasekera	SLIA
Rolan Silva	SLIA
K.R.S. Peiris	SLIA

There were two informal meetings, one at York in September 1976 and one at Kuala Lumpur in June 1977.

The sixth meeting of the Executive Committee only took place in Singapore between the 2nd and 4th of March 1979 some four and a half year after the fifth meeting with only four member institutes represented. The Meeting mainly discussed the draft of the Constitution which was circulated to all member Institutes in December 1974. No comments were received and the draft was first discussed and amended at the Colombo meeting in October 1975. Amendments raised at the Meeting together with the revised draft were sent to all member Institutes again after the Meeting, with a two month deemed approved condition.

Those attended the meeting were as follows :-

Chairman	Kington Loo	PAM
Hon. Secretary	Lai Lok Kun	PAM
Darmawan Prawirohardjo		IIA
I Hisham Albakri		
(Asian Representative)		
(CAA Executive Committee)		
Parid Wardi		PAM
Szeto Koh Yin		PAM
Felipe M Mendoza		UAP
Antonio S Dimalanta		UAP
Lim Chong Keat		Chairman CBAE
Sim Hong Boon		SIA

In 1970 the Ikatan Arsitek Indonesia (IAI) participated as observers and became a member in 1977. The Institute of Architects Bangladesh (IAP) became a member in 1971 and the Association of Siamese Architects became a member in 1972 and in 1974 the United Architects of the Philippines (UAP) participated as an observer and in 1977 as a full member.

The inaugural meeting of ARCASIA took place between the 17th and 18th of September, 1979 at the Executive Club, Jakarta Hilton, Jakarta. Ten years after the First Foundation Council meeting in Hong Kong. The Constitution of ARCASIA and the ARCASIA Board of Architectural Education (ABAE) were signed at the meeting.

ARCASIA and ABAE inaugural Meeting held in Jakarta, Indonesia in 1979

The meeting also received verbal reports from all representatives on individual member institute and allocated various projects to be carried out by member institutes such as standard form of building contract, sub-contract and supporting documents, limited liability and corporate practice; finance; directory; inter-country working system and mobility of architects; newsletter; role of architects & codes of practice and research organisation and research projects.

Those attended the inaugural meeting were as follows :-

Chairman	Kington Loo	PAM
Hon. Secretary	Lai Lok Kun	PAM
Szeto Koh Yim		PAM
Sim Hong Boon		SIA
Tan Choo Guan		SIA
Darmawan Prawirohardjo		IAI
Adhi Moersid		IAI
Felipe M Mendoza		UAP
Victor N Tiotuyco		UAP
Ronald Poon		HKIA
Jon Prescott		HKIA
Tham Chonchoo		ASA
I Hisham Albakri (Vice President CAA)		
Datuk Lim Chong Keat (Chairman CBAE)		

The Architects Regional Council for Asia became formally established and the acronym "ARCASIA" was retained for the new council which was held as a council of Presidents (or their representatives) of the various national institutes of architects in Asia.

8) ARCASIA (1980-1992)

By 1980, ten institutes had endorsed the constitution and in August that year the first council meeting of ARCASIA and ABAE took place in Bangkok. This set the precedence for the convening of regular annual meetings at the various countries of the member institutes.

At the first meeting of ARCASIA, Mr. Ronald Poon (HKIA) was elected the first Chairman of ARCASIA. Mr. Darmawan Prawirohardjo (IAI), Prof. Krisda Arunvongse (ASA) and Mr. L.A. Adithiya (SLIA) were elected Deputy Chairmen. Mr. Sim Hong Boon (SIA) was elected Chairman of the ARCASIA Board of Architectural Education (ABAE).

The meeting proceeded to receive individual member institute's report and discussed problems related to ARCASIA directory, ARCASIA handbook, communications and ARCASIA accounts.

First Meeting of Arcasia and ABAE in Bangkok, Thailand in August 1980.

The ABAE also met and discussed matters related to staff exchange, student exchange, travelling exhibition, publication and membership. A sub-committee was formed for publication projects chaired by Prof. Eric Lye with L.A. Adithiya, Syed Akeel Bilgrami, Victor Tiotuyco, Prof. Krishda Arunvongse and Parid Wardi as members.

Those attended the First meeting of ARCASIA & ABAE were as follows :-

Chairman	Kington Loo	PAM
Datuk Lim Chong Keat (First Founding Chairman)		
Darmawan Prawirohardjo		IAI
Prof. Krisda Arunvongse		ASA
L.A. Adithiya		SLIA
Ronald Poon		HKIA
Edwin Choo		SIA
Sim Hong Boon		SIA
Cheong Koh Cheow		PAM
Felipe Mendoza		UAP
Rear Admiral Prachuab Polakla		ASA
R.S. Khambatta		IIA
Victor N. Tiotuyco		UAP
Prof. Parid Wardi Sudin		PAM

Syed Akeel Bilgrami	IAP
Tham Chon Choo	ASA
Observers:	
Pree Buranasiri	ASA
Torpong Yomnak	ASA
Mok Yew Fun	SIA
M.L. Tri Devakul	ASA
I. Hisham Albakri (Vice President CAA)	

In March 1981, the 1st ARCASIA Workshop took place in Manila. During this meeting, an ARCASIA document for the international Code of Ethics on Consulting Services was proposed by UAP and by 1983, all ten member institutes endorsed the Code. This code was tabled at the 1981 UIA Congress in Warsaw, the CAA Nairobi Assembly 1981 and was presented to the Asian Federation of Engineering Organisations and the international Federation of Consulting Engineers. The Asian Association for Planning and Housing had already approved the document. The workshop also discussed matters related to the role and duties of Deputy Chairmen, the establishment of the History of Asian Architecture project, system of operation for future meetings, programmes for newsletter, pamphlet on ARCASIA, ARCASIA magazine, ideas for ABAE, ARCASIA travelling exhibition and visiting lecturership.

ARCASIA Workshop held in Manila in March 1981 where the ARCASIA-Manila Declaration was formulated which eventually became the International Code of Ethics on Consulting Services.

Those attended the First ARCASIA Workshop meeting were as follows :-

Chairman	Ronald Poon	HKIA
Hon. Secretary	Jon Prescott	HKIA
Kington Loo		PAM
Parid Wardi Sudin		PAM
Felip M. Mendoza		UAP
Victor N. Tiotuyco		UAP
Leandro V. Locsin		UAP

The 2nd ARCASIA Council Meeting was held in Hong Kong in August 1981. Architects from Beijing and Taipei met for the first time since 1949 and attended as invited observers. At this meeting, Target Publication and Investment Ltd. provided a sum to establish a travelling scholarship fund for architectural students. This was further supported by a donation from a prominent architect in the Philippines. The scholarship was intended to enable graduate students to travel and work in other Asian countries.

The 2nd ARCASIA Meeting and Workshop held in Hong Kong on August 6, 1981.

At the meeting each delegate made a report of activities of their Institute during the past year and raised problems for discussion. Jon Prescott tabled a report on communication within ARCASIA, Felipe Mendoza reported on the International Code of Ethics on Consulting Services. The meeting further discussed matters related to ARCASIA pamphlet, document exchange, ARCASIA publication company, ARCASIA newsletter, travelling exhibition, membership of architects from Beijing and Taipei. At the request of the ASA, ARCASIA supported the ASA's firm recommendation to the Thai Government against the repeal of the Architects profession act in Thailand.

Rear L to R: Sim Hong Boon, Visva Selvaratnam, Sumet Jumsai, Felipe Mendoza Front L to R: Darmawan Prawirohardjo, Ronald Poon (chairman), Leandro Loscin, Parid Wardi Sudin.Signing the "International Code of Ethics on Consulting Services".

The ABAE also held its meeting and the topics discussed included member institutes report, financing of ABAE meetings, inventory of academic institutions, recognition of ARCASIA schools, listing of research organisations within the area and History of Asian Architecture project.

Those attended the second meeting of ARCASIA were as follows :-

Chairman	Ronald Poon	HKIA
Hon. Secretary	Jon Prescott	HKIA
Hon. Treasurer	Ng Yook Man	HKIA
A.K. Rafique Uddin Ahmed		IBA
Darmawan Prawirohardjo		IAI
Yuswadi Saliya		IAI
Datuk Lim Chong Keat (1st Founding Chairman)		
Leandro V. Locsin		UAP
Felipe Mendoza		UAP
Lala Adithiya		SLIA
Dr. Sumet Jumsai		ASA
Paul Yuen		HKIA

Observers :

Maria Christina G.Esckerbana	UAP
Stuart King	HKIA
Tseng Chien	Beijing
Shu Jong Chuan	Taipei
Hsu Kun Nan	Taipei
Tsai Po An	Taipei

Those attended the ABAE meeting were as follows :-

Chairman	Sim Hong Boon	SIA
Barry Will		HKIA
Ir Suhartono Susile		IAI
Parid Wardi Sudin		PAM
Ahmed Rafique Suddin		IAP
Victor N. Tiotuyco		UAP
Pro. Krisda Arunvongse		ASA
Visva Selveratnam		SLIA

Following the Council meeting there was the second ARCASIA Workshop meeting held in the Royal Hong Kong Yacht Club on the 8th and 9th August 1981. The papers presented were practice with limited liability by Barry Will, Graham Campbell and Dennis Cheng, Code of Practice - role of the architect in society by Felipe Mendoza and Victor N. Tiotuyco, inter-country practice and mobility of architects by Dr. Tao Ho.

Those attended the workshop meeting were as follows :-

Felipe Mendoza
Victor Tiotuyco
Visva Selveratnam
Lala Adithiya
Ronald Poon
Jon Prescott
Barry Will
Paul Yuen
Tao Ho
Graham Campbell

The lighting of the ceremonial lamp at the opening of the 3rd ARCASIA Council Meeting.

The 3rd ARCASIA Council meeting took place in Colombo, Sri Lanka in October 1982. The opening ceremony was officiated by the Hon. Anandatissa de Alwis, Minister of States. In the opening speech, the Chairman Ronald Poon summed up the reasons for the existence of ARCASIA (see appendix 1). The meeting proceeded to dealt with the fee structure of architects, limited liability practice, individual member institute's report and travelling fund for architectural students. The meeting elected Mr. Lala Adithiya as the new Chairman of ARCASIA, Mrs. Yasmeen Lari (IAP), Felipe Mendoza (UAP) and Dr. Kenneth Yeang (PAM) as deputy chairmen.

Delegates to the 3rd ARCASIA Council Meeting at Colombo, 1982.

Those present at the meeting were as follows :-

Chairman	Ronald Poon	HKIA
Hon. Secretary	Jon Prescott	HKIA
Hon. Treasurer	David Lung	HKIA
Deputy Chairman	Lala Adithiya	SLIA
Ruslan Khalid		PAM
Ejaz Ahed		IAP
Arshad Abdulla		IAP
Chok Yan Hoi		SIA

Lim Chong Keat
(1st Founding Chairman)
Bunchong Sontitim   ASA
Surath Wickremasinghe   SLIA

Observers:
Kenneth Kan   HKIA
John Peh-Nien Lynn   Taipei
Lloyd Lee-Teh Lo   Taipei
Kao Erh Pan   Taipei
Kamil Khan Mumtaz   IAP
Lakshman Alwis   SLIA
Gemumo Fernando   SLIA
Ranjan Gopallawa   SLIA
Lochi Gunaratna   SLIA
H. Karaunadasa   SLIA
M. Krishnapillai   SLIA
J.C. Nilgiria   SLIA
Jayantha Perera   SLIA
Dennis R. Saunders   SLIA
G.S. de Silva   SLIA
Lalith de Silva   SLIA
Minnette de Silva   SLIA
Tudor W. Sirisena   SLIA
Raju Silvaraman   SLIA
Ashley de Vos   SLIA

Owing to the lack of a quorum, the ABAE only held an informal meeting at Colombo with its Chairman, Mr. Sim Hong Boon, Mr. Lochi Gunaratna and Mr. Parid Wardi Sudin attending.

Following the ARCASIA Council meeting, the first ARCASIA Forum, the brainchild of Mr. Ronald Poon was launched. The ARCASIA Forum is a platform for the ongoing discussion of current architectural ideas pertinent to the region. This is a gathering of prominent Architects and thinkers intended to serve as an Asian "catapult of minds" on matters concerning architecture and environment in the Asian context.

The 1st ARCASIA Forum.

The 4th Council meeting was held in Lahore, Pakistan in August 1983. The meeting discussed financial matters affecting ARCASIA, the matters discussed at the informal ABAE meeting in Colombo, individual Institute's report and two papers presented by Felipe Mendoza and SLIA on planning for the future.

The 4th ARCASIA Meeting at Lahore, Pakistan on August 26-28, 1983.

The meeting elected Felipe Mendoza (UAP) as the new ARCASIA Chairman, Dan Wongprasat (ASA), Barry Will (HKIA) and Victor Tiotuyco (UAP) as Deputy Chairmen.

Those attended the meeting were as follows :-

Chairman
L.A. Adithiya   (SLIA)
Hon Secretary
Lochi Guanaratna   (SLIA)
Hon Treasurer
Lalith de Silva   (SLIA)
Deputy Chairman
Yasmeen Lari   (IAP)
Deputy Chairman
Felipe Mendoza   (UAP)
Adviser & Forum Director
Ronald Poon   (HKIA)
Jon Prescott   (HKIA)
Misbath Najmi   (IAP)
Manuel Manosa   (UAP)
L.A. Adithiya   (SLIA)
S. Wickremasinghe   (SLIA)
Bunchong Sontitim   (ASA)
Dan Wongprasat   (ASA)

Observers :
P.G.K. Fernando   (SLIA)
Shahab Ghani Khan   (IAP)
Ejaz Ahed   (IAP)
Syed Akeel Bilgrami   (IAP)

In October 1984, the 5th Council meeting and the 2nd ARCASIA Forum were held in Manila to coincide with its first major international conference called ACA-1 or the first Asian Congress of Architects. The conference became an important milestone for ARCASIA. The unique format of the conference provided an extraordinary mechanism for the intellectual and cultural exchange for the architects in the region.

ACA-I Opening ceremony entrance of colors.

Delegates to the 5th ARCASIA Council Meeting at Manila, 1984.

The Council meeting discussed the printing of the International Code, the history of Asian Architecture project, the ARCASIA Target Scholarship, The work of the ARCASIA Board of Architectural Education, amendments to the ARCASIA Constitution, listing of architectural firms of ARCASIA member institutes for World Bank, Asian Development Bank and other international agencies and ARCASIA Architectural Competition.

ACA-I opening ceremony Manila, 1984.
Dato Kington Loo (V.P.-UIA Region IV), Felipe M. Mendoza (ACA-I Secretary General and Deputy Chairman Arcasia at the Rostrum), Yasmeen Lari (Deputy Chairman Arcasia), Manuel T. Manosa (ASA-I Congress Chairman), Leandro V. Locsin (ASA-I Keynote Speaker), L.A. Adithiya (Chairman Arcasia), Mrs. Inday Manosa, Ronald C.Y. Poon (Past Chairman Arcasia, ACA-I Rapporteur General), and Kenneth Yeang (Deputy Chairman Arcasia).

Those attended the meeting were as follows :-

Chairman
Felipe Mendoza (UAP)
Hon Secretary
Victor Tiotuyco (UAP)
Hon Treasurer
Rebecca Tobia (UAP)
Past Chairman
Lala Adithiya (SLIA)
Deputy Chairman
Rusi Khambatta (IIA)
Deputy Chairman
Barry Will (HKIA)
Deputy Chairman
Dan Wongprasat (ASA)
Adviser & Forum Director
Ronald Poon (HKIA)
2nd Founding Chairman
Data Kington Loo (PAM)
Sha Alam Zahiruddin (BIA)
A.H. Abduz Rahman (BIA)
Dr. Kenneth Yeang (PAM)
David Teh (PAM)
Syed Akeel Bilgrami (PIA)
Yasween Lari (PIA)
Mannel Manosa (UAP)
Chia Kok Leong (SIA)
Tan Cheng Siong (SIA)
Lochi Guanaratna (SLIA)
S. Wickremasinghe (SLIA)
Bunchong Sontitim (ASA)
Dan Wongprasat (ASA)
Observers :
P.G.K. Fernando (SLIA)
Ladith de Silva (SLIA)
Ejaz Ahed (PIA)

The 6th Council meeting took place in Lonavla, India on 16-19 November 1985. The meeting discussed updated report on the History of Asian Architecture project, the ARCASIA Target Scholarship, individual member Institute's report, the work of the ARCASIA Forum, ARCASIA exchange of faculty between countries, ARCASIA architectural competition, UAP's proposal for a foreign-local consulting architects relationship on projects funded from abroad, ACA-2 workshop, ARCASIA Architects directory, report of the ABAE, amendments to the ARCASIA constitution, the membership applications from architects from Beijing and Taipei and the Korea Institute of Registered Architects (KIRA). The meeting accepted in principle the application from KIRA and accepted the two Chinese group as observers until their differences are resolved by themselves. The meeting appointed Mr. Ronald Poon as Chairman of the ARCASIA Board of Advisers with Dato Kington Loo as member. The meeting elected Dr. Kenneth Yeang (PAM) as the new ARCASIA Chairman, Song Koon Atthakor (ASA), David Teh (PAM) and Syed Akeel Bilgrami as the Deputy Chairmen.

Delegates to the 6th ARCASIA Council Meeting at Lonavia, 1985.

Those attended the meeting were as follows :-
Chairman
Felipe Mendoza (UAP)
Deputy Chairman
Rusi Khambatta (IIA)
Past Chairman
L.A. Adithiya (SLIA)
ABAE Chairman (acting)
L. Gunaratna (SLIA)
Adviser
Ronald Poon (HKIA)
Hon Treasurer
Rebecca Tobia (UAP)
Songkorn Atthakor (ASA)
Pracha Suk Udom (ASA)
Gang Eng Don (SIA)
Chia Kok Leong (SIA)
Gemuno Fernando (SLIA)
Laksman Aliors (SLIA)
David Teh (PAM)
Dr. Kenneth Yeang (PAM)
Shahab Ghani Khan (IAP)
Ejaz Ahed (IAP)
Rusi Khambatta (IIA)
Yatish Kini (IIA)

Observers :
Mihindu Keerthiratne (SLIA)
Surah Wickremasinghe (SLIA)
Dennis Sannders (SLIA)
Minette de Silva (SLIA)
Fay Cheah (PAM)
Madham Ganesh
Deobhakta (IIA)
Gusmate Keshan (IIA)
Joshi Digambar (IIA)
Ogale Arum Vinayak (IIA)
Joe Ming Chu (Taipei)
Joshua Jih Pan (Taipei)
Ming Kuo Yu (Taipei)
Tse Hwan Chung (Taipei)
Chi Tok Kim (KIRA)
Wong Suk Oh (KIRA)
Kee Duk Song (KIRA)

The 7th Council meeting took place in Kuala Lumpur, Malaysia between 9th and 10th October 1986. The matters discussed were ARCASIA exchange faculty, ARCASIA architectural competition, special committee on ARCASIA publication, foreign-local consultants' relationship on project funded from abroad, criteria for membership, SLIA history book series, travelling fund for architectural students, ARCASIA forum, ARCASIA international code of ethics, ARCASIA newsletter, architects' directory, funding of Forum publication, status of Chinese Architect and the ABAE report. The meeting officially admitted KIRA as a member Institute of ARCASIA.

Delegates to the 7th ARCASIA Council Meeting at ACA-2 at Kuala Lumpur, 1986.

Those attended the meeting were as follows :-
Chairman
Dr. Kenneth Yeang (PAM)
Hon Treasurer
P.Kasi (PAM)
Deputy Chairman
David Teh (PAM)
Deputy Chairman
Songkoon Atthakor (ASA)
Deputy Chairman
Akeel Bilgrami (IAP)
Past Chairman
Felipe Mendoza (UAP)
Past Chairman
Lala Adithiya (SLIA)
Adviser
Datuk Kington Loo (PAM)
Adviser & Forum Director
Ronald Poon (HKIA)
C. Canchela (UAP)
Mauro C. Simpliciano (UAP)
Mochamad Daniswaro (IAI)
AdhiMoersid (IAI)

Masihuddin Shaker (IAB)
Bashirul Haq (IAB)
James Kinoshita (HKIA)
Edwin Wong (HKIA)
Chia Kok Leong (SIA)
Yang Soo Suan (SIA)
Rusi Khambatta (IIA)
Shirish Desphande (IIA)
Haji Hajeldon (PAM)
John Koh (PAM)
Dr. Pervaiz Vandal (IAP)
Woong Suk On (KIRA)
Song Kee Duk (KIRA)
Mrs. Apsorn Kuromarohit (ASA)
P.G.K. Fernando (SLIA)
Lochi Gunanatna (SLIA)

Observers:
Dai Nian Ci (Beijing)
Wu Liang Yong (Beijing)
ChenZhenXiang (Beijing)
Kun-Nan Hsu (Taipei)
Yu Ming Kuo (Taipei)
Tze Hwaa Chung (Taipei)
Laksman Alivis (SLIA)
Shahab Ghani Khan (PIA)
Joseph Kim (KIRA)
Young Bae Ahu (KIRA)
Pracha Suk Udom (ASA)
Rebecca Tobia (UAP)
Mihindu Keerthiratne (SLIA)
Young I Lee (KIRA)
LiuKaiji (Beijing)
Zhang Qinzhe (Beijing)
Tang Yiqing (Beijing)
Joe Ming Chu (Taipei)
Tsi Kwong Lan (Taipei)
Peh Nieu Lyun (Taipei)
Peng Yau-Hwa (Taipei)

There was a meeting of the ABAE on the 9th October 1986 and the matters discussed were, ABAE funding, Architectural Student Jamboree, exchange of information on recognised architectural schools, multi-lateral co-operation between member Institutes, student competition, individual member institutes' reports, sub-regional activities and future role of co-ordinators. The meeting elected Lochi Gunaratna (SLIA) to be the next chairman ABAE.

Student representatives to the ARCASIA Student Jamboree with the delegates to the ARCASIA ABAE meeting at Kuala Lumpur, 1986.

The Council meeting was followed by the second ARCASIA Congress of Architects (ACA-2) and ARCASIA Forum 3.

Those attended the ABAE meeting were as follows :-

ABAE Chairman
Lochi Gunaratna (SILA)
Prof. S.A. Deshpande (IIA)
P. Kasi (PAM)
Chia Kok Leong (SIA)
Zaenudin Kartadiwiria (IIA)
Chi Tok Kim (KIRA)
Bashirul Hag (IAB)
James Kinoshita (HKIA)
Mauro C. Simpliciano (UAP)
Dr. P. Vandal (IAP)
Mrs. Apsorn Kuromarohit (UAP)

Observers :
Prof. Wu Liang Yong (Beijing)
Yu Ming Kuo (Taipei)
Yung I.L. Lee (KIRA)
Chan Sui Him (SIA)
John Koh Seng Siew (PAM)
SuhardiHajiNoordin (PAM)

The 8th Council meeting took place in Bali, Indonesia between the 22nd and 23rd October 1987. The Meeting deliberated on the member Institutes' reports, ARCASIA Forum 4, ACAE

The 8th ARCASIA Council Meeting at Bali, 1987.

report, publication, constitution amendments, policy and programme for 1987 to 1988, 9th Council meeting and ACA-3, ARCASIA national projects, membership of architects from Beijing, Taipei and RAIA, ARCASIA awards, permanent secretariate proposal, ARCASIA think tank No. 1 "Urban future of Asia" and exhibition. Forum 4 took place after the Council meeting.

Those attended the meeting were as follows :-

Chairman
Dr. Kenneth Yeang (PAM)
Hon Secretary
Esa Mohamed (PAM)
Hon Treasurer
P. Kasi (PAM)
Deputy Chairman
Songkorn Atthakor (ASA)
Deputy Chairman
Syed Akeel Bilgrami (IAP)
Deputy Chairman
David Teh (PAM)
Past Chairman
Felipe Mendoza (UAP)
Adviser & Forum Director
Ronald Poon (HKIA)
Adviser
Dato Kington Loo (PAM)
Mati Tungpanich (ASA)
Pracha Suk-Udom (ASA)
Patrick Lau Sau Shin (HKIA)
Jon Prescott (HKIA)
Adhi Moersid (IAI)
Michael Sumaryanto (IAI)
Rusi Khambatta (IIA)
David Teh (PAM)
Hussein Hamzah (PAM)
Ejaz Ahed (IAP)
Shahab Ghain Khan (IAP)
Foilan L. Hong (UAP)
Majorie Musni (UAP)
Yang Soo Suan (SIA)
Chok Yan Hoi (SIA)
Lala Adithiya (SLIA)
Song Kee Duk (KIRA)
Kim Chi Tak (KIRA)

Observers :
Syaiful Arifin (IAI)
Soewondo B. Soetejo (IAI)
Zaenudin Kartadiwira (IAI)
Shamsuddin Mohammand (PAM)
Mohamad Farooq (IAP)
Husnain Lotia (IAP)
Ahn Byoung Yoon
Lee Jeong Keun

Delegates to the 9th ARCASIA Council Meeting at Seoul, 1988.

The 9th Arcasia Council Meeting at Seoul, 1988.

Opening ceremony at ACA-1, Seoul, 1988.

The 9th Council meeting took place in Seoul, Korea between the 7th and 8th November, 1988. The meeting deliberated on the members institutes' report, ARCASIA Forum 4 & 5, ACAE report, the work of the publication committee including Forum 4 book (R. Poon), ACAE booklet (SLIA), handbook (Chairman), practice survey (PAM), ARCASIA cities (Hon Sec./PAM), ARCASIA's Architecture (KIRA) and Habitat special issue, Constitution amendments, policy and programme for 1987 to 1988, ACA-3, ARCASIA national projects, membership of architects from Beijing and Taiwan, ARCASIA award, ARCASIA Think Tank No.1 "Urban Future in Asia", exhibition.

Those attended the meeting were as follows :-

Chairman
Dr. Kenneth Yeang   (PAM)
Hon. treasurer
P. Kasi   (PAM)
Deputy Chairman
Songkron Atthakor   (ASA)
Deputy Chairman
David Teh   (PAM)
Deputy Chairman
Syed Akeel Bilgrami   (IAP)
Adviser & Forum Director
Ronald Poon   (HKIA)
Adviser
Datuk Kington Loo   (PAM)
Past Chairman
Felipe Mendoza   (UAP)
Laksman Alwis   (SLIA)
Lal Balasuriye   (SLIA)
Tay Kheng Soon   (SIA)
Mati Tungpanich   (ASA)
Apson Kurmarohita   (ASA)
Lee Se Hoon   (KIRA)
Koui Chi Tok   (KIRA)
James Kinoshita   (HKIA)
Adhi Moersid   (IAI)
Michael Sumarijanto   (IAI)
Rusi Khambatta   (IIA)
N.A. Bachlheka   (IIA)
Hussain Hamzah   (PAM)
Zaigham Jaffery   (IAP)
M. Mahinddin Khan   (IAB)
John Koh   (PAM)

Observers :
Ahmad Pervez Mivza   (IAP)
Chouvabit Chonovatana   (ASA)
Lilith de Silva   (SLIA)
Mihindu Keerthiratua   (SLIA)
Elmo de Silva   (SLIA)

The ABAE meeting took place concurrently with the Council meeting and the 3rd ARCASIA Congress of Architects took place after the Council meeting.

Delegates to the 10th Council Meeting.

The 10th Council meeting took place in Bangkok, Thailand between the 19th and 20th October 1989. The meeting discussed individual member institutes' report, the report of the practice work group, report on out reach programme, the report of architects selection work group, the ARCASIA national projects, ARCASIA awards, ARCASIA city directory, architects' research programme, the ACAE report and the Forum report. The meeting admitted the Architectural Society of China as a member institute.

The ACAE also met during the Council meeting and deliberated on Asian student jamboree, ICAAP project by SLIA and data on Asian schools by HKIA.

After ARCASIA Forum 5 took place the Council meeting.

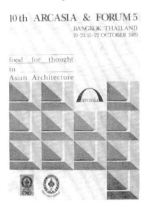

Mr. Ronald Poon at the opening of ARCASIA Forum 5 at Bangkok, 1989.

Those attended the Council meeting were as follows :-
Chairman
Rusi Khambatta (IIA)
Hon Secretary
N. Badheka (IIA)
Hon Treasurer
Ms. Meera Deobhakta (IIA)
Deputy Chairman
Adhi Moersid (IAI)
Deputy Chairman
Song Kee Duk (KIRA)
Past Chairman
Dr. Kenneth Yeang (PAM)
Chairman ACAE
Prof. Shirish Deshpande (IIA)
Adviser & Forum Director
Ronald Poon (HKIA)
Adviser
Datuk Kington Loo (PAM)
Iftekharuddin
   M. Choudhury (IAB)
M. Mahinddin Khan (IAB)
Xu Ronglie (ASC)
Gu Ziping (ASC)
Kit Haffner (HKIA)
Rocco Yim (HKIA)
Madhav Deshhakta (IIA)
Akhtar Chamhan (IIA)
Syahne Syarif (IAI)
Bombang Adi Setiose (IAI)
Jeong Keon Lee (KIRA)
Chitok Kim (KIRA)
Hussein Hamzah (PAM)
P. Kasi (PAM)
Ejaz Ahed (PIA)
Kalim A. Siddiqui (PIA)
Felipe Mendoza (UAP)
Angel Lazaro Jr. (UAP)
Mati Tungpanich (ASA)
Bunchong Sontitim (ASA)
Cha Kok Leong (SIA)
Edward H.Y. Wong (SIA)
Lalith de Silva (SLIA)
P.G.K. Fernando (SLIA)

Observers:
Apsorn Kumarohita (ASA)
Robtuk Hussain (IAB)
Fuji Xiang (ASC)

Jimmy Lim (PAM)
Hisham Albakri (PAM)
Shahab Ghani Khan (PIA)
Ahmad Pervez Mirza (PIA)
Muhammad Farooq (PIA)
Parvez L. Qureshi (PIA)
E.d. Aquilar (UAP)
Franciso T. Manosa (UAP)
Ms. Majorie L. Musni (UAP)
Ms. Sestoque (UAP)
Ms. Lizeta Uy (UAP)

The 11th Council meeting took place in Agra, India between the 1st and 2nd October 1990. The meeting dealt with the reports of the practice workshop, outreach programme, Architect selection working group and the ARCASIA think tank, the ARCASIA city directory, ARCASIA information

Mr. Ronald Poon, W. Badheka, Rusi Khambatta (Chairman), Ms Meera Deobhakta.

centre, ARCASIA awards, Forum 6, ARCASIA bye-laws, ACAE report, individual member institute's report, membership extension, ACA-5 and elected Syed Akeel Bilgrami(IAP) to be the next ARCASIA Chairman and Richeto Alcordo (UAP), Chia Kok Leong (SIA) and Rabiul Hussain (IAB) the deputy chairmen.

The ACAE met concurrently with the Council meeting and deliberated on data on Asian schools (HKIA), multi-lateral co-operation (IIA) and asian architectural students jamboree. The Meeting also elected Mr. P. Kasi (PAM) as the new ACAE Chairman, IAB the co-ordinator for south asia, IAI for South-East Asia and KIRA for East Asia.

ACA-4 took place at Delhi before the Council meeting.

Those attended the Council meeting were as follows :-
Chairman
Rusi Khambatta (IIA)
Hon Secretary
W. Badheka (IIA)
Hon Treasurer
Ms. Meera Deobhakta (IIA)
Deputy Chairman
Song Kee Duk (KIRA)
ARCASIA Forum Director & Adviser
Ronald Poon (HKIA)
Chairman ACAE
Prof. Shirish Deshpande (IIA)
Rabiul Hussain (IAB)
Zhang Baiping (ASC)
Xu Ronglie (ASC)
Christopher Haffner (HKIA)
Edward Shen (HKIA)
Madhav Deobhakta (IIA)
Akhtar Chauhan (IIA)
Syahrul Syarif (IAI)
Julius Hassan (IAI)
Kim Chi Tok (KIRA)
Lee Jeong Keun (KIRA)
Hussain Hamzah (PAM)
Jimmy Lim (PAM)
Syed Mohammad Irfan (IAP)
Kalim A. Siddiqui (IAP)
Richeto C. Alcordo (UAP)
M.A. Cristine V. Turalba (UAP)
Mati Tungpanich (ASA)
Apson Kuromarohit (ASA)
Chia Kok Leong (SIA)
Edward Wong (SIA)
Obervers:
Xu Anzhi (ASC)
Prof. Wu Liang Yong (ASC)
Ms. Shao Huyam (ASC)
Yan Xing Hua (ASC)
Bu Zhengwai (ASC)
Pisit Rojanavanich (ASA)
H.G. Malkani (IIA)
Atul Desai (IIA)
Ms. Neena Rangnekar (IIA)

The Council at the 11th Meeting, Agra.

Yayant Tipnis	(IIA)
Ms. Pragnya Chauhan	(IIA)
Jeon Bong Soo	(KIRA)
P. Kasi	(PAM)
Syed Akeel Bilgrami	(IAP)
Zaigam S. Jaffery	(IAP)
Ejaz Ahed	(IAP)
Mohammad Arshad	(IAP)
Mohammad Farooq	(IAP)
Shabit Ghani Khan	(IAP)
Felipe Mendoza	(UAP)
Ms. Rebecca Tobia	(UAP)
Ms. Angeline T. Chua Chiaco	(UAP)
Koichi Nagashima	(Japan)
Jose C.S. Marreiras	(Macau)

The 12th Council meeting took place in Beijing, China between the 24th and 25th September, 1991. The meeting discussed the individual member institute's reports, reports of the workgroups including practice workgroup, outreach programme workgroup, Architect selection workgroup, ARCASIA city directory and ARCASIA information centre, constitutional amendments, adoption of the bye-laws, ARCASIA Forum, ARCASIA awards, ACAE report ACA-5, and new membership accepting the Associacao de Arquitectos de Macau (AAM), the Japan Institute of Architects (JIA) and the Union of Mongolian Architects (UMA) as new member institutes of ARCASIA. Mr. Ronald Poon, founder and ARCASIA Forum director since 1982 offered his resignation and the Meeting accepted his wishes and commended his efforts for the founding of the Forum and his work as director for the past nine years. The Meeting appointed Mr. Jimmy Lim (PAM) as the co-ordinator for the next two years.

The ACAE met concurrently with the Council and discussed student jamboree, student competition (IAP), data on asian schools of architecture (HKIA), ARCASIA research journal (IIA), exchange of information, student/teacher exchange programme, accredition of schools and ARCASIA magazine. The meeting appointed Mr. Jimmy Lim as the first editor for the ARCASIA Magazine.

The Meeting was followed by ARCASIA Forum 6.

Those attended the Council meeting were as follows :-

Chairman
Syed Akeel Bilgrami	(IAP)

Hon Secretary
Syed Zaigham S. Jaffery	(IAP)

Hon Treasurer
Shahab Ghani Khan	(IAP)

Deputy Chairman
Chia Kok Leong	(SIA)

Deputy Chairman
Rabiul Hussain	(IAB)

Deputy Chairman
Richeto C. Alcordo	(UAP)

Past Chairman
Rusi Khambatta	(IIA)

Adviser
Felipe Mendoza	(UAP)

Adivser
Ronald Poon	(HKIA)

Chairman ACAE
P. Kasi	(PAM)
Uttam Kumar Saha	(IAB)
Mohiuddin Khan	(IAB)
Xu Ronglie	(ASC)
Gao Yilan	(ASC)
Stephen Poon	(HKIA)
Edward Shan	(HKIA)
Anil Nagrath	(IIA)
Syahrul Syarif	(IAI)
Bian Poen	(IAI)
Shoji Hayashi	(JIA)
Kuniabi Ito	(JIA)
Oh Woon Bong	(KIRA)
Joseph Chitok Kim	(KIRA)
Jon Prescott	(AAM)
Jimmy C.S. Lim	(PAM)
S.C. Vadiveloo	(PAM)
Gombyn Mygmar	(UMA)
Olziburengiin Ganboed	(UMA)
Pervez L. Qureishi	(IAP)
Kalim Siddiqui	(IAP)
Richeto C. Alcordo	(UAP)
Aurelio T. Juguillon	(UAP)
Tay Kheng Soon	(SIA)
Edward de Silva	(SIA)
Mihindu Keerthiratne	(SLIA)
M. Krishnapillai	(SLIA)
Mati Tungpanich	(ASA)
M.R. Chanvudhi Varavam	(ASA)

Delegates to the 12th Council Meeting at Beijing, 1991.

Delegates to ARCASIA Forum 6 at the garden of Fragrant Hill Hotel, Beijing 1991.

The 13th ARCASIA Council Meeting at Lahore, 1992.

The 13th ARCASIA Council Meeting.

The opening ceremony at ACA 5, Lahore, 1992.

Observers:
Edward Wong	(SIA)
Edgards A. Aguila	(UAP)
Mohammad Farovq	(IAP)
Dr. Lee Jeong Keun	(KIRA)
Koichi Nagashima	(JIA)
Michael Sumarijanto	(IAI)
Zhong Qinan	(ASC)
Uttam Kumar Saha	(IAB)

The 13th Council meeting took place in Lahore, Pakistan between the 24th and 25th October, 1992. The meeting discussed the individual member institute's report, Forum 6, ARCASIA awards, ACAE report, ARCASIA zones, 14th Council meeting and Forum 7 and ACA 6. The meeting elected Richeto Alcordo (UAP) the new ARCASIA Chairman.

ACA 5 took place after the Council meeting.

9) PAST OFFICIALS

1969-1974 (1st Foundation Council)
Chairman	Datuk Lim Chong Keat	(SIA)
Hon Secretary	Sim Hong Boon	(SIA)
Adviser	Kington Loo	(PAM)

1974-1980 (Executive Committee)
Chairman	Kington Loo	(PAM)
Hon Secretary	Lai Lok Kun	(PAM)

1980-1982 (1st ARCASIA Council)
Chairman	Ronald Poon	(HKIA)
Hon Secretary	Jon Prescott	(HKIA)
Hon Treasurer	Ng York Man (1980-81)	(HKIA)
	David Lung (1981-82)	(HKIA)
ABAE Chairman	Sim Hong Boon	(SIA)

1982-1984 (2nd ARCASIA Council)
Chairman	L.A. Adithiya	(SLIA)
Hon Secretary	Lochi Gunaratna	(SLIA)
Hon Treasurer	Lalith de Silva	(SLIA)
Deputy Chairman	Mrs. Yasmeen Lari	(IAP)
Deputy Chairman	Felipe Mendoza	(UAP)
ABAE Chairman	Parid Wardi Sudin	(PAM)
Forum Director	Ronald Poon	(HKIA)

1984-1986 (3rd ARCASIA Council)
Chairman	Felipe Mendoza	(UAP)
Hon Secretary	Victor Tiotuyco	(UAP)
Hon Treasurer	Rebecca Tobia	(UAP)
Deputy Chairman	Rusi Khambatta	(IIA)
Deputy Chairman	Barry Will	(HKIA)
Deputy Chairman	Dan Wongprasat	(ASA)
Adviser & Forum Director	Ronald Poon	(HKIA)
Adviser	Datuk Kington Loo	(PAM)
ABAE Chairman	Lochi Gunaratna	(SLIA)

1986-1988 (4th ARCASIA Council)
Chairman	Dr. Kenneth Yeang	(PAM)
Hon Secretary	Esa Mohammed	(PAM)
Hon Treasurer	P. Kasi	(PAM)
Deputy Chairman	David Teh	(PAM)
Deputy Chairman	Song Koon Atthakor	(ASA)
Deputy Chairman	Akeel Bilgrami	(IAP)
Adviser & Forum Director	Ronald Poon	(HKIA)
Adviser	Datuk Kington Loo	(PAM)
ABAE Chairman	Lochi Gunaratna	(SLIA)

1988-1990 (5th ARCASIA Council)
Chairman	Rusi Khambatta	(IIA)
Hon Secretary	N. Badheka	(IIA)
Hon Treasurer	Ms. Meera Deobhakta	(IIA)
Deputy Chairman	Adhi Moersid	(IAI)
Deputy Chairman	Song Kee Duk	(KIRA)
Adviser & Forum Director	Ronald Poon	(HKIA)
Adviser	Datuk Kington Loo	(PAM)
ACAE Chairman	Prof. Shirish Deshpande	(IIA)

1990-1992 (6th ARCASIA Council)
Chairman	Syed Akeel Bilgrami	(IAP)
Hon Secretary	Syed Zaigham S. Jaffery	(IAP)
Hon Treasurer	Shahab Ghani Khan	(IAP)
Deputy Chairman	Richeto Alcordo	(UAP)
Deputy Chairman	Chia Kok Leong	(SIA)
Deputy Chairman	Raburl Hussain	(IAB)
Adviser & Forum Director	Ronald Poon	(HKIA)
Adviser	Felipe Mendoza	(UAP)
ACAE Chairman	P. Kasi	(PAM)

1992-1994 (7th ARCASIA Council)
Chairman	Richeto Alcordo	(UAP)

10) PAST & CURRENT ACTIVITIES

Activities are mainly divided into two types. Those that occur during a two year period of each Chairman and those which spans the chairmanships becoming a kind of permanent fixtures of ARCASIA. For the former type we have, in the past, dealt with the following subjects :-

10.1 ARCASIA directory
10.2 Role of architects and code of practices
10.3 Research organisation and research projects
10.4 Limited liability and corporate practice
10.5 Inter-country working and mobility of architects
10.6 Student travelling fund
10.7 ARCASIA schools "recognition" listing
10.8 International code of ethics on consulting services
10.9 History of Asian Architecture
10.10 ARCASIA exchange of faculty between countries
10.11 ARCASIA architectural competition
10.12 Foreign-Local consulting architects relation on projects funded from abroad
10.13 Exchange of information on recognised schools of architecture
10.14 Information centre for asian architectural publications
10.15 Multi-lateral cooperation between institutions of architectural education in ARCASIA counties
10.16 Academic staff exchange programme
10.17 Student competition
10.18 ARCASIA newsletter
10.19 Practice survey
10.20 ARCASIA cities directory
10.21 ARCASIA architecture
10.22 ARCASIA think tank
10.23 Practice work group
10.24 Outreach programme
10.25 Architect selection working group
10.26 ARCASIA information centre

11) THE ARCASIA FORUM

For the permanent fixtures, one of the key activities of ARCASIA is the ARCASIA FORUM which is a platform for the ongoing discussion of current

WHAT IS AN ARCASIA FORUM?

architectural ideas pertinent to the region. This is a gathering of prominent architects and thinkers intended to serve as an Asian "catapult of minds" on matters concerning architecture and environment in the Asian context.

Forum 1 tool place in the SLIA premises in Colombo, Sri Lanka on 9-10 October 1982. The theme was "Innovations in Architecture". The speakers were Datuk Lim Chong Keat, Dr. Kenneth Yeang, Miss Minnette de Silva, Kamil Mumtaz, Lochi Gunaratna, Jon Prescott, Ruslan Khalid and Dr. Justin Samarasekera.

ARCASIA Forum 1, Colombo, 1982.

Ruslan Khalid, Datuk Lim Chong Keat & Jon Prescott at the ARCASIA Forum 1.

Forum 2 took place in the Philippines International Convention Centre in Manila on 26th October 1984 during ACA-1. The theme was "Asian Identity", the speakers were Ir Robi Sulato Sastrowardoyo, Koichi Nagashima with Bobby Manosa, Tao Ho, Dan Wongprasat and Ruslan Khalid as panelists.

A page from the ARCASIA Newsletter (October-December, 1986) on Arcasia Forum 3.

Presenters and panelists of the ARCASIA Forum 3.

Ronald Poon speaking at ARCASIA Forum 4.   *Geoffrey Bawa*

Forum 3 took place in Kuala Lumpur coinciding with ACA-2. The theme was "Design Directions in Asia" with the sub-theme "The State of the Art", there were two theme papers and eleven sub-theme papers given by representatives from member and observer institutes. The main speakers were Hiroyasu Higuchi and Reiko Hayashi of Team Zoo and Dr. Sumet Jumsai.

After the 1986 ARCASIA Council, it was decided that the Forum would be held on the alternate years to ACA to enhance the council meetings in between the congresses.

Forum 4 took place in Bali, Indonesia on the 24th October 1987 with the theme "My architecture". The speakers were William Lim, Mohamad Daniswaro, Kazuhiro Ishii and Geoffrey Bawa.

The venue of the Forum was Puri Ubud, a traditional Balinese castle in the hill area of Bali. The castle has a central courtyard with a stage platform in the middle and a large raised terrace on one side leading to the main entrance to the domestic quarters. The central stage served as the speakers' platform and the audience was seated or comfortably spreading themselves on the marble floor of the terrace looking across to the stage. It was indeed the nearest setting we had for an ARCASIA Forum resembling the ideal environment of "meeting under a tree". The surrounding had a very relaxed air about it and yet the discussion which occured during the Forum was highly exciting.

Forum 5 took place in Bangkok, Thailand on the 21st and 22nd October, 1989. The theme was "Food for Thought in Asian Architecture". The speakers included Dr. Surapol Virulrak, Beng Huat Chua, Fumihiko Maki and Rocco Yim.

Felipe Mendoza, Ronald Poon, Ejaz Ahed, Dr. Ken Yeang, Prof. Maki & Apsorn Kuromarohit at ARCASIA Forum 5.

ARCASIA Forum 5 in session.

Forum 6 took place in the Fragrant Hill Hotel, Beijing on the 27th and

Fragrant Hill Hotel, site for the ARCASIA Forum 6.

28th September, 1991. The theme was "New Thoughts in Mass Housing in the Asian Context". The speakers were Tay Kheng Soon, Prof. Lin Zhiqun, Lee Kwong Yan, Akira Ozawa and Raj Rewal.

Mr. Ronald Poon founded the ARCASIA Forum in 1982 during his chairmanship of ARCASIA and has been acting as its Director and Session Chairman for all six Forum. He has also published records of Forum 1 to 4. In 1991 after Forum 6 Mr. Poon offered to relinquish his directorship and Council decided to nominate a coordinator for each of the future forum.

Raj Rewal & Ronald Poon at the ARCASIA Forum 6, Beijing, 1991.

Forum 7 will take place in Hong Kong in September, 1993 with the theme "The Vanishing Asian City". The speakers will be Mr. B.V. Doshi, Mr. Kiyonori Kikutake, Professor Tunney Lee, Mr. Manuel Vicente and Mr. Zhang Zugang.

Brochure for the ARCASIA Forum 7.

The papers for Forum 1 to 4 have been published by Mr. Ronald Poon who funded the publication for Forum 1 to 3 and papers for Forum 4 was funded by the Target fund. Papers for Forum 5 were not published due to lack of fund. Papers for Forum 6 were published by ASC.

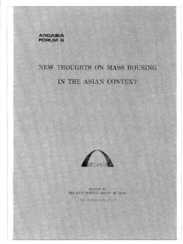
Papers for Forum 6 published by ASC

12) ASIAN CONGRESS OF ARCHITECTS (ACA)

Another key activity of ARCASIA is the Asian Congress of Architects (ACA) which is major regional conference that is held every two years and coinciding with a meeting of the Council.

ACA is more than a conference for the intellectual exchange of current architectural ideas. It is also a memorable convivial occasion that combines cultural activities of the participants. The host institute for the congress is selected from the member institutes. The theme is chosen amongst pressing major issues concerning architecture and environment in the Asian region.

ACA-1 took place in October 1984 at the Philippines International Convention Center, Manila. The theme was "Strategies for Asian Architects". The speakers were Robi Sularto Sastrowardoyo, Koichi Nagashima, Tay Kheng Soon, Eric Lye, Yasmeen Lari, Charles Correa, Leandro Locsin and Felipe Mendoza.

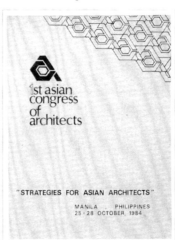

Papers for ACA 1 published by UAP.

Opening Ceremony of ACA 1, Manila, 1984.

Opening banquet at ACA-1

Session Chairman, speakers and panelists of ACA 2. (a page from the ARCASIA Newsletter)

ACA-2 was held in Kuala Lumpur in October 1986. The theme was "Strategies for Professional Practice in Asia". The speakers were A. Eugene Kohn, Takekuni Ikeda, Kington Loo, Yang Soo Suan, A.P. Kanvinde, Froilan L. Hong, Hiroyasu Higuchu, Reiko Hayashi and Sumet Jumsai.

Session Chairman, speakers and panelists of ACA 2. (a page from the ARCASIA Newsletter)

The PAM delegation at the Friendship night of ACA 3 at Seoul.

ACA-3 took place in Seoul, Korea in November 1988. The theme was "The Asian Spirit – Architecture for the Community". The speakers were Han Tai Dong, C.B. Wilson, Kang Hong Bin, John V. Daly, N.J. Habrakan, Rod Hackney and Tsutomu Shigemura.

ACA-4 took place in New Delhi, India with the theme "Architecture, Development and Environment in Asian Context". The speakers were Hans Bjoriess, Romi Khosla, Ken Yeang, B.V. Doshi, Rod Hackney, Mrs. A. Baig, Ms. Barayu Akiya, Soumyendra Ray, David Jackson, Kubbhuskan Jain, A.K. Maitro, K. Bhattacharjee, FuJi Xiang, Wu Liang Yong, Raj Rewal, Zaijham Jaffrey, Miki Desai, Jeet Malkotra, Bir Zhenwei and Zu Anzhi.

The ARCASIA group visiting historical buildings at Agra, India.

ACA-5 took place in Lahore, Pakistan in October 1992 with the theme "Towards a new Direction in Architecture". The speakers were Pervez Vendal, Sang H. Lee, Tay Kheng Soo, John Wells Thorpe, T. Shigermura, Jennifer C. Habib, Sajjad Kanson, Huang Wei, Christine Vadaez, Akhtar Chauhan, Kamil Khan Mumtaz, Philip Cox, Koichi Nagashima, Jia Sheng Bao and Fawah Ali Butt.

Programme for ACA 4.

Advertising card box for ACA 5.

HKIA NEWSLETTER

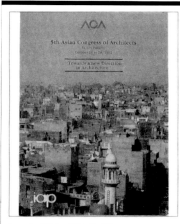

Papers for ACA 5 published by IAP

All ACA papers were published by the host Institute and distributed at the Congresses.

13) ARCASIA STUDENT JAMBOREE

The Student Jamboree is another ARCASIA fixture. Accompanying each Asian Congress of Architects is the gathering of architecture students from member countries of ARCASIA under the auspices of the ARCASIA Committee for Architectural Education (ACAE). the objects of the Student Jamboree are as follows :-

- To facilitate talented Asian students of architecture to travel in the region and experience the architecture of different but related environments from their own;

- To foster among the up-and-coming generation of architects in the region a better understanding of each other's countries and their contemporary conditions and problems;

- To encourage the building of interpersonal contacts between the architectural students in the region;

- To promote an awareness among Asian architectural students of the role and work of ARCASIA and ACAE and to engender their support to and future participation as architects in these institutions; and

- To enliven ARCASIA deliberations by the presence of students.

14) ARCASIA AWARDS

ARCASIA in its endeavous to raise the standard of the built enviornment throughout Asia in general and in its member countries in particular, had instituted the arcasia award for architecture in order to encourage and recognize exemplary work done by architects working in Asia. The awards are given every two years at the ASIAN CONGRESS OF ARCHITECTS organized by ARCASIA.

An independent panel of jury members consisting of leading architects of Asia, nominated by the ARCASIA Council, shall assess and announce the AWARDS.

The aim of the ARCASIA AWARD is to acknowledge exemplary architectural work and in doing so encourage the sustenance of the Asian spirit, the development and improvement of the Asian built environment and enhancement of the awareness of the role of architecture and architects in the socio-economic and cultural life of Asian countries.

With this Award, the intention is also to demonstrate that good architecture is a major component of the positive influence on the human environment, and that physical development in Asia need not be in disharmony with the cultural values, national identity or the natural environment of developing countries in Asia.

The Award will be given in the following categories :

Category A
RESIDENTIAL PROJECTS (single storey, multi-storey, individual units, complexes)

Category B
PUBLIC AMENITY BUILDINGS (office buildings, institutional & religious buildings, recreational facilities, shopping centers, health care facilities, etc)

Category C
INDUSTRIAL BUILDINGS (factories, warehouses, etc)

Category D
CONSERVATION PROJECTS (projects which conserve or restore the architectural heritage of asian countries)

Entries selected by the Jury will be awarded a GOLD Medal as the ARCASIA AWARD FOR ARCHITECTURE in each category for the year 1992 and subsequently every two years.

The ARCASIA Award for architecture ICI Dulux Gold Medal 1992 was made at the 13th Council meeting at Lahore, Pakistan in October 1992 as follows :

Category A (residential buildings)
The Courtyard House, Beijing
by Prof. Liang-yong Wu

Category B (public amenity buildings)
Al-Hamra Art Centre, Lahore
by Nayyar Ali Dada

Category C (industrial buildings)
Kanagawa Science Park, Japan
by Takekuni Ikeda

Category D (conservation projects)
Ahsan Manzil Dhaka, Bangeadesh
by Shan Alam Zahiruadin

15) ARCASIA PUBLICATIONS

1) ARCASIA leaflet (1987-1988) published by PAM.

ARCASIA Forum leaflet and ARCASIA leaflet.

2) The ARCASIA Forum leaflet published by Ronald Poon.

3) ARCASIA Handbook 1986 edition published by PAM.

ARCASIA Handbook 1986

4) ARCASIA Handbook 1992 edition published IAP.

ARCASIA Handbook 1992

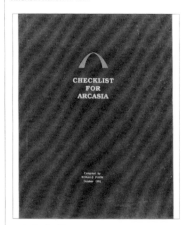

ARCASIA Checklist

5) ARCASIA Checklist by Ronald Poon, published by IAP.

6) ARCASIA Forum 1-4 papers published by Ronald Poon.

7) International Code of Ethics on Consulting Services published by UAP.

ARCASIA Forum papers

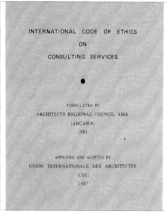

International Code of Ethics on consulting services

8) ARCASIA Forum 6 papers published by ASC.

9) ACAI papers published by UAP.

10) ACA3 papers published by KIRA.

11) ACA5 papers published by IAP.

12) ARCASIA Newsletters published by UAP.

One of the ARCASIA Newsletter published by UAP.

13) ARCASIA magazine published by PAM. (Jimmy Lim, editor-in-chief)

The ARCASIA magazine

16) CONCLUSION

ARCASIA has gone a long way since its first Foundation Council meeting in December 1969 in Hong Kong. It started with six founding member institutes and presently it has fifteen member institutes covering a wide spread over Asia. The organisation proved to be highly influential in the shaping of the architecture and environment in Asia and it remains the most respected regional body of architects in the world.

ARCHITECTS REGIONAL COUNCIL ASIA (ARCASIA)

THE OPENING SPEECH BY RONALD POON
FOR THE THIRD ARCASIA MEETING
AND FORUM ON
INNOVATION IN ARCHITECTIRE COLOMBO – SRI LANKA 1982

APPENDIX 1

**OPENING ADDRESS BY
CHAIRMAN, MR. RONALD POON**

The Hon. Minister of State, Mr. Lala Adithiya, delegates, fellow architects, ladies and gentlemen, it is my great pleasure to welcome you to the 3rd meeting of ARCASIA.

As I was walking down the never-ending airport corridor in Hong Kong at 7 a.m. last Wednesday to board the plane for this meeting, I started to think aloud about why architects like us leave our families and work behind to attend this sort of meetings half way round the world and maintain honourary positions in international professional institutions such as ours. Surely to many people and our wives in particular, this sort of pursuit in the abstract world of international cross fertilisation, regional professionalism etc. is bordering on insanity and yet many of us are addicted to be involoved in such pursuits. Though I must admit it is rather an acquired taste, very much the same as being addicted to the durian fruit, I suppose. Incidentally, I am not particularly fond of the fruit, but I was told that once you liked it, you cannot do without it and when you eat it, you are either envied by those who like it or not appreciated by those who don't. Either way you lose. Nevertheless, we keep on hearing the same old names being mentioned in the ARCASIA circle, such as Arunwongse, Loo, Albakri, Lim, Sim, Hong, Selvaratnam, Prescott, Adithiya, Bhalla, Prawirohardjo, Sudin, Lai, Szeto, Mendoza, Tiotuyco etc. The fact that their names keep on re-appearing can only confirm that the philosophy and work of ARCASIA is worthy of their continuous pursuit and association, not to mention that without fail all these men (and I regret not yet a woman though Mrs. Lari of Pakistan nearly made it to our Hong Kong meeting last year), all these men are also highly successful in their profession. This fact alone will put to shame the constant challenges and ridicule that we receive from amongst our local institute membership that involvement in ARCASIA or all other international bodies is a waste of time and effort. There is a recent exception, of course, and that is the Commonwealth Association of Architects where I wasted 4 years of involvement without a single sense of achievement. No regrets though as others have wasted more and will waste more time and effort.

All of us can survive, of course, on just rice and water. In our part of the world, we call this combination, plain congee, but sooner or later you would need a little spice, a drop of French maggie or worcester sauce, or even a piece or two of beef and so on. As modern design philosophy continues to be evolved and as modern technology advances in ever increasing scale and complexity and as foreign consultants keep on coming, we need the spices and meat of our profession i.e. the international cross fertilisation to keep us properly fed to face the challenges of our profession.

We have in the past dwelled in management refinement of ARCASIA, over ambitious projects, practice problems etc. and for this meeting in Colombo, we hope we can efficiently deal with our business sessions and settle down to start the first ARCASIA Forum, a happening which can have tremor-like effects on our profession in the region. I called it the Asian catapult of new architectural philosophies.

On behalf of ARCASIA I would like to take this opportunity to thank the Hon. Mr. Ananda-tissa de Alwis for taking time off in his busy schedule especially prior to the election to come and officiate at our opening ceremony. I would like to thank the SLIA, Lala Adithiya and his Council and his energetic organising Committee for making our meeting possible in such magnificent settings. Their efficiency in putting the organisation of this meeting on the road in such a short time under world recession is to be highly commended.

It now gives me great pleasure in declaring the 3rd ARCASIA meeting open and wish all ARCASIANs a successful meeting.

Thank you.

ARCHITECTS REGIONAL COUNCIL ASIA (ARCASIA)

SNAPSHOTS OF ARCASIA

2nd Council Meeting, Hong Kong, 1981

Datuk Lim Chong Keat, Ronald Poon (Chairman), Jon Prescott.

Dr. Sumet Jumsai, Tao Ho and Visva Selvaratnam.

Ruslan Khalid confronted by the Pakistan delegation.

Lala Adithiya in happy mood.

Felipe Mendoza, Leandro Loscin, Victor Tiotuyco, Miss Esckerbana.

Visua Selvaratnam singing with Lala Adithiya and Jon Prescott clapping.

The ARCASIA Forum 1, 1982.

Ashad Abdulla, Ejaz Ahed and Ronald Poon.

ACA 1

The architects from Beijing and Taipei.

Ronald Poon in the midst of architects from Beijing, Taipei and Hong Kong.

3rd Council Meeting, Colombo, 1982.

The evening reception at Colombo, 1982.

Jon Prescott, Dr. Samarasekera, Visva Selvaratnum, Mrs. Adithiya, Lala Adithiya & Surah Wickremasinghe.

Mrs. Inday Manosa, Felipe Mendoza, Ronald Poon & Leandro Loscin at the opening ceremony of ACA 1, Manila, 1984.

Datuk Lim Chong Keat giving a talk.

The opening ceremony at 3rd Council Meeting David Lung, Jon Prescott, Ronald Poon, Minister of State, Lala Adithiya.

Ronald Poon, John Lynn, Kao Erh Pan & Lloyd Lee Teh Lo.

*Front L to R : Ken Yeang, Ronald Poon, Dan Wongprasat, Sumet Jumsai.
Rear L to R : Bunchong Sontitim, William Lim.*

Front L to R : Leandro Loscin, Ms. Esckerbana, Darmawan Prawirohardjo. Rear L to R : Lala Adithiya, Felipe Mendoza, Ronald Poon, A.K. Rafique Uddin Ahmed, Tseng Chien & Andrew Lee.

Lala Adithiya, the Hon. Auandatisia de Alwis, Ronald Poon.

Ruslan Khalid & Kamil Mumtaz.

Opening luncheon at the Philippine International Convention Center.

5th Council Meeting at Manila, 1984

Mrs. Yasmeen Lari, Victor Tiotuyco, Felipe Mendoza, Manuel Manosa, Lala Adithiya.

Ken Yeang, Lalith de Silva, Barry Will, Ronald Poon.

Surath Wickremasinghe, Syed Akeel Bilgrami, Bungchong Sonititim.

Evening gathering at 6th Council Meeting, Lonavia, India, 1985.

Distributing momentos at the 7th Council Meeting, Kuala Lumpur, 1986.
Ronald Poon, Rebecca Tobia, Felipe Mendoza, Datuk Kington Loo.

The evening gathering at ACA 2, Kuala Lumpur

From the ARCASIA Newsletter published by UAP.

From the ARCASIA Newsletter published by UAP.

From the ARCASIA Newsletter published by UAP.

From the ARCASIA Newsletter published by UAP.

David Teh in the centre of the crowd at ACA 2 friendship night.

Ronald Poon, Dr. Kenneth Yeang and Adhi Moersid welcoming the Chief Guest at Forum 4 opening ceremony, Bali, 1987.

Dr. Ken Yeang and Adhi Moersid chatting to Ronald Poon at the evening reception at Bali, 1987.

9th Council Meeting & ACA 3 at Seoul, 1988

Ronald Poon (Past Chairman & Adviser), Felipe Mendoza (Past Chairman), Dr. Ken Yeang (Chairman), Datuk Kington Loo (Past Chairman & Adviser).

Speakers' briefing session at ACA-3, Seoul.

Exhibition at ACA-3.

The delegates at ACA-3.

Some of the delegates and ARCASIA officials at ACA 3 in Seoul.

The delegates from Taipei at ACA-3.

Jim Kinoshita and one of the speaker N.J. Habrakan at ACA-3.

The Hong Kong delegation at friendship night, ACA-3.

The opening act of the friendship night, ACA-3.

10th Council Meeting at Bangkok, 1989.

Student's cartoons at 10th Council Meeting, Bangkok, 1989.

Student's cartoons at 10th Council Meeting, Bangkok, 1989.

Dr. Kenneth Yeang

The Friendship night on a boat at 10th Council Meeting.

Prof. & Mrs. Fumihiko Maki.

The ACA 5 at Agra, 1990.

The three past Chairmen Ronald Poon, Datuk Kington Loo & Dr. Kenneth Yeang.

P. Kasi, Bobby Manosa, Jimmy Lim and friends.

Arcasia delegates touring historical site at Agra, 1990.

Ipe Mendoza making a point.

Song Kee Duk, Beng Huat Chua & Dr. Surapol Virulrak singing.

The cocktail reception at ACA 5.

Chia Kok Leong

Dr. Surapol Virulrak, Jeong Keon Lee & Chitok Kim singing.

Prof. Wu Liang Yong sketching.

Sharing a joke at Agra.

Four Past Chairmen

Datuk Kington Loo

Edward Shen with the architects from Beijing.

Prof. Wu explaining a point.

Ronald Poon

P. Kasi, Ejaz Ahed, Felipe Mendoza and the Pakistan delegation.

Ronald Poon having a smoke & whiskey at Chairman Rusi Khambatta's room.

Bottoms up.

Felipe Mendoza

P. Kasi, Ronald Poon, Majorie Musni, Chris Haffner and Bobby Manosa.

The ladies at Arcasia, Agra, 1990.

Rabiul Hussain, Felipe Mendoza, Xu Ronglie, Lee Jeong Keun.

The garden reception at ACA 5.

Rod Hackney, Rusi Khambatta, Ms. Angeline T. Chua Chiaco & Ms. Rebecca Tobia.

Richeto Alcordo, Ms. Angeline T. Chua Chiaco, Ronald Poon, Felipe Mendoza and Rod Hackney.

Delegates from China, Taipei, Hong Kong and Sri Lanka at the 12th Council Meeting

Kim Chi Tok with the ladies

Jon Prescott looking at Ms. Jose C.S. Marreiros

Arcasia Forum 6 at Beijing, 1991.

Koichi Nagashima

Dr. Lee Jeong Keun

Mohammad Farooq, Syed Akeel Bilgrami & Ejaz Ahed having a meal

The Speakers: Tay Kheng Soon, Ken Yeang, B.V. Doshi, Lee Kwong Yan.

Richeto Alcordo

Meng Ta Cheong

Mr. & Mrs. Akeel Bilgrami & Mr. & Mrs. Rusi Khambatta

Tay Kheng Soon speaking at Forum 6.

The participants of Forum 6.

Arcasia delegates with the performers at Friendship night at ACA 5.

The ARCASIA delegates with Mr. G.C. Zhou, Deputy minister of Construction

The friendship night at Forum 6.

Mr. & Mrs. Adhi Moersid

Koichi Nagashima singing.

The Hong Kong presidential delegation
Front L to R :
Chris Haffner (past President HKIA)
Ronald Poon (past President)
Rear L to R :
Stephen Poon (past President)
Jon Prescott (past President)

The Korean delegation performing.

The Hong Kong delegation at the historical site, Lahore.

The Hong Kong delegation performing.

The 13th Council Meeting & ACA-6 at Lahore, 1992.

The student work exhibitions.

Zaigam S. Jaffery

Datuk Hisham Albakri and friend.

Robert Lam, Ronald Poon, Rita Cheung and Li Sau Lung at a historical site, Lahore.

Ronald Poon, Raj Rewal, Koichi Nagashima

Tay Kheng Soon making a point to Ronald Poon.

The exhibition of architects' work from member Institutes.

Another Maria from Macau.

Angel Lazaro Jr.

The singing session.

Arcasia delegates being transported by horse carts to the venue of the cultural evening at ACA-6, Lahore.

The evening entertainment at IAP member's home.

Architects Regional Council Asia

30 HKIA NEWSLETTER

附錄 (8)
Appendix (8)

百花齐放的中国建筑学理论

—— 现代中国建筑创作研究小组第四届年会后记 ——

一九八八年十月十八日早上，五十多位来自五湖四海的建筑设计院领导人、高级建筑师及大学建筑系教授，齐集在杭州花港宾馆大礼堂上，展开了一连五天的「现代中国建筑创作研究小组第四届年会（杭州）」。与会者还有中国建筑学会副秘书长张祖刚等国内资深的建筑界知名人士及浙江省内多位单位领导人以及来自各地的新闻出版单位。

本刊编辑部蒙创作研究小组邀请出席这次盛会，亲身体验与会人士的积极参与、融洽相处的过程，共同探讨现代建筑中面对的环境、建筑与文化等突出问题。本刊工作人员着实得益不少。

会议编排十分紧凑，五日内共有三十多位人士作专题发言，然后由台下与会者回应，大家毫无隔阂的提出个人见解，既有意见相近者，亦有持不同观点的看法。除此之外，为了使会议形式更生动，大会安排了分组讨论、作品展览及参观。他们透过不同场景启发对建筑理论的思考。据部分人士反映，他们晚间也争取时间，与同房的会员就当日讨论内容或某些问题交流意见，从而提高中国建筑学的理论与实践水平，并促进了彼此之间的友谊。

经过五天不同形式的讨论，本刊特地访问了杭州市建筑设计院院长兼总建筑师程泰宁及潘祖尧顾问有限公司董事长潘祖尧，两位从不同角度就会议的整体情况发表了个人意见，甚具参考价值。另外，本刊亦摘录了是次会议十多位发言人士的论文内容，以飨读者。

虽然读者在其中不难发现他们之间的理论论据是不尽相同、各有特色的，但仍能在会上公开讨论，这正符合程泰宁院长形容的「平等、民主、坦诚、热情」，中国建筑理论亦是甚具发展潜质的。

会议报导

潘祖尧
潘祖尧顾问有限公司董事长

国内建筑理论具发展潜质

目前国内建筑设计创作处于过渡时期，多模仿古代建筑及国外建筑形式，暂时仍未建立独特的中国建筑特色。但因中国建筑师理论思想发达，有儒、释、道哲学等思想薰陶，甚具发展潜质。

是次会议，与会者都是国内建筑专业的第一线人物，对国内建筑设计发展甚具影响力，今次能聚首一堂，公开交流讨论，拓阔彼此视野，能藉此探寻一条长远及具影响力的建筑理论。

类似这样公开讨论的会议形式，在国内及国外均甚少举行，多是事后以文章批评，不及即时现场作出反应般具启发性。许多香港建筑师也甚少出席国际性建筑创作会议，但随着中、港建筑师合作日频，香港建筑师应多参与此类会议，增加对国内建筑业的认识，对工作有裨益。

对于国内建筑业以追上国际水平为当前工作目标，我个人认为是不合国情的。建筑师们应考虑社会实际情况，设计适合人民生活和工作需求的建筑，香港建筑师在施工技术和材料运用方面较有认识，可以与国内建筑师互相补足。

纵观整个会议，若能增加对与会者作品的公开评论，可收互相观摩、吸收和融合的效果。此外，创作小组计划组织「华人建筑师学会」，吸纳海外会员，共同推动世界建筑创作设计。

附錄 (9)
Appendix (9)

科協明年辦連串活動 促進中港台科技交流

明報 11/12

【本報專訊】香港中國通訊社十二月十二日電：香港科技協進會新任會長、建築師潘祖堯日前出任後，使香港科技和工商業界獲得更實質性的收穫，並藉此機會擴大中港科技交流合作的領域。

潘祖堯介紹說，上述活動包括將於明年初組織一個由香港科技和工商界人士組成的考察團，前往台灣和南韓實地考察兩地的科技和工業的發展，希望能夠借鑒兩地的經驗，並尋求合作的機會。

明年五月，舉辦一系列科技活動，為提高香港的科技水平和促進香港與中國大陸、台灣及國際間科技交流合作出新的努力。

明年九月在港舉辦一個題為「亞洲四小龍的新科技發展」國際研討會，與會的除來自「四小龍」的科技、工商界人士之外，還計劃廣邀中國大陸及其他地區的學人參加。

潘祖堯透露，「亞洲四小龍的新科技發展」國際研討會結束之後，將組織與會的學者訪問中國北京、上海和西安等城市，如果屆時台灣學者也能夠成行的話，將會成為海峽兩岸學者的一次直接交往。

該會計劃在新的一年裏，舉辦一系列中港科技交流合作的領域活動，表示，為提高香港的科技水平和促進中國大陸、台灣及國際間科技交流合作出新的努力。同時亦希望繼今年多個中國國民辦科技實業家協會聯合舉辦「科技交易會」，希望繼今年多個中國中聯合舉辦科技實業家協會國民辦「科技交易會」。

■香港科技協進會會長潘祖堯認為政府應成立科技發展局。

科技協進會籲港府應設立科技發展局

明報 15/7/88

【本報專訊】香港科技協進會會長潘祖堯昨日指出：港府應考慮成立香港科技發展局，積極推動香港的新科技發展。

他昨日在一個午餐會上表示：港府最近成立了「科技委員會」，是對解決香港科技發展中的種種問題的第一步。但他認為委員會的權力有限，只限於諮詢性質。眾所週知，香港科技發展遠不如亞洲其他三小龍，香港政府在推廣其他事業有很積極的機構，如貿易發展局、建造業訓練局、生產力促進局等方面就沒有同類推動機構，他認為政府應該成立科技發展局，贊助科技界推動香港的新科技。

他指出：韓國、新加坡、台灣等地都有政府在科技發展上支持。他舉例說，台灣政府給予工業不足以上香港在與台灣、新加坡及韓國競爭。

他說：香港科技協進會是由各界專業代表成立於一九八五年，主要協調科技界各行專業之間的業務關係，使本港科技界能和諧合作。該會將於今年十一月舉辦一個國際性會議，主題是亞洲四小龍之工業發展，該會希望藉此會議加強與亞洲區內各國家與地區能會的密切聯繫與交流。

府提供金錢及便宜地租作支持，以吸引外商投資。

他表示，香港雖然只是一個地區而非國家，但事實上香港是非發展在科技發展上支持，但事實上香港是非發展工業不足以上香港在與台灣、新加坡及韓國競爭。

科技協進會明年大計 國際交流活動連串 與內地合辦交易會及國際研討

文匯 11/12/87

【香港中國通訊社十日電】香港科技協進會新任會長、著名建築師潘祖堯接受本社記者訪問時表示，該會計劃在新的一年裏，舉辦一系列科技活動，為提高香港的科技水平和促進香港與中國大陸、台灣及國際間科技交流合作出新的努力。

潘祖堯介紹說，上述活動包括將於明年初組織一個由香港科技和工商界人士組成的考察團，前往台灣和南朝鮮實地考察兩地某些成功的經驗，並尋求合作的機會。明年五月，計劃與中國國民辦科技實業家協會聯合舉辦「科技交易會」，希望繼今年多個中國科技成果展在港展出後，使香港科技、工商界人士獲得更多實質性的收穫，並藉此機會擴大中港科技交流合作的領域。明年九月在港舉辦一個題為「亞洲四小龍的新科技發展」國際研討會，與會的除來自「四小龍」的科技、工商界人士之外，還計劃廣邀中國大陸及其他地區的學者參加。

香港科技協進會 計劃舉辦一系列活動

專訪

趙遠光

香港科技協進會新任會長、著名建築師潘祖堯接受本社記者訪問時表示，該會計劃在新的一年裏，舉辦一系列科技活動，為提高香港的科技水平和促進香港與中國大陸、台灣及國際間科技交流合作作出新的努力。

潘祖堯介紹說，上述活動包括將於明年初組織一個由香港科技和工商界人士組成的考察團，前往台灣和南韓實地考察兩地的科技和工業的發展，希望能夠借鑒兩地某些成功的經驗，並尋求合作的機會。明年五月，計劃與中國民辦科技實業家協會聯合舉辦「科技交易會」，希望繼今年多個中國科技展展出後，使香港科技、工業界獲得更多實質性的收穫，並藉此機會擴大中港科技合作。

明年九月在港舉辦一個為「亞洲四小龍」的科技發展」國際研討會，與會的除來自「四小龍」的科技、工商界人士之外，還計劃廣邀中國大陸及其他地區的學者參加。

潘祖堯透露，「亞洲四小龍的新科技發展」國際研討會結束之後，將會組織與會的學者訪問中國北京、上海和西安等城市，與中國科技、工商界探討交流合作的途徑。他表示，如果屆時台灣學者也能夠成行的話，將會成為海峽兩岸學者的一次直接交往。

由一群資深的科技、工商界和專業人士組成的香港科技協進會，成立兩年多來會舉辦和參與多項科技活動。其中至今仍為人們所津津樂道的是該會與中國科研機構聯合舉辦的「核技術展覽」。由於當時不少市民擔心大亞灣核電站的安全問題，該項展覽吸引了逾十萬市民參觀。展覽資料豐富，內容深入淺出，對市民起到了一定的釋疑解惑作用。

潘祖堯表示，雖然香港科技界各行業早有專業學會的存在，但是綜合性「香港科技協進會」成立之後，使香港科技界線狀的個別功能進一步成為網狀的、全面性的整個功能。他介紹說，該會主要工作包括整體和全面地建立科技聯絡網，搜集和提供科技訊息和資料；致力於提高香港的科技水平，促進香港與中國大陸、台灣及國際間的科技交流合作，擬訂香港長遠的科技發展方案，供港府有關部門參考。

（香港中國通訊社十二月十日電）

新聞通訊

香港中國通訊社編印

第173期
1987年12月11日

香港軒尼詩道342號十樓
電話：5-728069

Urgent demand for steps to promote industrial research

15/7/87 SCMP

HONGKONG'S prosperity could be threatened if the Government failed to promote research in the industrial sector.

The president of the Hongkong Association for the Advancement of Science and Technology, Mr Ronald Poon Cho-yiu, said the establishment of a science and technology council would be the first much needed step to promote research.

Speaking at a Rotary Club meeting yesterday, Mr Poon said Hongkong was the least developed member of the four dragons of Asia (Hongkong, Taiwan, South Korea and Singapore) in terms of science and especially in industrial technology.

The problem was serious as Hongkong thrived on its textiles, electronics, toys and plastic industries, he said.

"What Hongkong's industries were doing before was just copying products already on the market, instead of coming up with new ideas and products," he said.

"In the long run, it is not the road to survival, especially for the electronics industry."

However, only the Government could make it possible for Hongkong to engage in research, he said.

Quoting the successful cases of Taiwan and Singapore, Mr Poon said individual companies received financial aid such as low interest rates, long term loans, and reduced prices on land from their governments.

"Government subsidies in land are especially important in Hongkong since land prices are high," he said.

Mr Poon said Government support would attract foreign investors who would import technology, and also bring back the manpower that had been lost in the immigration surge.

The Government's recently established Science and Technology Committee was not sufficient, Mr Poon said.

"The committee has very limited financial and decision making powers. It should only be the stepping stone to the formation of the science and technology council, which should be the goal," he said.

為提高香港工業產品競爭力 港宜設科技發展局

潘祖堯指港府發展科技欠積極

【本報訊】香港科技協進會會長潘祖堯認為香港政府對於推動本港科技發展有欠積極，應考慮成立一個科技發展局，以免削弱本港工業產品與亞洲其餘三小龍的競爭力。

潘祖堯昨日在北區扶輪會午餐例會上表示，目前新加坡、台灣和南韓的官方部門都積極鼓勵科學研究工作，且提供低息貸款，所以他們的科技能在日新月異的時代，有全面性發展。而香港在相比之下，科研工作十分缺乏，只有個別廠家有進行一些小型的研究工作，例如正研究紡織程序中繼線步驟的自動化。

雖然香港殷近成立了科技委員會，但潘祖堯覺得這只是一個諮詢機構，沒有資金和實權。因此，他希望政府可以設立一個科技發展局，類似現時的生產力促進局，以能有更大權力和效率發展科技，例如可以建設與新加坡和台灣相若的科學園地，能吸引不少人才回流。他認為假如成功，可能會提高港的競爭力。

文滙報 15/7/1988 (Fri)

香港科技協進會 擬舉辦連串活動

【電】香港中國通訊社：新任會長、著名建築師潘祖堯接受本社記者訪問時表示，該會計劃在新的一年裡，舉辦一系列科技活動，為提高香港的科技水平和促進香港與中國大陸、台灣及國際間科技交流合作作出新的努力。

潘祖堯介紹說，上述活動包括將於明年在港展出後，使香港科技成果展覽會，供港府有關部門和工商界人士組成的考察團，前往台灣和南朝鮮實地考察兩地某些成功的經驗，希望能夠借鑒兩地的科技和工業的發展，搜集和提供科技信息和資料，致力於提高香港的科技水平，明年五月，計劃與中國國民辦合辦科技實業家協會聯合舉辦「科技交易會」，希望繼今年多個中國科技交流會成立之後，「香港科技協進會」將成為全面性的整個功能。

但是綜合性「香港科技協進會」成立之後，個別功能進一步成為專業學會的存在，故在推動科技發展方面，亦應考慮成立類似的組織。

香港科技協進會 16/2/87報

港應積極發展科技 加強產品競爭能力 專業者促設局負責

【本報專訊】一位科技界的專業人士昨日批評港府未有積極推動本地的科技發展，導致香港產品的競爭力日漸削弱，他建議當局考慮成立一個科技發展局，專責推動本港的科技發展。

香港科技協進會會長潘祖堯昨日出席扶輪社午餐會時指出，本港科技發展在亞洲四小龍中排行最尾，但港府迄今顯然仍未有意積極改善這些差距。

他強調，雖然政府較早時已成立科技委員會協助當局制訂長遠發展科技的政策，但由於該委員會的權力不大，要全面地推廣發展科技，將會涉及很多問題，加上港方向來被評為若當局再不正視的話，將會進一步擴大與海峽兩岸學者的一次直接交往。

潘祖堯表示，雖然香港科技界各行業早期有專業學會的存在，

科技、工業界獲得更多實質性的收穫，並藉此機會擴大中港科技合作的新領域，與會的除未明年九月在港舉辦一個題為「亞洲四小龍」的國際科技研討會，與會的除自「四小龍」的科技、工商界人士之外，還計劃廣邀中國大陸的學者與會，屆時台灣學者也能夠成行的話，將會組織與會成員，如北京、上海和西安等城市的學者訪問中國北京、上海和西安等城市，與中國科技、工商界探討交流合作的途徑。他表示，如果屆時台灣學者也能夠成行的話，將會成為海峽兩岸學者的一次直接交往。

星島日報 15/7/88 (Fri)

附錄 (10)
Appendix (10)

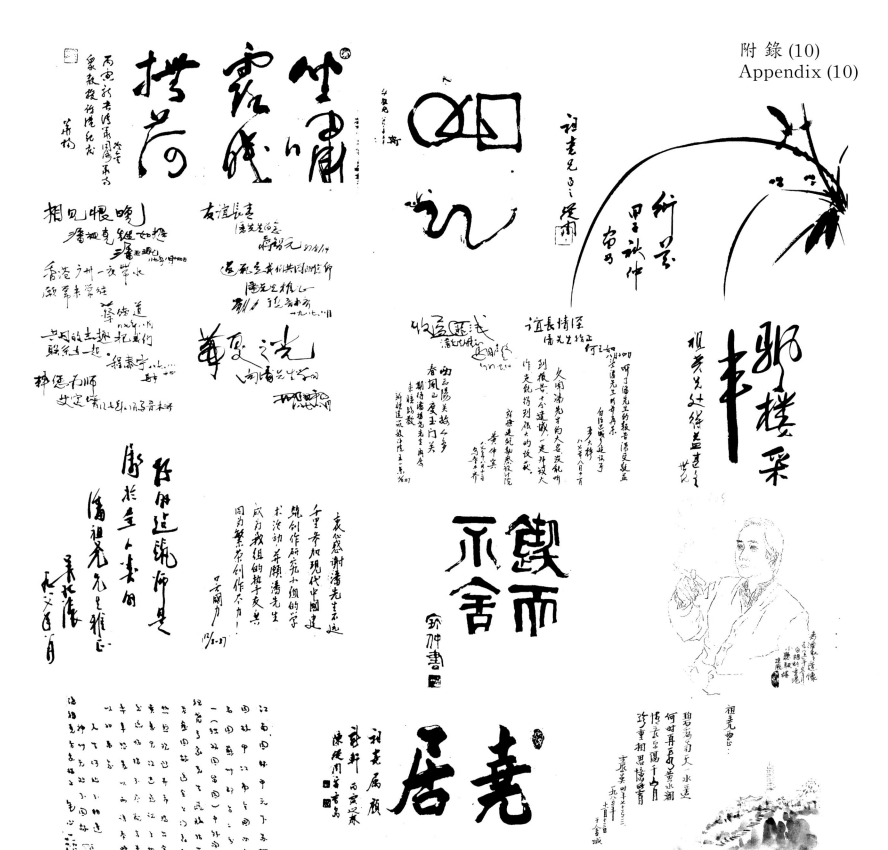

西山神駿通佳賓
同行聚首喜春城
杜鵑花開齊歡笑
駕道彩虹到昆明

敬贈
潘祖堯先生　　毛朝屏 八五三

珠江亂入戰池塘
梅新噴火傲骨霜
廣美理石名竹又
皮板搞摹羊毫球
祖堯仁兄正　素古老師戊辰秋

遠湘兄先生再博　奉傳

魚上龍門
源溯九隆

蘭亭韻事傳
千秋
祖堯先生留念　戴念慈 一九八九

人生不相見 動如参与商
今夕是何夕 共此灯烛光
珍惜可知对此日難遇突
祖堯先生過宁共此多
　楊廷寶手于南京

雨中嵐山所作
綠扁磐石銘
乏澤及天下
遠望一水情

山不在高有仙則名
水不在深有龍則靈
為祖堯先生李建華書
戴念慈 一九八九年十二月吉日

孟夏
宗孝筆 一九八九年四月廿六日

冬去祖生春又生
勞燕繞書天更潤
洗滿玉壺冰
祖堯大才留念
　三月省于上海 一九八七

祖堯畫竹扇面
參竹畫祖堯同畫

祖堯兄指正
設門有待來賓
留徑可通涵室
竹修林戌柳暗
花明玉敵何拘
且效溫公之獨樂
四時不謝
祖明書於
香港

祖堯兄指正
香城添香
建築傳基
友誼傳情
九九待來
祖堯先生正　美代達

附錄(11)
Appendix(11)

SUNDAY MORNING POST, JANUARY 19, 1992

When debating points deteriorate

WHAT do you give the man who has everything for Christmas? Well, copies of all 34 speeches from last year's Urban Council debates if your name is Roy Spencer and the recipient is to be Urban Councillor representative Ronald Poon Cho-yiu.

In a delightfully worded speech to the Urban Council's annual conventional debate on Wednesday, Mr Poon described how "after a sumptuous Christmas lunch in the terrace of my home with a vintage port in my glass and my family around me, I decided to read all the 34 speeches from last year's annual debate — a Christmas gift from Roy Spencer".

Sounds like a perfect way to wind up the festive period but, in a speech that proves the art of public speaking is not dead, he goes on to point out that the speeches contained a six-page report on a "small but important part of the council's work — public toilets".

Mr Poon then takes the council to task for the poor quality of debates during the monthly open meetings, and suggests they are responsible for

Ronald Poon: strange seasonal reading.

"deteriorating audience rating in the gallery".

"For those of us who have not yet acquired the art of appearing attentive while taking forty winks, it is sheer torture to sit through two hours of meeting — particularly after a few glasses of wine during the monthly council lunch."

Nice work if you can get it . . .

附錄 (12)
Appendix (12)

大公報　一九九七年十二月五日

▶ 在一個人大候選人座談會上，曾憲梓向選舉會議成員講述自己的參政方針

港區全國政協委員中的選舉委員會委員昨日舉辦候選人論壇。

1997年11月26日 星期三　明報

「心水選擇」 政協施子清拿出了他的「心水選擇」，惺惺地給了鄰座的幾名政協，說讓他們作參考，而這幾名政協都把施氏的「心水」一一記錄下來。（黃俊帽攝）

「選賢不避親」Vs「忽然愛國」 投票心得

一個座談會，真正愛國愛港的人士並不太多。

昨日在「國」者，而主持人劉宇新亦說，公開講述自己的投票心得。港進聯監委會主席楊西表示，他首先辦會考慮姜恩柱，其次是現任人大，在投票時亦會考慮到各界別的代表性，着重均衡代表。本身是福建人的楊氏坦言「選賢不避親」，親朋戚友，籍貫亦是考慮之列，但會放在最後一個博士學位，所以同是來自福建籍的王紹爾則強調，不要投票支持哪些「忽然愛」

會上，出席的「選舉會議」成員公開講述自己的投票心得，港進聯政協馬志民認為，要挑選哪些有原則、真正而又敢於表達意見的人士。不希望選出的代表只懂看風頭火勢，彷如「牆頭草」。此外他認為，另一名政協潘祖堯則認為，形容全國人大的工作量迫如考取一個博士學位，所以，若功續越多，就會分身乏術。

港區政協委員辦座談會
選會界別候選人談政綱

【本報訊】港區全國政協委員中的首屆立法會選舉委員會的委員，昨晚舉辦選委座談會，邀請選委會界別候選人介紹競選政綱。發起人之一的李祖澤表示，政協中的各位選委會委員將按照有利於實行「一國兩制」、「港人治港」和有利於保持香港繁榮穩定的標準，自由選擇人選，並不存在協商名單之事。

李祖澤在回應一些輿論指選舉委員會選舉是「小圈子選舉」時表示，民主要循序漸進地發展，基本法規定第一屆立法會選委會有十個議席，第二屆有六個議席，第三屆以後由港人自行決定，這是走向立法會全部直選的必不可少的步驟，是特區政府可做到的最民主的選舉方式。而八百名選委是有公信力、有代表性的，絕不是什麼「小圈子選舉」。他說，選委中的港區全國政協委員也是香港永久性居民，舉辦座談會是履行基本法賦予選委的權利，這是天經地義的。

這次的座談會由十二位身兼選委的全國政協委員發起，他們是：李祖澤、余國春、阮北川、施子清、施展熊、施祥鵬、高敬德、張兩葡、梁欽榮、陳瑞球、劉浩清、潘祖光、應邀出席的選委逾五十人，除政協委員之外，還有紡織製衣界和出版界人士。由於這次二十三位候選人到會發表了政綱。

會選舉日前最後一次論壇，每位候選人又獲得較其他論壇時限為多的五分鐘時間，因此大家都盡力利用這個機會「推銷」自己。不少候選人都提到，在一個月的競選過程裡，他們通過與各界選委的接觸聽到了很多意見和建議，無論日後能否當選，這都是一次有益的經驗。

245 文匯

附錄 (13)
Appendix (13)

SOUTH CHINA MORNING POST
DECEMBER 2, 1991

香港房屋协会主席潘祖尧说，内地应加强具体住房对策

中新社北京三月二十五日电 （记者 刘兆义）香港地区政协委员、香港房屋协会主席潘祖尧在接受记者采访时表示，内地各有关部门应制订具体可行的住房政策，其主旨不该是"炒"而是"住"。

潘祖尧以其自身在香港进行房屋计划和经营的经验，在谈及内地的一些住房方面问题时坦言，内地首先需要加强具体住房政策、法规的制订，以制止某些不正的"炒风"。他认为，每个城市必须制订好总体的长远的城市规划，要充分考虑到城市的发展和交通的负荷能力诸因素，同时亦需有完整配套的基础设施。

他表示，应该鼓励个人购房，但制订政策时要符合实际，需顾及到各阶层人士的实际收入水平。他认为，鉴于目前内地的实际情况，不宜建设过多的高档次住房，应以"居者有其屋"的方针，建设一些普通的住宅，使大家均有房可住。

潘祖尧还说，对于一些先富起来的人，他们可能会有更高层次的住房需求，但毕竟是少数。对大多数人来说，首要的是"住"房而不是"炒"房，有关部门在制订政策时亦需顾及此点。

潘祖尧还建言，有些地方，一些人先富起来，但不文明的东西也随之而来。所以，也要注意精神文明的建设，因为它是一个民族素质的重要标志之一。

Ronald Poon

THE executive committee of Hongkong Housing Society has elected Mr Ronald C. Y. Poon, an appointed Urban Councillor, its chairman.

A chartered architect and a past president of Hongkong Institute of Architects, Mr Poon has served on the Housing Society Executive Committee since 1982.

三度蟬聯執委會主席
潘祖堯談房協工作
房受忽視階層建屋
未來八年耗資三百億建二萬七千單位

【本報訊】剛當選連任的香港房屋協會主席潘祖堯昨日表示，房協計劃在未來八年耗資三百億元，興建二萬七千個住宅單位，並將極為受社會忽視的「特別階層」，例如老年人、軍身人士等提供住屋。

他表示，房協為今後將積極為受社會忽視的若干「特別階層」提供住屋。例如房協已在六個屋邨興建七百四十五個老人單位，為單身人士提供樓房之所。此外，房協亦開創年「單身人士住屋計劃」，為單身人士提供樓房之所。房協方便會在明年中設立一房屋政策組，他表示歡迎，認為這是解決香港日益複雜的住屋問題的明智之舉。

潘祖堯在房協週年大會後表示，房協於93/9選為執行委員會主席，前義務司庫陳文裘連任為副主席，劉達全則獲選為義務司庫。

四財政年度內大力動用房屋發展項目資本達港幣三十八億元，在未來八年，將斥資港幣三百億元，增建二萬七千個住宅單位，但現在這方面的出租房屋，但房協以往主要為普羅大衆興建出租房屋的角色主要由房屋機構所肩負。

房協擬在港及內地
興建高檔長者居所

【本報專訊】香港房屋協會準備在本港及內地興建高檔長者住屋，為中等收入階層的康健老人提供可負擔的居所和全面舍監服務，讓老長者可安享晚年。

連續第三年當選房協執行委員會主席的潘祖堯接受本報訪問時說，房協目前正在尋找市區土地興建類長者住屋，一俟資料研究結果可望在六個月內完成。

潘祖堯接納新華社香港分社提出有關構思，初步接受有關委託，並向房協提供擬於深圳等地興建的可行性初步研究結果。不過他強調，有關機構思是否落實有待房協執行委員會通過。

他說，房協現時已有為長者提供優惠租金的住屋服務。但對象是低收入老人。鑑於高齡人士日漸增多，當中不乏經濟能力較佳的中等階層，該類人士要求的是更好住屋質素、樓宇設計和管理服務，故房協亦有意向有關機構擴闊該階層老人。

他表示，在內地提供長者居所的構思，亦是針對愈來愈多本港老人返回內地安渡晚年的趨勢而設。他認為，除基於文化因素外，內地生活指數較低可令老人生活得較舒適亦是一個主要原因。

另外，潘祖堯表示，在未來八年，房協將斥資三百億港元，興建二萬七千個住宅單位，而為應付有關建屋計劃，該會於明年中將朝向外舉債，確保有關建屋計劃如期開展。

公僕建屋合作社物業
房協斟介擬携手重建

香港房屋協會主席潘祖堯指出土地資源、除打算向私人發展商購入土地外，很多時需向私人發展商購入土地，除打算向私人發展商購入土地外，慈善機構及私人發展商提加強合作，使房協每年提供的居住單位可維持在三千個或以上的水平。

房協主席潘祖堯對本報記者表示，房協最近的周年大會上，潘祖堯和其他委員曾就今後大計的探討。在討論上，初步確定今後除向政府尋覓土地外，若與慈善機構或私人發展商合作建屋，對房協有好處。

他說，一些慈善機構輒下有不少物業，以擴大每年的建屋計劃。

尋覓土地資源已成為其中研究與討究。

潘祖堯說，由於港府每年沒有規定撥地給予房協，故房協助該機構推行建屋計劃。

潘祖堯

長久以來欠缺興建屋邨以提供廉宜租值住宅單位予香港市民的資源，加上進度緩慢，他表示房協在興建住宅的餘，政府能每年給予一定數量的土地予房協，經房協發展成土地緊絀問題。

他補充，在該個計劃中，與會委員亦同意檢討一項五年建屋綱領，初步建議目標，希望能於一九四八至一九五一年或其法定年期內，提供三千個或以上住宅單位的建屋量。

較現時每年提供二千個單位為多。他又說，房協正在研究興建「長者住屋」，提供廉價出售單位予六十歲以上老人居住。房協打算與政府合作，以解決老人住屋需求。

房屋協會對社會貢獻極大，希望資源終有一日能向香港政府承認，並覽撥地給房協。

本報記者 陳偉強

明報・1992年12月14日 星期一

人物專訪
房協擬於粵省建高檔老人屋
中方反應積極提供磋商渠道

香港房屋協會繼取得夾心階層住屋計劃的「綠燈」後，正構思在本港及內地興建另類住屋，為中等收入階層的康健老人提供可負擔的居所及護理服務，可安享晚年。房協主席潘祖堯接受本報記者訪問時說，雖然該項構思最終能否落實推行仍未有定數，但不少長者都希望返回內地故鄉因在內地興建高檔老人屋的構思，除非有地皮問題，他們的腹笥已批出於內地享受不受港人歡迎的廉宜物業。

該批出於內地享受不受港人歡迎的廉宜物業，似乎已受到同鄉情意結的熟識，在廣東或福建地，房協亦曾為此物色居所。

潘祖堯指出，房協該項構思是與新華社香港分社會談後獲得鼓勵繼續研究。潘說，該構思大致上規範該類長者住屋之基本要求、樓宇設計管理及服務為有服務，故協方面現時亦有意把住屋服務擴展至該階層老人，迎合社會需求。

他解釋，不少人退休後都希望返回內地故鄉因在內地興建高檔老人屋的構思，除可滿足晚年生活所需外，他們的腹笥已批出於內地享受不受港人歡迎的廉宜物業。

老人有望鄉情意結

回顧房協過去一年的工作，潘祖堯說，房協形容為豐收的一年，經歷四年多協商後推行夾心階層住屋計劃，正式發展其他建屋計劃。

檢討目前架構運作

他表示，由於房協未來的建屋量有所增加，為應付龐大繁重的發展需要，該會已委聘專業管理顧問公司檢討協會現時的架構及研究有否需要一間非牟利法定機構，本身的架構，他說，結果是以自負盈虧形式運作，這種運作方針是否有所改變。

本報記者 陳偉強

潘祖堯說在內地另類長者住屋的構思，初步反應良好。

Sandwich class housing contracts worth $1.72 mb

BY JOSHUA FELLMAN

Standard 3/6/93

CONSTRUCTION contracts worth more than $1.72 billion will be tendered during the first quarter of next year for 5,000 units of sandwich class housing, the chairman of the Housing Society says.

Ronald Poon said the design and approval process for the project was under way, and should be completed by the end of this year.

The Society had already identified six sites, four of which totalling 3.85 hectares had been finalised, with enough space for 3,600 units of between 50 and 80 square metres, according to Poon.

Exact locations of confirmed sites have not been officially finalised, but Poon said that they would be located in or near Tseung Kwan O, Tuen Mun, Ma On Shan, and one possible site near Sha Tin.

RONALD POON: End of year completion date.

"The government is looking for enough space to accommodate the further 1,400 units," Poon said.

The 5,000 units were in addition to any units purchased with the help of government subsidies before the purpose-built flats were ready in 1996.

Construction costs for the 5,000 units, average size of which was 65 square metres, had been set at $5,000 per square metre.

Contractors would be chosen from the government approved list and Poon did not expect any difficulty to be created by the concurrent demands of the new airport project.

"It probably won't effect our costs, as our contractors are not likely to be involved in the airport project," Poon said.

He added that the formula by which the sale price of the flats would be set had not been settled.

The Housing Society usually sold at 60 per cent of market value, but was considering charging more because the buyers had more disposable income, and the society had not yet been given land prices.

"But we will stay with our principle of setting prices at an affordable level.

"Any surplus from the scheme will be ploughed back into other projects."

When near completion, the two and three bedroom homes, would be allocated by ballot.

Sites had to be finalised, design work completed and construction under way as soon as possible, if the Society was to have any hope of completion near its deadline, Poon said.

The housing society had originally promised the government to complete the 5,000 sandwich class units by March 1996. However, there was a delay in receiving approval of the land grant by the Sino-British Land Commission because of political difficulties.

"We will finish the units sometime in 1996...but we have already had a five month delay, as we expected the land grant to be approved in January, rather than May."

房協決撥出青衣第三區地盤
發展95年首批夾心家庭住屋

記者李榮強/專訪

香港房屋協會主席潘祖堯指出,該會已決定撥出屬下青衣第三區地盤,改作發展首批在九五年後完成的夾心家庭住屋用途,而這項動工中的項目,料為最先落成的未來夾心家庭住屋。

房協在未來幾年,因承擔龐大的夾心家庭住宅計畫,九六年後每年比均生產住屋單位將達一萬個(數字包括房協本身發展的住屋)。潘氏認為,這不會構成財政壓力,或分散協會資源運用。過往,房協每年生產單位平均則只維持二千至三千個單位的水平。

另一方面,房協亦推行市區重建計畫,自行收購物業樓,向政府申請改用途並隨市價出售,最近便推出上環的尚雅苑,雖然銷情平平,所致。

他又謂,房協在明年將首次向銀行貸款,但比例會極低,這次動用貸款是基於該會有多個大型屋邨落成,加上各項開銷,形成消費赤字所致。

潘氏透露,房協亦展開與公務員建屋合作社,磋商合作重建計劃,房協亦積極找開發商洽談屬下公務員宿舍的可能性,他說,雖然不少私人發展商有意欲,但房協可提供公務員具重建宿舍等所無的,是房協可提供按揭。

潘氏認為不會構成該會財政太大影響,他說,這些中收購發展佔比例極少,普遍而言,房協推出物業,皆提供合資格低於市價四成的住屋,故管銀行最近收緊貸款政策,但樂於對房協合資格買家提供按揭。

他說,假如未來樓價上升,而引致申請夾心家庭置業貸款反應冷淡的話,該會主要是盡量有效用盡申請名額的辦法,他說,政府撥出二十億元予夾心階層置業貸款。

本身已發展物業作選拆用途。目前房協持有的發展中地盤共十多個,已作出發展的單位約三萬個,以低於市價租金為十四萬居民提供居所。

另一方面,首批由房協負責的夾心家庭住屋貸款計畫正透過遴選程序,潘氏表示對三千四百人申請一千個限額,可謂反應不俗,他稱,現正以每個申請者調查其背景四至六周的進度進行,估計十一月初便完成首批審批工作。

他續說,在四千六百四十一份申請書中,百分之七十三點三來自平均月入二萬三千零一元至二萬三千元的家庭,而百分之八十一則屬於第一優先類別的家庭,即申請人加配偶及一名子

房屋協會擴展物業計劃
發行廿億浮動利率票據

【本報訊】獲多利融資有限公司消息,期為兩年之年息為三個月港元存款加○.○八厘。發行人可於發行日之後的兩年零三個月行使購權利。該等票據透過香港金融管理局債務工具中央結算系統進行結算,不會於任何證券交易所上市。

香港房屋協會於一九四八年成立,並於一九五一年根據香港房屋協會立案法團條例註冊成立。房屋協會的宗旨和目的是為香港市民提供價格相宜、高質素和切合需要的住宅單位,推廣和實行有效率的屋價。

據悉,香港房屋協會現正安排在港元資本市場發行二十億港元浮動利率票據,以滿足其一般資金需要,並於昨日推出,由獲多利亞洲有限公司、法國國家巴黎銀行香港分行、德意志銀行亞洲有限公司及奧地利銀行、法興業銀行香港分行全數包銷;它們並將成為該項發行的牽頭經理。在牽頭經理之下,將邀請其他財資機構出任聯席經理。

據透露,該等票將按面值發行。

第二期夾心階層貸款買樓
房協昨抽籤提供八億款項

【簡報訊】香港房屋協會昨日為夾心階層住屋貸款計劃第二期收到的四千六百四十一份有效申請書進行抽籤。房協預計會在本月中公布,合格證書將在六月底前發出,抽籤結果將於五月十三日張貼於房協辦事處。

房協執行總幹事蘇慶和稱,將按抽籤的次序約見申請者,直至第二期所提供的一千五百個名額額滿為止,第一批合格證明書,預計在六月底前發出,現時共有八間銀行同意參與該計劃。

房協將在住屋貸款計劃第二期抽籤儀式昨日在房協總部舉行,並由房協主席潘祖堯及夾心階層住屋貸款計劃圖案協助委員會主席彭榮按分別抽出十個號碼,順序為零、八、六、五、零、三、九、一、七、二號。房協並將這六、七、八個號碼輸入電腦,按申請人的條件,編排出四千六百四十一個編號。抽籤結果將於五月十三日張貼於房協辦事處。

潘祖堯稱,第二期貸款計劃在五月三日截止,共有四千七百五十六份申請書,其中一千四百零八因重複申請或逾期遞交原因而被取消資格。在申請期間,房協並接獲超過一萬八千個電話查詢,比第一期接獲三千四百三十五份申請書及一萬八千四百六十六個電話查詢有顯著增加。他稱,這與放寬申請條件有關。

女。

「房協,也初步入選申請人已完成審批過程,其中共有七百一十四名合資格申請人,平均貸款額為四十萬,總數約一億元,他們大部分均購置居所,在該批合資格申請人中,百分之二十九在九龍,百分之五十位於新界,百分之二十的住宅區。王麗珍謂,整個夾心階層住屋貸款計劃(物業管理)二十億元,餘下款項將動用於第三期貸款計劃中,第二期則為八億,則希望盡量安排在第三期撥出。

房協高層抽出夾心階層申請貸款號碼

房協主席潘祖堯表示該會正磋商與公務員合作社共同重建屬下物業。

建築設計最忌好看不中用
潘祖堯望創現代中國流派

本報記者 丁永家

秋冬之際，是郊遊的黃金季節。大家到新界郊區，不難見到一些紅牆綠瓦的西班牙式別墅，與古舊樸實的村莊相混而立。儘管有不少人稱讚這些建築美輪美奐，別具特色，但是建築師潘祖堯則對它們有這樣的評價：「好看不好吃。」

「建築設計最重要的是符合人的要求，在建築形式、材料運用，與周圍環境配合這幾方面如果不考慮得徹底，美感便自然浮現出來。」潘祖堯日前在接受記者訪問談到建築設計時便強調：「建築物的美感，其實只是一件副產品。」

潘祖堯自六六年便開始建築設計工作，他的「建築作品」包括翡翠明珠寬銀幕戲院、蕭明中學等。他雖然只有四十多歲，但在建築界已工作了二十多年，算是「老行尊」了。

年輕的建築老行尊

潘祖堯非常強調「建築設計主要是創造適當環境給人類享用」的看法。而這個以人為本的思想，與他的一言一行都很脗合。他怎樣評價日本的傳統建築、「日本建築從中國傳統而來，再加上自己的傳統，簡化了於簡單，好像炒菜一樣，要做得好吃並不是簡單地加點鹽就好了。」他說：「有時覺得日本建築過於簡單，好像炒菜一樣，適當地加點鹽，適當地把它炒熱算數，」

日本建築源於中國

他又說，日本的金閣寺線條已開始複雜，但它的幾何方式和線條還是很簡單，好在其坐落的周圍工地處理及池湖配合得當，像炒菜加點鹽一樣，令人覺得好吃。

潘祖堯批評紅牆綠瓦的西班牙別墅是「好看不好吃」，而金閣寺則簡單得來「好吃」，這種以實用為主的觀點，又怎樣應用於本港的建築上呢？

「能交代地方氣候，反映住戶的生活方式，便是實用而且美觀的建築。」他說：「香港算是熱帶地方，所以需要有遮擋太陽的設計。通風也是需要的。」他又以自己設計的「豐樂閣」為例解釋：「這棟樓座南向北的一層四個單位，全部可見海景及山景，又有季候風吹入房內。」「這棟樓的客廳都是三角形的，住客除風涼，不過住落就覺地方好用。」「走火樓梯沒有牆，走火時可直接吸收新鮮空氣，而且又省地方。」

香港建築受限制

他又說，香港地少，道路縱橫，限制很多，所以，一般樓宇都不理會氣候，靠冷氣機調節溫度，盡量爭取平面空間，而且差不多完全沒有建築裝飾，只求自己看出是港式建築。這些建築，有些內行建築師一眼便看出是磚頭托，但做建築師則為之之外，還要考慮別人的需要。

潘祖堯說起他的入行經過，也是與「以人為本」的實用觀點息息相關。他在英國唸完中學後進入大學修讀美術，到處繪畫不問世事。過了差不多一年的悠閒生活，他毅然轉修建築系。他說：「做藝術家，多求自己人設計房子，要了解別人的生活，因為房子的設計對家庭生活是很有影響的。」他又說：「設計公共建設，例如大會堂，如果建築氣氛營造成功，可以鼓勵交流，這些都給自己和別人很大的滿足感。」

對於中國的建築，潘祖堯自己也有不少心得。他在自己寫的一篇《香港建築師眼中的現代中國建築》中曾提出「……建築師的想法，只集中於外型和技術方面，忽略了生活哲學及生活方式的根底，等於是人失去了靈魂，……如果只顧實事求是，以實用經濟為原則，就會創出很單調的建築環境，如果只是集中用傳統建築中的特徵來交代地方特色，那麼建築只會停頓在後現代派的水平。」

建築忌「食古不化」「食洋不化」

潘祖堯最近被選為香港科技協進會會長，工作外，他致力於發展本港工作高科技化和科普少期望，希望新一代的建築學者寄予不現代中國的流派，及早革除過往的「食古不化」，食洋不化」通病。我們使用潘祖堯的兩句四言警句，作為結語：「食古不化，食洋早化」。

潘祖堯與內地建築界人士常有交流。

三千萬建香港廳 中港美攜手設計

[本報北京專電] 耗資三千萬港元，位於北京人民大會堂內的香港廳設計圖則已經定案，由中方設計師參考了香港建築師潘祖堯和一位美籍設計師的意見綜合而成。港區全國人大代表陳永棋形容設計「清淡、高貴和雅潔」。

人大代表陳永棋昨日在北京與香港一批港區人大代表會晤後，視察香港廳設計的最後圖則，現有人員並帶來一些用作廳內地面的白色雲石比較。陳永棋形容「大娘」，是令人滿意的。記者看到，圖則是以大廳彩色照片、大廳設計好像酒店禮堂，廳內牆身以淺色為主，中間會擺放一個圖片。

陳永棋表示，他們將成立一個委員會，研究將來在大廳內擺放哪些可代表香港的物品。

1996年8月13日　明報

設計清淡高貴雅潔為原則
人民大會堂香港廳明年五月竣工

【本報記者范傑揮北京專電】北京人民大會堂香港廳的設計圖則經已最後確定，設計以達到「清淡、高貴、雅潔」為原則。香港廳總面積達二萬五千平方多呎，是人民大會堂內最大的一個廳。

一九九七年後將作為香港區人大代表在人大會議期間專用的開會地方。

港區人大代表陳永棋昨天介紹香港廳的圖則。陳永棋表示，香港廳由清華大學的建築系教授王煒鈺負責劃則設計，經政協委員潘祖堯建築師與另一名美籍華人建築師作最後定稿。王氏亦是澳門廳的設計者，澳門廳已於今年二月全部竣工。

香港廳整個改建裝修工程預期在明年五月一日完成，但港區人大代表爭取在明年三月人大會議期間可以初步使用。

陳永棋稱，未修改前的香港廳圖則未能捉摸切合香港人的心理，現在拍版的圖則能夠體現「清淡、高貴和雅潔」原則。

據悉，未修改前的圖則過分強調裝潢、華麗，予人印象是一如某些香港酒樓，因此被部分港區人大代表要求修改。

香港廳分「主廳」和「展覽廳」，主廳呈長方形，展覽廳則除可用作展覽用途外，更可擺設花圃而裝修成為一個花園。

香港廳的特色是有兩道門作出入用途，以體現香港作為中國門戶、及特別行政區的獨立特色。一道門從南門入，另一道則從西南門入。大會堂內的其他省、市廳均只有一道門。香港廳裝修費達三千萬，在香港富商兼全國政協副主席「扯頭纜」下，經已接近籌得三千萬。港區人大代表已組織了一個小組，研究如何擺設陳設。

蘋華日報 13/8/96

文匯報　1996年7月31日　星期三

北京西單商辦建築 獲選首都十佳設計
潘祖堯公司設計年底開工造價十四億

本報記者 蘭婉婷

由本港一間建築公司負責設計的北京西單北大街西側商場、五號地盤商場及辦公樓，最近獲選為首都十佳設計。

潘氏公司表示，該項目—北京西單北大街北側的四號、五號地盤（西工程），為該區最重要的綜合商業發展項目之一。佔地二十畝。五公頃總建築面積約二十六萬平方米，由辦公樓、零售中心、娛樂場所、配套公寓式酒店公寓，及地下停車場等設施組成，項目分兩階段施工，首期（一九九七至二○○○年），包括三號地及八號地互相通，車道穿梭各街道。

潘氏表示，二號與四、五號地進行設計工作外，並同時為二、三號地及八號地進行設計。二、三號地及八號地盤已由中國銀行買下使用，供中國銀行使用，而現時已動工。地則仍未可以動工。

至於其他工地的地則仍未可以動工興建。

本公司所設計的北京西單北大街西工程四、五號地地盤商場及辦公樓，佔地面積約一萬七千多平方米，總建築面積一十二萬平方米。地盤分為南北兩部分，南部地下三層，地上十二層商業樓，以商場及辦公為主；北部地下四層，地上九層，建築造價二億五千萬元。

據悉，西工程之三號地西單國際大廈，由潘氏公司及中國建築師事務所聯合設計，其旗下銀行設計，並已於日前動工。

潘祖堯建築師

中国国家大剧院第二轮设计方案竞赛拉开帷幕

世界建筑 5/1998

中国国家大剧院是中国最高表演艺术中心，建筑面积12万m²，占地3.89hm²。1998年4月13日—7月13日举办了第一轮建筑设计方案邀请竞赛，共收到来自36家设计单位的44个方案，其中来自国外设计方案20个，国内设计方案24个（含香港特别行政区方案4个）。

11人组成的评委会由吴良镛教授担任主席，委员是埃里克森（Arthur Erickson）、傅熹年、何镜堂、潘祖尧、彭一刚、包费尔（Ricardo Bofill）、宣祥鎏、芦原义信、张锦秋和周干峙。评委会认为，全部参赛方案中，没有一个方案能较综合地、圆满地、高标准地达到设计任务书提出的要求。决定举行中国国家大剧院第二轮建筑设计方案竞赛。第二轮参赛单位除了第一轮中得票过半数的5家设计单位：法国巴黎机场公司、英国塔瑞法若建筑设计公司、日本矶崎新建筑师株式会社、中国建设部建筑设计院和德国HPP国际建筑设计有限公司外，还邀请了北京市建筑设计研究院、香港王欧阳建筑师事务所、清华大学和深圳大学。第二轮竞赛已于1998年8月24日拉开帷幕，方案提交日期为1998年11月10日，最终评选结果预定在11月20日发布。□（凡夫）

"建筑师杯"全国中小型建筑优秀设计评选揭晓

1992 建筑师（48期）111

由中国建筑工业出版社《建筑师》编委会主办的"建筑师杯"全国中小型建筑优秀设计评选活动，于9月11日在哈尔滨市圆满结束。共评选出优秀奖8项，表扬奖19项。

这次参评的工程都是1986～1991年底6年期间建成投入使用的。全国有122家设计院送来209项工程图纸参评。

评选工作由张开济主持。除《建筑师》编委会成员外，还聘请了戴复东、钟训正、常怀生等专家教授参加评选。香港著名建筑师潘祖尧也应邀前来参加评选活动。

这次活动得到潘祖尧先生赞助，并得到哈尔滨市建委、哈尔滨建工学院、哈尔滨市建筑设计院的大力支持。获奖项目将在《建筑师》第49、50期上详细介绍（分别于今年12月、明年2月出版）。

桂林晚报
GUILIN EVENING NEWS 1998年5月

借名师之手 画名城之美
潘祖尧先生受聘为秀峰区发展顾问

本报讯（记者 罗素玲）我市作为旅游名城，旧城改建不仅需要加快步伐，更需要高起点、高水准建设。为此，5月2日，秀峰区政府真诚聘请全国政协委员、中国建筑学会名誉理事、香港著名建筑师潘祖尧先生为发展顾问，以加速秀峰区的经济发展和旧城改建步伐。

潘先生在建筑业上有过突出贡献，1994年曾应邀担任北京西单北大街商业开发区总顾问，他所负责总规划的方案，获"首都十佳建筑设计方案"；1996年底又应邀担任人民大会堂香港厅设计顾问。1996年到广西考察时，就对桂林的发展产生了浓厚兴趣。秀峰区得知这个消息后，为加快辖区内的经济发展和加快桂林市繁华地段的旧城改建步伐，聘请潘先生为秀峰区发展顾问。

潘先生对这个"顾问"的名誉很看重，他表示今后要加强合作，尽力搞好秀峰区的开发工作和桂林市城市建设。

北京城市规划信息
BEIJING CITY PLANNING INFORMATION

香港潘祖尧谈古都风貌

香港建筑师潘祖尧在1995年第二期建筑学报中载文论"建筑风格与古城风貌"。他认为：提倡"古都风貌"是十分恰当的，也是有积极意义的。"风貌"主要是涵盖城市的建筑，但又不但是建筑，一个城市的风貌还涵盖有城市的精神文明这样一个层面，也就是说既指"硬件"，也指"软件"。他说：作为泱泱大国的首都北京，理应有文化的独特风貌，就其大部分街区来说，它不应该是仿古的、复古的建筑的堆砌，也不应该是迪斯尼乐园的翻版或者是纽约的再现，而应该是具有五千年文明的中华民族现代社会的风貌，体现中国的四个现代化，蒸蒸日上的民族心态和社会进步，体现出十亿人口大国政治文化中心的风貌。在分析了一些建筑屋顶仿效古亭的做法后说："古都风貌"也好，"首都风貌"也好，唯一能适应现代化社会环境的，就是从功能、规模、地方色彩、地方气候方面去发挥，不能以表面去哗众取宠，在"内应外合"的外型设计上能充分发挥古都建筑的特点。在具体体现上述原则时就自然会产生一个合理的"古都风貌"。

（各示）

東方日報
一九九六年七月二十四日　星期三

● 入選北京首都十佳建築設計的建築物模型，建成後將座落北京西單北大街。 24/7/96

北京舊城西單大街重建計劃
港建築設計首獲十佳榮譽

【本報訊】本港一間建築顧問公司負責的北京舊城西單大街北側兩個地盤重建計劃，最近獲揚威北京，其所負責的北京西單北大街西側兩個地盤重建計劃，為北京舊城西單北大街重建計劃之一部份，獲選為此一重建計劃中九個地盤中首個試行重點重建區，整個項目包括北京西單北大街，為一綜合性大型重建，特色是此重建計劃所得出的總效果為北京首都十佳建築設計方案之一，亦為香港建築公司首次得此殊榮。

該項目的總顧問及總規劃師是由身為香港建築師學會前任會長潘祖堯先生主理，潘氏指出該公司所負責地盤面積約三十一萬五千平方呎，整個項目包括一座十二層樓高辦公大樓及四層商場、地庫兩層停車場，首期工程造價達七億港元。

他表示，整個計劃方案分二期進行，首期已動工，預算工程需時兩年竣工。

該公司負責地盤建築設計方案，他指出，該建築設計採用米黃色、雅緻的鋁合金及大幅玻璃作為物料，整個外形充分發揮現代化商業大樓的後現代建築風格。

潘氏指出，他本身是香港人，主要參與他比較喜歡的中港兩地的市場發展階段重建計劃，念頭仍可給予建築師更大的發揮空間，不像香港一般處處容易接受新意念；於內地工作滿足感較大，但他亦表示內地在解釋活性及解釋上仍有限制。

"建筑论坛"第四次研讨会召开
主题：建筑与社会

本报讯 由香港建筑师学会前任会长、全国政协委员潘祖尧先生倡议并赞助的"建筑论坛"第四次研讨会于近日在重庆召开。在杨永生、潘祖尧主持下，与会者认真讨论了建筑创作与社会环境的关系。

彭一刚院士、钟训正院士、建设部总工姚兵以及程泰宁、崔愷、吴耀东、张为耕、梁晓歧、张兴国、戴志中等出席研讨会。齐康院士向会议提交了论文。与会者从总结我国当代建筑创作的经验教训和分析西方建筑发展史入手，深入地讨论了当前建筑创作的社会环境、建筑文化的导向、建筑美学知识的社会普及以及建筑师如何走出困境等问题。

"建筑论坛"每年举办一次，前三次的主题分别是"建筑与评论"、"比较与差距"和"现状与出路"，论文已分别由天津科技出版社出版发行。本次研讨会由重建大建筑与城规学院和天津科技出版社主办。

（永生）

98.11.3 中国主流报刊版

圖片來源

潘衍壽　　　—潘衍壽土木工程師事務所的作品
大衛羅素　　—合作設計項目
拿士敦　　　—城市理工大學
黑川紀章　　—PAV6/8105系列的幻燈片
東方日報、華南早報、香港建築雜誌、英國建築師
學會雜誌、明報、大公報、星島日報及商報
等報刊上的照片。

Illustration credits

Peter Y.S. Pun for all projects executed at Peter Y.S. Pun & Associates
David Russell for all jointly designed projects
Denis Lasdun Redhouse and Softley for City Polytechnic University
Kisho Kurakawa for the use of the slides in his serie PAV6/8105
Oriental Daily, South China Morning Post, Building Journal Hong Kong, Wen Wei Po, RIBA Journal, Ming Pao, Ta Kung Pao, Sing Tao Jih Pao and Hong Kong Commercial Daily for the photographs in the newspaper clippings.